SOPHIE & TOMMY
HAVE COPIES

EXPLORING THE SOLAR SYSTEM

Palgrave Studies in the History of Science and Technology

James Rodger Fleming (Colby College) and Roger D. Launius (National Air and Space Museum), Series Editors

This series presents original, high-quality, and accessible works at the cutting edge of scholarship within the history of science and technology. Books in the series aim to disseminate new knowledge and new perspectives about the history of science and technology, enhance and extend education, foster public understanding, and enrich cultural life. Collectively, these books will break down conventional lines of demarcation by incorporating historical perspectives into issues of current and ongoing concern, offering international and global perspectives on a variety of issues, and bridging the gap between historians and practicing scientists. In this way they advance scholarly conversation within and across traditional disciplines but also to help define new areas of intellectual endeavor.

Published by Palgrave Macmillan:

Continental Defense in the Eisenhower Era: Nuclear Antiaircraft Arms and the Cold War
By Christopher J. Bright

Confronting the Climate: British Airs and the Making of Environmental Medicine
By Vladimir Janković

Globalizing Polar Science: Reconsidering the International Polar and Geophysical Years
Edited by Roger D. Launius, James Rodger Fleming, and David H. DeVorkin

Eugenics and the Nature-Nurture Debate in the Twentieth Century
By Aaron Gillette

John F. Kennedy and the Race to the Moon
By John M. Logsdon

A Vision of Modern Science: John Tyndall and the Role of the Scientist in Victorian Culture
By Ursula DeYoung

Searching for Sasquatch: Crackpots, Eggheads, and Cryptozoology
By Brian Regal

Inventing the American Astronaut
By Matthew H. Hersch

The Nuclear Age in Popular Media: A Transnational History
Edited by Dick van Lente

Exploring the Solar System: The History and Science of Planetary Exploration
Edited by Roger D. Launius

Exploring the Solar System

The History and Science of Planetary Exploration

Edited by
Roger D. Launius

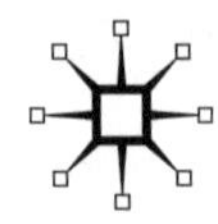

First published in 2013 by
PALGRAVE MACMILLAN®
in the United States—a division of St. Martin's Press LLC,
175 Fifth Avenue, New York, NY 10010.

Where this book is distributed in the UK, Europe and the rest of the world, this is by Palgrave Macmillan, a division of Macmillan Publishers Limited, registered in England, company number 785998, of Houndmills, Basingstoke, Hampshire RG21 6XS.

Palgrave Macmillan is the global academic imprint of the above companies and has companies and representatives throughout the world.

Palgrave® and Macmillan® are registered trademarks in the United States, the United Kingdom, Europe and other countries.

ISBN: 978–1–137–27316–1

Library of Congress Cataloging-in-Publication Data is available from the Library of Congress.

A catalogue record of the book is available from the British Library.

Design by Newgen Imaging Systems (P) Ltd., Chennai, India.

First edition: January 2013

10 9 8 7 6 5 4 3 2 1

Contents

Figures and Tables

Figures

Tables

Introduction

Roger D. Launius

Without question, the solar system exploration program has become the stuff of legends and myths in some measure because of its rich harvest of knowledge about Earth's neighboring planets, a transformation of our understanding of the solar system's origin and evolution, and a demonstration of what might be accomplished using limited resources when focusing on scientific goals rather than large human spaceflight programs aimed at buttressing American prestige.[1] That is the purpose of this collection of essays about episodes in the history of solar system exploration. It seeks to illuminate a broad set of perspectives on this unique topic, and to pose questions about the trajectory of planetary exploration from the beginning of the space age to the present.

Most assuredly, success in opening a window to the solar system effort did not take place by magic. It required considerable effort. The foundation for this was laid in the 1950s, when space science first became a major field of study. During the decade of the 1960s both the United States and the Soviet Union began an impressive effort to gather information on the planets of the solar system using ground-, air-, and space-based equipment.[2] Especially important was the creation of two types of spacecraft, one a probe that could be sent toward a heavenly body and the second an Earth-orbiting observatory that could gain the clearest resolution available in telescopes because it did not have to contend with the atmosphere. The studies emanating from this new data revolutionized humanity's understanding of Earth's immediate planetary neighbors. These studies of the planets, perhaps as much even as Project Apollo, captured the imagination of people from all backgrounds and perspectives. Photographs of the planets and theories about the origins of the solar system appealed to a very broad cross-section of the public. As a result, NASA had little difficulty in capturing and holding a broad interest in this aspect of its efforts.

This story has been told largely as a set of flight projects from the 1950s to the present. During the decade of the 1960s, as a direct outgrowth of the Apollo mandate to land Americans on the Moon by the end of the decade, NASA space science focused much of its efforts on lunar missions with projects Ranger, Surveyor, and Lunar Orbiter.[3] Even so, a centerpiece of NASA's planetary exploration effort in this era was the Mariner program, originated by NASA in the early part of the decade to investigate the nearby planets. Built by Jet Propulsion Laboratory scientists and technicians, satellites of this program proved enormously productive throughout the 1960s in visiting both Mars and Venus. Mariner made a huge impact in the early 1960s as part of a race between the United States and the Soviet Union to see who would be the first to reach Venus. It was also the closest planet to Earth, and a near twin to this planet in terms of size, mass, and gravitation. Later missions would later expand on the knowledge of the planet.[4]

At the same time Mars attracted significant attention, an attraction it has yet to relinquish for most planetary scientists, prompting missions there as well. In July 1965 Mariner 4 flew by Mars, taking 21 close-up pictures. Mariners 6 and 7, launched in February and March 1969, each passed Mars in August 1969, studying its atmosphere and surface to lay the groundwork for an eventual landing on the planet. Their pictures verified the Moon-like appearance of Mars and gave no hint that Mars had ever been able to support life. Among other discoveries from these probes, they found that much of Mars was cratered almost like the Moon, that volcanoes had once been active on the planet, that the frost observed seasonally on the poles was made of carbon dioxide, and that huge plates indicated considerable tectonic activity. Mariner 9, scheduled to enter Martian orbit in November 1971, detected a chilling dust storm spreading across Mars; by mid-October dust obscured almost all of Mars. Mariner 9's first pictures showed a featureless disk, marred only by a group of black spots in a region known as Nix Olympia (Snows of Olympus). As the dust storm subsided, the four spots emerged out of the dust cloud to become the remains of giant extinct volcanoes dwarfing anything on the Earth. Mons Olympus, the largest of the four, was 300 miles across at the base with a crater in the top 45 miles wide. Rising 20 miles from the surrounding plane, Mons Olympus was three times the height of Mt. Everest. Later pictures showed a canyon, Valles Marineris, 2,500 miles long and 3.5 miles deep. As the dust settled, meandering "rivers" appeared indicating that, at some time in the past, fluid had flowed on Mars. Suddenly, Mars fascinated scientists, reporters, and the public.[5]

While successes in planetary science have been very real all was not rosy with the politics of planetary exploration. In many respects the 1960s proved a training ground for how to envision, develop, and gain approval for planetary science missions. These political realities were played out thereafter. The labyrinth of modern science policy ensures that those engaged in government-funded science must play a savvy game of bureaucratic politics that is at once both insightful and extreme. A variety of strategies arose to succeed at this game. These included keeping individual projects small so as to avoid serious scrutiny, bringing aboard the project as many scientific disciplines as possible to ensure that everyone has a stake in the effort, developing large partnerships with multifaceted research and educational institutions in numerous congressional districts, and creating international coalitions, to name only a few.

One issue constantly debated, and never fully resolved, was the tradeoff resulting from the balance of cost, scale, and schedule for space probes. A perennial source of debate in planetary exploration, those engaged in deciding on planetary missions ask whether or not NASA should build a large number and variety of small, inexpensive probes or consolidate many kinds of experiments onto a few large, expensive spacecraft? Both sides have valid rationales. Small, inexpensive satellites could not accomplish a great deal at any one time and had limited scientific value but their smallness made them less conspicuous in the political process and perhaps many could be built and flown and thereby overcome the limitations of any one probe. Also, if one or more of them failed, the entire planetary program would not suffer as much. Large, costly satellites, on the other hand, were a scientist's (but not an accountant's) dream provided they worked properly, but if any component failed the returns could be greatly diminished. They also attracted more scrutiny in Washington, and had to be astutely managed to ensure funding. Finally, they took much longer to shepherd to completion. It was not uncommon for huge projects to take more than a decade for research, development, and launch.

Between the 1960s and the present various NASA leaders have swayed back and forth on this question, much of the time advocating, but not always able to deliver, a mixture of large and small spacecraft to avoid the long hiatus that came if a mission failed. Such an approach, while also having drawbacks, was designed to minimize the potential difficulties envisioned in a spacecraft's failure.

One overwhelmingly significant incident in this story is the long shadow cast by the cancellation of a Mars lander in 1967. No event was more significant in the first quarter century of planetary exploration than the political debacle of losing that mission. It was an enormously important object lesson, and its legacy is everywhere apparent.

In the summer of 1967, even as the technical abilities required to conduct an adventurous space science program were being demonstrated, the planetary science community suffered a devastating defeat in Congress and lost funding for a satellite lander to Mars. No other NASA effort but Project Apollo was more exciting than the Mars program in the middle part of the 1960s, yet this enormous setback took place. The planet had long held a special attraction for Americans, so much like Earth and possibly even sustaining life, and the lander would have allowed for extended robotic exploration of the red planet. A projected $2 billion program, the lander was to use the Saturn V launch vehicle being developed for Apollo.

The problem revolved around the lack of consensus among scientists on the validity of this Mars exploration initiative. Some were excited and supported the mission; most opposed it as too risky and too expensive. Without that consensus in 1967 and with other national priorities for spending for "Great Society" social programs, combating urban unrest, and for the military in Vietnam, the Mars lander was an easy target in Congress. It was the first space science project ever killed on Capitol Hill. The NASA administrator, James E. Webb, frustrated by congressional action and infuriated by internal dissension among scientists, stopped all work on new planetary missions until the scientists could agree on a planetary program. As 1968 began, the entire US planetary exploration program consisted of two Mars flybys scheduled for 1969.[6]

The scientific community learned a hard lesson about the pragmatic, and sometimes brutal, politics associated with the execution of "Big Science" under the suzerainty of the Federal government. Most important, scientists realized that strife within the scientific community had to be kept within the community in order to put forward a united front against the priorities of other interest groups and other government leaders. They learned that they had to resolve internal differences inside their community, not in complaints to the media or in testimony before Congress. While imposing support from the scientific community could not guarantee congressional support for a mission, without it virtually any initiative would not be funded. They also learned that while a $750 million program found little opposition at any level, a $2 billion project crossed an ill-defined but very real threshold triggering intense competition for those dollars.[7] Having learned these lessons, as well as some more subtle ones, the space science community regrouped and went forward in the latter part of the decade with a trimmed-down Mars lander program called Viking, which was funded and eventually provided important scientific data in the mid-1970s.

To avoid future imbroglios, NASA formed a Lunar and Planetary Mission Board and an Astronomy Mission Board to assist in planning future missions and to provide a forum to identify and resolve differences among the scientists. In 1967 and 1968, space scientists hammered out a mutually acceptable planetary program for the 1970s. Although this program continued to emphasize the exploration of Mars by recommending what became Project Viking, a scaled-back mission to attempt a soft landing on Mars, it also included two Mars orbiters and other initiatives.[8]

In addition, the planetary science community developed a set of "decadal surveys" beginning in 1968 that developed a set of questions concerning lunar and planetary exploration, as well as options for answering them and missions for conducting scientific research. A succession of seven reports extending from 1968 to the most recent in 2011 have charted a comprehensive science and mission strategy for planetary science based on extensive review and input from a broad swath of planetary scientists in the United States. They have served for some 45 years to identify the most important scientific questions to be tackled by the scientific community, broadly considering the planets, moons, small and icy bodies, comets, and asteroids. Collectively, the planetary science community has been able to rally around the decadal survey thereafter to win political support for many of its priorities in scientific investigation.

At sum, these surveys have, according to Wesley T. Huntress, a former head of space science at NASA, succeeded in achieving the following:

1. Creating a revolution in space science by building a consensus for change among all stakeholders.
2. Working with divergent people and institutions, many of whom viewed others as rivals or even threats, to undertake some of the most spectacular robotic space science missions in NASA's history.

3. Negotiating the shoals of difficulties within the science community, its various components, and its relationships with other communities.
4. Working with the Office of Space Science staff, the NASA administrator and his staff, the administration's OMB and OSTP, and the members and staff of the Congress to obtain the resources and authority to carry out this exploration program.
5. Facilitating the often problematic relations between the National Academy of Science, NASA, other federal agencies, and international partners in space science.
6. Pursuing the strategic planning, management, and road-mapping necessary to undertake this exploration program.[9]

The various planetary decadal surveys, therefore, have provided a national plan for developing a stepwise exploration agenda and fostering associated scientific discoveries.

Relations between NASA's human and space science entities have been strained from the very beginning of the space age, although an uneasy existence has persisted to the present. Space scientists resented the priorities and media attention enjoyed by the human spaceflight programs, especially Apollo. They complained about the lack of plans or funding in these programs for scientific research in general and about the manner in which planetary science went lagging with the budgetary priorities of the piloted spaceflight effort. So intense were rivalries that these organizations contended for control of the Apollo science program. An uneasy sharing of power emerged in which the NASA associate administrator for space science, Homer E. Newell, created a Manned Space Science Division and required that the head of the division report to him on scientific issues and to the head of the Apollo program on technical and funding issues. One could argue that these measures led to remarkable scientific returns from Apollo, clearly never envisioned as a science program, but not without a fair measure of controversy and in-fighting among representatives of these two unique facets of NASA's overall mission.[10]

Similar challenges of negotiating the priorities of space science with the human spaceflight effort occurred during the Space Shuttle program. For example, only slowly and reluctantly did the shuttle program management adjust to the use of the shuttle for planetary and other space science activities.[11] For their part, the scientists had to modify many projects—including its planetary probes to Jupiter, Saturn, and Venus—so that they could be launched aboard the shuttle. Accordingly, after the launch of the Voyagers in 1977, NASA canceled the Titan-Centaur launch vehicle program and made plans to phase out the Delta and Atlas class expendable launch vehicles. Since the performance of the shuttle and its then-planned upper stage were less than that of the Titan-Centaur, the cancellation decreased the size of the payload that NASA could send to the planets. Subsequently, to restore this capability, NASA decided to develop a shuttle-compatible version of the Centaur. Later, it also canceled these plans for Centaur, then reinstated them, and finally canceled them for good after the *Challenger* accident in 1986.[12]

As a result of these decisions the planetary science community often found itself whipsawed between launch vehicle priorities and shuttle prerequisites. As only one example of the effect these issues had on the planetary science program, the team responsible for Galileo, NASA's Jupiter probe, spent several frustrating years and many millions of dollars trying to adjust the spacecraft's configuration and trajectory to accommodate the capabilities of the shuttle. Originally scheduled for launch in 1982 as an extended follow-on to the Voyager probe, NASA finally launched Galileo in 1991. Over that period, the cost of Galileo increased by about $1.3 billion. The *Challenger* accident delayed this project, and perhaps the entire planetary science program, by five–ten years. The *Challenger* accident had one salutary effect; scientists no longer had to use the shuttle for all of its launches. It could then purchase expendable launch vehicles from commercial vendors and use the shuttle only when required by the mission.[13]

Throughout the space age robotic exploration of the planets took second stage to the human effort, but there were notable successes. One of them was the Viking mission to Mars. After a succession of missions that pulled back the curtain on the red planet, the first long-duration lander reached the Martian surface in 1976. Launched in 1975 from the Kennedy Space Center, Florida, *Viking 1* spent nearly a year cruising to Mars, placed an orbiter in operation around the planet, and landed on July 20, 1976, on the Chryse Planitia (Golden Plains), with *Viking 2* following in September 1976. These were the first sustained landings on another planet in the solar system. While one of the most important scientific activities of this project involved an attempt to determine whether there was life on Mars, the scientific data returned mitigated against the possibility. The two landers continuously monitored conditions at the landing sites and found both exciting cyclical variations and an exceptionally harsh climate that prohibited the possibility of life. The failure to find any evidence of life on Mars, past or present, devastated the optimism of scientists and led to a 20-year hiatus in the exploration of Mars.[14]

Likewise, the outer planets were opened to discovery by a set of daring missions in the 1970s. During the early 1960s, G. A. Flandro and Michael Minovitch, from the Jet Propulsion Laboratory, discovered that once every 176 years both the Earth and all the giant planets of the solar system gathered on one side of the Sun. This geometric line-up made possible close-up observation of all the planets in the outer solar system (with the exception of Pluto) in a single flight, the "Grand Tour." The flyby of each planet would bend the spacecraft's flight path and increase its velocity enough to deliver it to the next destination. This would occur through a complicated process known as "gravity assist," something like a slingshot effect, whereby the flight time to Neptune could be reduced from 30 to 12 years. Such a configuration was due to occur in the late 1970s, and it led to one of the most significant space probes undertaken by the United States.

To prepare the way for this outer planetary mission, NASA conceived Pioneer 10 and Pioneer 11 to visit Jupiter and Saturn. Both were small, nuclear-powered, spin-stabilized spacecraft that Atlas-Centaur sent beyond Earth. The first of these was launched on March 3, 1972, traveled outward to Jupiter, and in May 1991 was about 52 Astronautical Units (AU), roughly twice the distance from Jupiter to the Sun, and still transmitting data. In 1973, NASA launched Pioneer 11, providing scientists with their closest view of Jupiter, from 26,600 miles above the cloud tops in December 1974. The close approach and the spacecraft's speed of 107,373 mph, by far the fastest ever reached by a an object from Earth, hurled Pioneer 11 1.5 billion miles across the solar system toward Saturn. It was expected that as Pioneer 11 passed beyond Saturn it would continue to return data to Earth through the year 2000, in the process extending its original 30-month design life to 28 years.[15]

Meantime, NASA technicians prepared to launch what was called Project Voyager. While the four-planet mission was known to be possible, it was quickly deemed too expensive to build a spacecraft that could go the distance, carry the instruments needed, and last long enough to accomplish such an extended mission. Thus, the two Voyager spacecraft were funded to conduct intensive flyby studies only of Jupiter and Saturn, in effect repeating on a more elaborate scale the flights of the two Pioneers. Even so, the spacecraft builders designed as much longevity into the two Voyagers as possible with the $865 million budget available. NASA launched these from Cape Canaveral, Florida: Voyager 2 lifting off on August 20, 1977, and Voyager 1 entering space on a faster, shorter trajectory on September 5, 1977. Both spacecraft were delivered to space aboard Titan-Centaur expendable rockets.

As the mission progressed, with the successful achievement of all its objectives at Jupiter and Saturn in December 1980, additional flybys of the two outermost giant planets, Uranus and Neptune, proved possible—and irresistible—to mission scientists and engineers at the Jet Propulsion Laboratory in Pasadena, California. Accordingly, as the spacecraft flew across the solar system, remote-control reprogramming was used to reprogram the Voyagers for the greater mission. Eventually, between them, Voyager 1 and Voyager 2 explored all the giant outer planets, 48 of their moons, and the unique systems of rings and magnetic fields those planets possess.

The two spacecraft returned to Earth information that has revolutionized the science of planetary astronomy, helping to resolve some key questions while raising intriguing new ones about the origin and evolution of the planets in this solar system. The two Voyagers took well over one hundred thousand images of the outer planets, rings, and satellites, as well as millions of magnetic, chemical spectra, and radiation measurements. They discovered rings around Jupiter, volcanoes on Io, shepherding satellites in Saturn's rings, new moons around Uranus and Neptune, and geysers on Triton. The last imaging sequence was Voyager 1's portrait of most of the solar system, showing Earth and six other planets as sparks in a dark sky lit by a single bright star, the Sun. The Voyagers are expected to return scientific data until about the next decade since communications will be maintained until their nuclear power sources can no longer supply enough electrical energy to power critical subsystems.[16]

In the 1990s and the first decade of the twenty-first century a new enthusiasm for planetary exploration transformed our knowledge of the solar system. Numerous projects came to fruition during the period. For example, the highly successful Magellan mission to Venus provided significant scientific data about that planet.[17] Another such project was the Galileo mission to Jupiter, which even before reaching its destination had become a source of great concern for both NASA and public officials because not all of its systems were working properly, but it returned enormously significant scientific data.[18] The Cassini mission to Saturn enthralled scientists and the general public alike as a partnership between NASA and the European Space Agency successfully opened new vistas of understanding about the ringed planet and its fascinating satellite, Titan.[19]

Finally, Mars exploration received new impetus beginning on July 4, 1997, when Mars Pathfinder successfully landed on Mars, the first return to the red planet since Viking in 1976. Its small, 23-pound robotic rover, named Sojourner, departed the main lander and began to record weather patterns, atmospheric opacity, and the chemical composition of rocks washed down into the Ares Vallis flood plain, an ancient outflow channel in Mars's northern hemisphere. This vehicle completed its projected milestone 30-day mission on August 3, 1997, capturing far more data on the atmosphere, weather, and geology of Mars than scientists had expected. In all, the Pathfinder mission returned more than 1.2 gigabits (1.2 billion bits) of data and over ten thousand tantalizing pictures of the Martian landscape.[20]

Thereafter the strategy for much of Mars exploration has been built upon the motto "Follow the Water." In essence, this approach noted that life on Earth is built upon liquid water and that any life elsewhere would probably have chemistries built upon these same elements. Accordingly, to search for life on Mars, past or present, NASA's strategy must be to follow the water. If scientists could find any liquid water on Mars, probably only deep beneath the surface, the potential for life to exist was also present.[21]

Evidence of changes to the planet's surface from fast-flowing water has been collected by many space probes orbiting the planet since the latter 1990s. The spacecraft to open this possibility was Mars Global Surveyor, reaching the planet in 1998 and beginning a new and exciting era of scientific missions to study the red planet. Its recent discoveries offer titillating hints for learning about the possibility of life on Mars, at least in the distant past. Operating for several years, Mars Global Surveyor continued to send back views of the Martian surface that seemed to show evidence of dry riverbeds, flood plains, gullies on Martian cliffs and crater walls, and sedimentary deposits that suggested the presence of water flowing on the surface at some point in the history of Mars. This led scientists to theorize that billions of years ago, Earth and Mars might have been very similar places. Of course, Mars lost its water and the question of why that might have been the case has also motivated many Mars missions to the present. At that point, a consensus emerged that on any mission to Mars we should "follow the water" and seeking the answer to the ultimate question: "Are we alone in the universe?" Mars may well provide a definite answer.

There are even a few scientists who would go somewhat further and theorize that perhaps some water is still present deep inside the planet. In that case simple life forms might still be living beneath Mars's polar caps or in subterranean hot springs warmed by vents from the Martian core. These might be Martian equivalents of single-celled microbes that dwell in Earth's bedrock. Scientists are quick to add, however, that these are unproven theories for which evidence has not yet been discovered.

This strategy of "follow the water" has dominated all planning for Mars science missions for more than a decade and results thus far have been promising. A major step forward came with the Mars Exploration Rovers (MER), two extraordinary robots named Spirit and Opportunity that displayed noteworthy toughness and flexibility. Designed for 90-day missions to search for evidence of the effects of water in shaping Martian geology, Spirit and Opportunity remained operational more than six years after they first reached different parts of the planet in January 2004. During that time, they drove more than seven times the distance originally set as the criterion for a successful mission, with Spirit covering 3.4 miles and on February 6, 2007, Opportunity becoming the first spacecraft to traverse 10,000 meters—or 6.2 miles—on the surface of Mars. Both contended with hills and craters, escaped sandtraps, and overcame numerous technical problems.[22]

Spirit explored the Gusev Crater and revealed a basaltic setting, one not greatly suggestive of past water on Mars. It traveled to "Columbia Hills" and found a variety of rocks indicating that early Mars was characterized by impacts, explosive volcanism, and abundant subsurface water. Unusual-looking bright patches of soil turned out to be extremely salty and affected by past water. At "Home Plate," a circular feature in the "Inner Basin" of the "Columbia Hills" region, Spirit discovered finely layered rocks that are as geologically compelling as those found by Opportunity. A successful lander, Phoenix, in 2008 went even further and confirmed that the planet once offered a warm and watery habitat for life.[23]

Finally, the exploration of the Kuiper Belt has changed the perspective on the nature of the solar system. This region is a disk-shaped region of icy debris located about 50–100 AU from the Sun. Scientists only reached consensus about its existence in the 1990s; before that time they were uncertain if this was truly a cohesive entity. Since that time the Kuiper Belt has been an emerging area of research in planetary science. The detection of several large icy bodies at the outer reaches of the solar system kick-started the field. The planet Pluto, discovered in 1930 by Clyde Tombaugh, is only the largest member of the Kuiper Belt. Moreover, Pluto's largest moon, Charon, is half the size of Pluto and the two form a binary planet, whose gravitational balance point is between the two bodies. Other named objects soon joined Pluto, including 1992 QB_1, Orcus, Quaoar, Ixion, 90377 Sedna, and Varuna.

The discovery of these many objects, nearly as large as Pluto and occupying the range in the outer solar system, led the International Astronomical Union (IAU) in 2006 to redesignate Pluto from a planet—there would henceforth be eight of them in the solar system—and call it by the new title of "dwarf planet." The first members of the "dwarf planet" category were Ceres, Pluto, and 2003 UB_{313}. Scientists and astronomers deliberated on this long and hard before reaching a definition of planets that included the following criteria: (i) it is in orbit around the Sun; (ii) it has sufficient mass for its self-gravity to overcome rigid body forces so that it assumes a hydrostatic equilibrium (nearly round) shape; and (iii) it has cleared the neighborhood around its orbit. Its members also specifically commented that the "dwarf planet" status of Pluto would hereafter be recognized as a critical prototype of this new class of trans-Neptunian objects. While this decision remains controversial, it represents an important recent step in understanding the origins and evolution of the solar system. In the first decade of the twenty-first century two NASA missions—New Horizons to Pluto and Dawn to Vesta and Ceres—represented a new initiative to explore these outer bodies.[24]

Through these explorations knowledge of the solar system has been transformed. In 50 years we have redefined the planets, characterized their nature, and developed new, fundamentally significant perspectives on the nature of our place in the cosmos. Many specific instances of this dramatic transition may be cited; I will use only three. The first is the debate over the origins of the Moon that came in the aftermath of the focused exploration culminating in Project Apollo. Second, the discoveries concerning water in the solar system have profoundly affected human understandings of astrobiology. Finally, the 2011 planetary decadal survey of projects listed 12 major discoveries that came through this research.

In the first case, prior to the 1960s the origin of the Moon had been a subject of considerable scientific debate and careers had risen and fallen on championing one of three principal theories:

1. Co-accretion—a theory that asserted that the Moon and the Earth were formed at the same time from the Solar Nebula.
2. Fission—a theory that asserted that the Moon split off from the Earth.
3. Capture—a theory that held that the Moon formed elsewhere and was subsequently drawn into orbit around the Earth.

The data supporting these various theories had been developed to an amazingly fine point over time but none of these theories actually explained enough open questions to convince a majority of planetary scientists.

The answer to that question came only with the missions to the Moon, including landing missions made by the Apollo astronauts between 1969 and 1972, and exhaustive analysis by scientists over many years. While it took more than a decade, data from these explorations eventually led to a consensus on the origins of the Moon. Indeed, if there is one dramatic moment—as opposed to myriad important but mundane events—in the history of lunar science it is the 1984 conference in Kona, Hawaii, in which scientists around the world presented papers on the sole topic of how the Moon originated. The new and detailed information from the Moon's explorations pointed toward an impact theory—which suggested that the Earth had collided with a very large object (perhaps as big as Mars and named after the fact "Theia")—and that the Moon had formed from the ejected material of both Theia and Earth.[25]

This proved to be a theory that fit the fact that although the Earth has a large iron core the Moon does not, because the debris blown out of both the Earth and the impactor would have come from iron-depleted, rocky mantles. Also lending credence to this theory, although the Earth has a mean density of 5.5 grams/cubic centimeter the Moon's density is only 3.3 grams/cubic centimeter, which would be the case were it to lack iron, as it does. The Moon has exactly the same oxygen isotope composition as the Earth, whereas Mars rocks and meteorites from other parts of the solar system have different oxygen isotope compositions. While there were some details left unexplained by this conclusion, the impact theory came out as the scientific consensus and is now widely accepted.

This "big whack" theory, as it was called, explained well what was learned about the geology and selenogony of the Moon during the Apollo program. Lunar scientist Paul D. Spudis recently concluded on this issue:

> The giant-impact hypothesis appears to explain, or allow for, several fundamental relations—not just bulk composition, but also the orientation and evolution of the lunar orbit... Part of the reason for this model's current popularity is doubtless because we know too little to rule it out: key factors such as the impactor's composition, the collision geometry, and the Moon's initial orbit are all undetermined.

In the end, further research is required, for "as it turned out, neither the Apollo astronauts, the Luna vehicles, not all the king's horses and all the king's men could assemble enough data to explain circumstances of the Moon's birth" to everyone's satisfaction.[26] Thereafter, the impact theory made its way into the textbook arena. First it was mentioned as one of several theories, but by 1990 it had gained precedent as the preferred explanation in the undergraduate curricula. In this case, much has changed in terms of scientific knowledge about the Moon because of explorations in the space age.

In the second instance, prior to the space age knowledge about the presence of liquid water in the solar system was essentially without foundation. Speculations abounded, but little actual knowledge existed. Famously, Percival Lowell and others believed Mars was inhabited and had been losing water. Canals had been built by the inhabitants to take water from the poles to populated locations nearer the equator. None of this has proved out but scientists have discovered that water existed on Mars for lengthy periods of time, and perhaps some still remain deep in underground fissures. Likewise, speculations about Venus as a Precambrian rain forest harboring life have been dashed as spacecraft from Mariner to Magellan to MESSENGER have collected scientific data about this strange and forbidding planet. At the same time, based on the data coming from the Galileo and Cassini-Huygens probes, scientists now strongly suspect that there is water under Europa's icy surface and even at Ganymede and Enceladus.[27]

Finally, the National Academies' decadal study of planetary science offered 12 fundamental discoveries that have taken place recently ranging across a broad spectrum of fields and objects in the solar system:

1. *An explosion in the number of known exoplanets.* Confirmed examples have grown from a few dozen at the beginning of this decade to many hundreds, including numerous multiplanet systems.
2. *The Moon is less dry than once thought.* Evidence is mounting that the lunar surface and interior is not completely dry as previously believed. Observations from Lunar Prospector, LRO, LCROSS, Cassini, and Chandrayaan-1 also suggest small, but significant, quantities of water on the Moon.
3. *Minerals that must have formed in a diverse set of aqueous environments throughout Martian history.* Observations from multiple orbiters and rovers have identified a broad suite of water-related minerals including sulfates, phyllosilicates, iron oxides and oxyhydroxides, chlorides, iron and magnesium clays, carbonates, and hydrated amorphous silica.
4. *Extensive deposits of near-surface ice on Mars.* These deposits are a major reservoir of Martian water, and because of oscillating climate conditions, potentially lead to geologically brief periods of locally available liquid water.
5. *An active meteorological cycle involving liquid methane on Titan.* Observations from Cassini and Huygens have confirmed the long-suspected presence of complex organic processes on Titan.
6. *Dramatic changes in the atmospheres of the giant planets.* Notable examples include: observations of three impacts on Jupiter in 2009–2010; striking atmospheric seasonal change on Saturn and Uranus; evidence of vigorous polar vortices on Saturn and Neptune; and the discovery of rapid changes in the ring systems of Jupiter, Saturn, Uranus, and Neptune.
7. *Recent volcanic activity on Venus.* The European Space Agency's Venus Express spacecraft has found zones of higher emissivity associated with volcanic regions, suggestive of recent volcanic activity.
8. *Geothermal and plume activity at the south pole of Enceladus.* Observations by the Cassini spacecraft have revealed anomalous sources of geothermal energy coincident with curious rifts in the south polar region of Enceladus.
9. *The anomalous isotopic composition of the planets.* Analysis of data from the Genesis solar wind sample return mission has revealed that the Sun is highly enriched in oxygen-16.
10. *The differentiated nature of comet dust.* Analysis of samples returned by the Stardust mission revealed that cometary dust contains minerals that can only form at high temperatures, close to the Sun. This result has

changed ideas concerning the physical processes within the protoplanetary disk.

11. *Mercury's liquid core.* Radar signals transmitted from NASA's Deep Space Network station in California and detected by NRAO's Green Bank Telescope detected Mercury's forced libration and demonstration that the planet has a liquid core.
12. *The richness and diversity of the Kuiper belt.* A combination of ground- and space-based telescopic studies has revealed the diversity of the icy bodies forming the Kuiper belt. This diversity includes many objects as large as or larger than Pluto and, intriguingly, a large proportion of binary and multiobject systems.

A hallmark of these discoveries highlighted in the planetary science decadal survey has been their diversity. Chemistry, geology, astrobiology, climatology, geomagnetism, and a host of other scientific disciplines have contributed to this harvest of knowledge about the solar system. Collectively, they have transformed the human perspective on this corner of the universe and how it came to be.[28]

The forgoing discussion represents the relative state of the solar system exploration master narrative as it stands at present. Overall, from the beginning of the Space Age the United States and the Soviet Union, followed soon by other nations, began an impressive effort to learn about the planets of the solar system that has yet to abate. The data collected and analyzed by scientists revolutionized understanding of Earth's neighboring bodies in the solar system. These efforts also captured the imagination of people from all backgrounds like nothing else except the Apollo lunar missions. As a result, the cause of planetary exploration had little difficulty in capturing and holding widespread public support. Science textbooks have been rewritten multiple times because of these missions. This is fine as far as it goes, taking a progressive approach toward its subject and emphasizing both a single line set of accomplishments and too often leading to an exceptionalist perspective.[29]

Through a succession of broadly analytical essays on major aspects of the history of robotic planetary exploration, this book opens new vistas in the understanding of the development of planetary science in the Space Age. Its purpose is to consider, over the course of the more than 50-year history of the Space Age, what we have learned about the other bodies of the solar system and the process whereby we have learned it. This collection of essays pursues broad questions about this issue:

- The various flight projects and their broader implications for the exploration of other solar system bodies.
- The development of space science disciplines and institution building.
- The big questions of planetary science and what has been learned in the nearly 50 years of the Space Age.
- The relationships of organizations/international bodies, civil/military organizations, and so on, one to another.
- The uneasy alliance between robotic exploration and human spaceflight.

- Herding the space science community and setting priorities for missions, instruments, and knowledge generation.
- The manner in which scientific knowledge has been acquired, refined, analyzed, and disseminated over time.
- The development of theories about planetary science.
- The development of instruments and methodologies for scientific exploration.
- Broad recapitulations of the science of solar system origins and evolution.
- The significance of this work is twofold: (1) it captures in one place a set of broad studies of the history of robotic solar system exploration, and (2) it opens opportunities for greater involvement of social scientists, historians, and scientific and technical academics in considering a topic that has not attracted them heretofore. There is no question but that the greatest share of historical work has been on human spaceflight, and that story is very much in the public consciousness. This work is intended as a means of offering important insights about themes that might be productively explored by others in the future and as an opportunity to energize scholars to pursue this important topic.

The essays in this volume build on the history of planetary exploration, space science and technology studies, and larger themes in historical study to help reorient the discipline toward a wider view. At some level, these essays represent a contribution to the "new aerospace history." I believe there is an emergent "new aerospace history" of which this collected work is a part. It moves beyond a fetish for the artifact to emphasize the broader role of the spacecraft, and more importantly the whole scientific and technological system including not just the vehicle but also the other components that make up the aerospace climate, as an integral part of the human experience. It suggests that many unanswered questions are present in the development of flight, and perhaps inquisitive individuals might illuminate areas never even considered before. Many of the works reflecting this perspective are both insightful and provocative, and move beyond traditional realms and themes long the province of space history, to contribute to understandings of ideology, culture, and society.[30]

As only one example, several essays in this collection investigate the history of the heterogeneous management of planetary exploration programs. One of the fundamental tenets of successful spaceflight was the program management concept, originated in the ballistic missile development program of the 1950s and transferred to NASA almost with the creation of the agency in 1958. It posited that successful spaceflight missions rested on maintaining a balance between three critical factors—cost, schedule, and reliability. These three factors were inextricably interrelated and had to be managed as a group. Many also recognized these factors' constancy; if program managers held cost to a specific level, then one of the other two factors, or both of them to a somewhat lesser degree, would be adversely affected. This has held true for virtually every spaceflight program, including planetary science missions. The schedule for these missions was often less firm than human spaceflight activities. So it often took a backseat to cost and reliability.[31]

Getting all of the personnel elements to work together always challenged project managers, regardless of whether or not they were civil service, industry, or university personnel. This was especially true of planetary science missions, in which engineers designing and flying the spacecraft and scientists developing and using scientific instruments competed to ensure that their priorities found expression in the spacecraft. These groups had competing goals and cultures and always enjoyed an uneasy working relationship. As ideal types, engineers usually worked in teams to build hardware that could carry out the functions necessary to fly the spacecraft to its target and to perform its task once it arrived. Their primary goal involved building vehicles that would function reliably within the fiscal resources allocated to the project. Again as ideal types, space scientists engaged in pure research and were more concerned with designing experiments that would expand scientific knowledge about the planetary body to be explored. They also tended to be individualists, unaccustomed to regimentation and unwilling to concede gladly the direction of projects to outside entities. The two groups always contended with each other over a great variety of issues associated with every planetary mission throughout the history of the space age. For instance, the scientists disliked having to configure payloads so that they could meet time, money, or launch vehicle constraints. The engineers, likewise, resented changes to scientific packages added after project definition because these threw their hardware efforts out of kilter. Both had valid complaints and had to maintain an uneasy cooperation to accomplish all their objectives.

The scientific and engineering communities within NASA, additionally, were not monolithic, and differences among them thrived. Add to these representatives from industry, universities, and research facilities, and competition on all levels to further their own scientific and technical areas resulted. The NASA leadership generally viewed this pluralism as a positive force within the space program, for it ensured that all sides aired their views and emphasized the honing of positions to a fine edge. Competition, most people concluded, made for a more precise and viable space exploration effort. There were winners and losers in this strife, however, and sometimes ill-will was harbored for years. Moreover, if the conflict became too great and spilled into areas where it was misunderstood, it could be devastating to the conduct of the planetary science program. The head of the Office of Space Science worked hard to keep these factors balanced and to promote order so that NASA could effectively further knowledge about the solar system, but it was neither an easy task nor one accomplished expeditiously. What did this mean for the development of planetary science at NASA?[32]

The 14 essays contained in this volume might have been organized in several different ways. The structure I have chosen highlights certain basic themes emphasized throughout the book. In part one, "Managing Planetary Science," the three authors focus on the organizational bureaucracy that made planetary exploration possible. In the first essay John D. Ruley elucidates the central role played by Homer Newell in establishing the policies, procedures, and first missions sent out by NASA to explore other bodies in the solar system. Newell's background as a respected scientist, his demeanor as a consensus builder and effective negotiator, and his long view on the nature of planetary science in the United States enabled him to build over a period of more than a decade a very effective organization for accomplishing a succession of spectacular missions.

John M. Logsdon's contribution focuses on the near demise of NASA's planetary exploration program in the Reagan administration of the 1980s. Throughout the 1970s NASA had enjoyed remarkable success in this arena with the Pioneer and Voyager probes to the outer solar system and the Viking landings on Mars serving as flagships for an aggressive, effective science effort. Attempts to demolish this accelerating program was only staved off through effective political maneuverings. Even so, the proposed mission to Halley's Comet in 1986 suffered termination through the budgetary constrictions roiling Washington at the time.

NASA's "faster, better, cheaper" efforts undertaken to reduce the costs of planetary science in the 1990s led Amy Paige Kaminski to assess the push/pull of politics between larger, complex, expensive, and highly capable probes and the smaller, less expensive, and more modest probes. While "faster, better, cheaper" was not really a program, this approach to planetary exploration emerged in the aftermath of the loss of the multibillion dollar Mars Observer spacecraft in 1993 when NASA administrator Daniel S. Goldin announced that he could no longer abide the building of "Battlestar Galactica"–type space probes. NASA had to find a way to accomplish lower-cost missions to the planets. In this context, as Kaminski notes, the fundamental tenets of the program management concept was that three critical factors—cost, schedule, and reliability—were interrelated and had to be managed as a group. Managing for all three proved exceptionally difficult; if program managers held cost to a specific level, then one of the other two factors, or both of them to a somewhat lesser degree, would be adversely affected. The "faster, better, cheaper" concept tried to squeeze each of these issues—although many said you could successfully "pick two" to control rather than all three—but remarkably NASA project managers did so with some success.

In a second part concerning "Developing New Approaches to Planetary Exploration" the four authors explore the technologies and processes that facilitated successful planetary missions. Andrew J. Butrica's contribution explores the little-known history of space navigation and the manner in which mission controllers ascertained the location of their spacecraft and their target. W. Henry Lambright applies management theory to the processes used to bring to fruition the Viking, Cassini, and Hubble Space Telescope projects. Peter J. Westwick explores the fascinating relationship between Hollywood technicians and planetary exploration programs in the creation of visually stunning imagery for public consumption. Finally, Roger D. Launius explores the manner in which long-outdated reentry and landing technology was resurrected for the Genesis and Stardust sample return missions to Earth. Part three, "Exploring the Terrestrial Planets," focuses on the terrestrial planets of the inner solar system and critical elements in their exploration beginning with Earth. Recognizing the nature of our home planet as one of several planets in the solar system proved central to the advancing knowledge about the cosmos. Erik M. Conway relates planetary science to the discovery of the phenomenon of global climate change, while Andrew K. Johnston investigates the evolution of ever more sophisticated satellites collecting data about Earth and helping scientists to understand the many natural processes at work here and how they interrelate one to another. Roger D. Launius explores the manner in which Venus, Earth, and Mars have been linked in the search for life in the solar system. Finally, Robert Markley draws a complex but succinct portrait of the manner in which Mars was reinterpreted in the 1960s because of the discoveries made for the first probes there.

In part four, "Unveiling the Outer Solar System," two historians offer unique perspectives on Saturn and its moons and Pluto and its place in the pantheon of planets. Arturo Russo tells the fascinating story of the international mission, Cassini/Huygens, to Saturn. The European Space Agency built the probe that landed on the surface of Titan and revealed a world of remarkable beauty and sophistication. David H. DeVorkin relates the manner in which Pluto rose and fell as a planet and what this has meant for the larger objective of solar system exploration. A third scholar, William E. Burrows, draws together the larger meaning of the exploration of the solar system. This cornucopia of knowledge about worlds beyond ours has altered human understanding. He takes as his theme the remark of scientist Merton Davies who put it memorably: "The joy of exploration is finding answers for which there are no questions." Such activities have affected how humanity sees itself.

Acknowledgments

Whenever historians take on a project such as this they stand squarely on the shoulders of earlier investigators and incur a good many intellectual debts. The editor would like to acknowledge the assistance of several individuals who aided in the preparation of this book. For their many contributions in completing this project I wish especially to thank Jane Odom and her staff archivists at the NASA History Division who helped track down information and correct inconsistencies, as well as Bill Barry, Steve Dick, Steve Garber, Glen Asner, and Nadine Andreassen at NASA; the staffs of the NASA Headquarters Library and the Scientific and Technical Information Program who provided assistance in locating materials; Marilyn Graskowiak and her staff at the NASM Archives; and many archivists and scholars throughout NASA and other space organizations. Patricia Graboske, head of publications at the National Air and Space Museum, provided important guidance for this project. The National Aeronautics and Space Administration provided a generous grant to make this work possible, and we thank it for this assistance.

In addition to these individuals, we wish to acknowledge the following individuals who aided in a variety of ways: Debbora Battaglia, David Brandt, Bruce Campbell, Bob Craddock, Tom D. Crouch, Gen. John R. Dailey, Robert Farquhar, Jens Feeley, James Rodger Fleming, James Garvin, Lori B. Garver, Michael H. Gorn, John Grant, G. Michael Green, Barton C. Hacker, James R. Hansen, Wes Huntress, Peter Jakab, Dennis R. Jenkins, Violet Jones-Bruce, Sylvia K. Kraemer, John Krige, Alan M. Ladwig, Jennifer Levasseur, W. Patrick McCray, Howard E. McCurdy, Jonathan C. McDowell, Karen McNamara, Ted Maxwell, Valerie Neal, Allan A. Needell, Michael J. Neufeld, Frederick I. Ordway III, Scott Pace, Robert Poole, Anthony M. Springer, Alan Stern, Harley Thronson, and Margaret Weitekamp. I wish, especially, to thank the individual contributors to this volume. Finally, my greatest debt is to the women in my life, my partner Monique and my daughters Dana and Sarah. All of these people would disagree with some of the materials and observations made here, but such is both the boon and the bane of scholarly inquiry.

Notes

1. Indeed, Representative George E. Brown Jr. (D-CA) remarked in a speech to the National Academy of Sciences in 1992 that "[i]t is also important to recall that some of our proudest achievements in the space program have been accomplished within a stagnant, no growth budget. The development of... the Viking lander, Voyagers I and II, Pioneer Venus,... were all carried out during the 1970s when the NASA budget was flat. It would be wise to review how we set priorities and managed programs during this productive time" (George E. Brown, Remarks, February 9, 1992, copy in NASA Historical Reference Collection, NASA History Division, Washington, DC).
2. See Homer E. Newell, *Beyond the Atmosphere: Early Years of Space Science* (Washington, DC: NASA Special Publication-4211, 1980).
3. On lunar science, see these major histories: R. Cargill Hall, *Lunar Impact: A History of Project Ranger* (Washington, DC: NASA SP–4211, 1977); Bruce K. Byers, *Destination Moon: A History of the Lunar Orbiter Program* (Washington, DC: NASA TM X-3487, 1977); Linda Neuman Ezell, comp., *NASA Historical Data Book, Vol. II: Programs and Projects, 1958–1968* (Washington, DC: NASA SP-4012, 1988), pp. 325–31; David M. Harland, *Exploring the Moon: The Apollo Expeditions* (Chicester, England: Wiley-Praxis, 1999); W. David Compton, *Where No Man Has Gone Before: A History of Apollo Lunar Exploration Missions* (Washington, DC: NASA Special Publication-4214, 1989).
4. Jet Propulsion Laboratory, *Mariner-Venus 1962: Final Project Report* (Washington, DC: NASA SP-59, 1965), pp. 4–6; Ladislav E. Roth and Stephen D. Page, *The Face of Venus: The Magellan Radar-Mapping Mission* (Washington, DC: NASA Special Publication-520, 1995).
5. See William K. Hartman and Odell Raper, *The New Mars: The Discoveries of Mariner 9* (Washington, DC: NASA Special Publication–337, 1974).
6. For a discussion of this political debate, see Edward Clinton Ezell and Linda Neuman Ezell, *On Mars: Exploration of the Red Planet, 1958–1978* (Washington, DC: NASA SP-4212, 1984), pp. 83–120.

7. The average percentage of the total NASA appropriation allotted for space science in 1959–1968 was 17.6 percent; it was 17 percent in 1969–1978. By 1996 it was less than 10 percent but it rebounded thereafter and as of 2010 it was approaching 30 percent.
8. NASA, Office of Space Science and Applications, "Planetary Program Review," July 11, 1969, copy in NASA Historical Reference Collection; John E. Naugle, "Goals in Space Science and Applications," *Nuclear News*, January 1969.
9. Wesley T. Huntress, "Space Science in the 1990s: Bust to Boom," Lindbergh proposal, January 21, 2008, copy in possession of author.
10. For example, see A. A. Levinson, ed., Proceedings of the Apollo 11 Lunar Science Conference, Houston, Texas, January 5–8, 1970, vol. 1 (New York: Pergamon Press, 1970), pp. 1–54; Don E. Wilhelms, *To a Rocky Moon: A Geologist's History of Lunar Exploration* (Tucson: University of Arizona Press, 1993); and Paul D. Spudis, *The Once and Future Moon* (Washington, DC: Smithsonian Institution Press, 1996); Curator for Planetary Materials, Johnson Space Center, "Top Ten Scientific Discoveries Made During Apollo Exploration of the Moon," October 28, 1996, NASA Historical Reference Collection, NASA History Division, Washington, DC.
11. Homer E. Newell, associate administrator, NASA, to Dr. [James] Fletcher, administrator, NASA, "Relations with the Science Community and the Space Science Board," December 3, 1971, NASA Historical Reference Collection.
12. For a discussion, with documents concerning the expendable launch vehicle situation, see John M. Logsdon, gen. ed., with Ray A. Williamson, Roger D. Launius, Russell J. Acker, Stephen J. Garber, and Jonathan L. Friedman, *Exploring the Unknown: Selected Documents in the History of the U.S. Civil Space Program, Volume IV, Accessing Space* (Washington, DC: NASA Special Publication-4407, 1999).
13. US General Accounting Office, Space Exploration: Cost Schedule and Performance of NASA's Galileo Mission to Jupiter, May 27, 1988, GAO/NSIAD–88–138FS; US General Accounting Office, Space Exploration: Cost Schedule and Performance of NASA's Ulysses Mission to the Sun, May 27, 1988, GAO/NSIAD–88–129FS; US General Accounting Office, Space Exploration: Cost, Schedule and Performance of NASA's Magellan Mission to Venus, May 27, 1988, GAO/NSIAD–88–130FS; US General Accounting Office, Space Science: Status of the Hubble Space Telescope Program, May 2, 1992, GAO/NSIAD–88–118BR, copies available in NASA Historical Reference Collection.
14. Roger D. Launius, "'Not too Wild a Dream': NASA and the Quest for Life in the Solar System," *Quest: The History of Spaceflight Quarterly* 6 (Fall 1998): 17–27.
15. Richard O. Fimmel, William Swindell, and Eric Burgess, *Pioneer Odyssey* (Washington, DC: NASA SP-349, 1977); Richard O. Fimmel, James A. Van Allen, and Eric Burgess, *Pioneer: First to Jupiter, Saturn, and Beyond* (Washington, DC: NASA SP-446, 1980).
16. Henry C. Dethloff and Ronald A. Schorn, *Voyager's Grand Tour: To the Outer Planets and Beyond* (Washington, DC: Smithsonian Institution Press, 2003); Stephen J. Pyne, *Voyager: Seeking Newer Worlds in the Third Great Age of Discovery* (New York: Viking, 2010).
17. On Magellan, see Ladislav E. Roth and Stephen D. Wall, *The Face of Venus: The Magellan Radar–Mapping Mission* (Washington, DC: NASA Special

Publication–520, 1995); David Harry Grinspoon, *Venus Revealed: A New Look Below the Clouds of our Mysterious Twin Planet* (Reading, MA: Addison-Wesley, 1997).

18. On the Galileo mission, see Daniel Fischer, *Mission Jupiter: The Spectacular Journey of the Galileo Spacecraft* (New York: Copernicus Books, 2001); Michael Hanlon, *The Worlds of Galileo: The Inside Story of NASA's Mission to Jupiter* (New York: St. Martin's Press, 2001); David M. Harland, *Jupiter Odyssey: The Story of NASA's Galileo Mission* (Chicester, England: Springer-Praxis, 2000); Cesare Barbieri, Jürgen H. Rahe, Torrence V. Johnson, and Anita M. Sohus, eds., *The Three Galileos: The Man, The Spacecraft, The Telescope* (Dondrecht, The Netherlands: Kluwer Academic, 1997).
19. On Cassini, see Ralph Lorenz and Jacqueline Mitton, *Titan Unveiled: Saturn's Mysterious Moon Explored* (Princeton, NJ: Princeton University Press, 2008); Laura Lovett, Joan Horvath, and Jeff Cuzzi, *Saturn: A New View* (New York: Harry N. Abrams, 2006); David M. Harland, *Mission to Saturn: Cassini and the Huygens Probe* (Chicester, England: Springer-Praxis, 2003).
20. Price Pritchett and Brian Muirhead, *The Mars Pathfinder Approach to "Faster-Better-Cheaper"* (Dallas, TX: Pritchett and Associates, 1998); Judith Reeves-Stevens, Garfield Reeves-Stevens, and Brian K. Muirhead, *Going to Mars: The Untold Story of Mars Pathfinder and NASA's Bold New Missions for the 21st Century* (New York: Pocket Star, 2002); Brian K. Muirhead and William L. Simon, *High Velocity Leadership: The Mars Pathfinder Approach to Faster, Better, Cheaper* (New York: HarperBusiness, 1999); Andrew Mishkin, *Sojourner: An Insider's View of the Mars Pathfinder Mission* (New York: Berkeley Publishing Group, 2003).
21. A basic history of Mars exploration may be found in Joseph M. Boyce, *The Smithsonian Book of Mars* (Washington, DC: Smithsonian Institution Press, 2003).
22. Steve Squyres, *Roving Mars: Spirit, Opportunity and the Exploration of the Red Planet* (New York: Hyperion, 2005).
23. Andrew Kessler, *Martian Summer: Robot Arms, Cowboy Spacemen and My 90 Days with the Phoenix Mars Mission* (*New* York: Pegasus Books, 2011); Pat Duggins, *Trailblazing Mars: NASA's Next Giant Leap* (Gainesville: University Press of Florida, 2010); Fredric W. Taylor, *The Scientific Exploration of Mars* (New York: Cambridge University Press, 2010); Andrew Chaikin, *A Passion for Mars: Intrepid Explorers of the Red Planet* (New York: Harry N. Abrams, 2008).
24. Alan Boyle, *The Case for Pluto: How a Little Planet Made a Big Difference* (New York: Wiley, 2009); Mike Brown, *How I Killed Pluto and Why it Had it Coming* (Berlin, Germeny: Spiegel & Grau, 2010).
25. Dana Mackenzie, *The Big Splat: Or How Our Moon Came to Be* (Hoboken, NJ: John Wiley & Sons, 2003).
26. Paul D. Spudis, *The Once and Future Moon* (Washington, DC: Smithsonian Institution Press, 1996), p. 246.
27. This has been discussed in such works as G. Scott Hubbard, *Exploring Mars: Chronicles from a Decade of Discovery* (Tucson: University of Arizona Press, 2012); Frederic W. Taylor, *The Scientific Exploration of Mars* (Cambridge, UK: Cambridge University Press, 2010); Roger D. Launius, Erik M. Conway, Andrew K. Johnston, Zse Chien Wang, Matthew H. Hersch,

Deganit Paikowsky, David J. Whalen, Eric Toldi, Kerrie Dougherty, Peter L. Hays, Jennifer Levasseur, Ralph L. McNutt Jr., and Brent Sherwood, "Spaceflight: The Development of Science, Surveillance, and Commerce in Space," *Proceedings of the IEEE* 100(5) (2012): 5–7, DOI: 10.1109/JPROC.2012.2187143.

28. National Research Council, *Vision and Voyages for Planetary Science in the Decade 2013–2022* (Washington, DC: National Academies Press, 2011), chapter 1.
29. On American exceptionalism, see Kerwin Lee Klein, *Frontiers of Historical Imagination: Narrating the European Conquest of Native America, 1890–1990* (Berkeley: University of California Press, 1997); Richard T. Hughes, *Myths America Lives By* (Urbana: University of Illinois Press, 2003).
30. Roger D. Launius, "The Historical Dimension of Space Exploration: Reflections and Possibilities," *Space Policy* 16 (2000): 23–38; "National Aerospace Policy and the Development of Flight in America," in Tom D. Crouch and Janet R. Daly Benarek, eds., *1998 National Aerospace Conference: The Meaning of Flight in the 20th Century, Conference Proceedings* (Dayton, OH: Wright State University, 1999), pp. 280–85.
31. Aaron Cohen, "Project Management: JSC's Heritage and Challenge," *Issues in NASA Program and Project Management* (Washington, DC: NASA Special Publication-6101, 1989), pp. 7–16.
32. Howard E. McCurdy, *Inside NASA: High Technology and Organizational Change in the U.S. Space Program* (Baltimore, MD: Johns Hopkins University Press, 1993), pp. 11–98; Wesley T. Huntress Jr., V. I. Moroz, and I. L. Shevalev, "Lunar and Planetary Robotic Exploration Missions in the 20th Century," *Space Science Reviews* 107 (2003): 541–649.

Part I

Managing Planetary Science

Chapter 1

Homer Newell and the Origins of Planetary Science in the United States

John D. Ruley

The rise of planetary science as a recognized discipline became possible with the launch of the first planetary probes in the late 1950s and early 1960s.[1] In the United States, from the beginning of the Space Age until 1967, this effort blossomed at NASA under the leadership of a quiet former mathematics professor named Homer E. Newell Jr. who came to the American space agency from the Naval Research Laboratory (NRL). There, he had worked on upper atmosphere research using sounding rockets since the end of World War II, and then filled an important role as science coordinator for the Vanguard Earth satellite program of the late 1950s. He joined NASA less than a week after it opened its doors in 1958, as assistant director for space sciences in the new agency's Office of Space Flight Development (OSFD).[2]

At the time, there was no clear notion of planetary science as a separate scientific discipline—it was just one of several research areas that NASA expected to investigate under the mandate provided by the National Aeronautics and Space Act of 1958, whose provision to expand "human knowledge of phenomena in the atmosphere and space" covered a broad range of activities.[3] When considering planetary exploration, however, Newell realized that geology, meteorology, volcanism, climate and weather, and a range of other disciplines would be needed in NASA's effort to explore our solar system. Still, to the degree that any thought was given to a new discipline, it was the more general "space science." In his 1980 historical memoir *Beyond the Atmosphere: Early Years of Space Science*, Newell wrote that space science turned out to be inseparable "from the rest of science and the broad range of disciplines to which space techniques promised to contribute."[4] Nonetheless, as historian Joseph Tatarewicz pointed out in his 1990 book, a new discipline focused on study of the planets, variously called planetary astronomy, planetology, or planetary science, did emerge over time—as a direct result of NASA support through the various offices Newell headed.[5]

Newell's accomplishments during his time at NASA included:

- Creating the space science organization at headquarters, including a planetary science division.
- Establishing policies for robotic mission and experiment selection, and announcements of opportunities to include the academic science community.
- Successfully lobbying to create a planetary sciences section at the American Geophysical Union (AGU), becoming that section's first president.
- Representing NASA to the academic science community and the community to NASA.
- Exploiting the high priority granted to lunar projects after 1961 to accelerate development of launch vehicles and probes that benefitted both lunar and planetary programs.
- Identifying Mars as the planet in our solar system most likely to harbor life, and working to develop a capability to land scientific instruments there.
- Expanding NASA's astronomy program to include ground-based telescopes to support planetary missions, over the objections of his own staff.
- Supporting development of spin-scan imaging sensors—again, over the objections of his staff. Originally developed for weather satellites these were later used on deep-space missions to image Jupiter and Saturn.

These accomplishments were offset by one great failure: After his promotion to associate administrator in 1967, Newell became overfocused on internal NASA issues and failed to realize that the scientific community would not support a manned mission to Mars (figure 1.1). Relations between NASA and the planetary science community nearly came apart as a result, and took years to recover.

The story of how this happened involved a complex series of struggles between (and among) NASA's headquarters and field centers, the National Academy of Sciences, and the wider academic science community.

Planetary exploration became an issue for Newell almost immediately after he joined NASA; the first specific mission noted by Newell in one of his long series of green cloth-covered notebooks (now in the National Archives) was an early Mars probe. His major activity early on, however, was building a staff—he did so in large part by recruiting his former colleagues at NRL—and establishing NASA's policy for mission and experiment selection. This latter activity fundamentally affected planetary missions and all other space science activities at the agency from its origins to the present.[6] This almost immediately put him in conflict with the newly created Space Science Board (SSB) of the National Academy of Sciences.[7]

The SSB had been formed a few months earlier, when National Academy of Sciences president Detlev Bronk became concerned that the legislation authorizing NASA did not provide an adequate mechanism for scientific input. He discussed this with Hugh L. Dryden, research director of the National Advisory Committee on Aeronautics (NASA's administrative predecessor),

Figure 1.1 Homer E. Newell, NASA associate administrator of space science in 1967. (Credit: NASA.)

Advanced Research Projects Agency (ARPA) chief scientist Herbert York, and International Council of Scientific Unions president Lloyd V. Berkner during a meeting in June 1958. The following month, Berkner sent a telegram to scientists around the nation soliciting those interested in participating in space research, and received over two hundred replies.[8] Berkner then organized the board and a dozen committees covering topics including exploration of the Moon and planets.[9]

According to Newell's subordinate and eventual successor, John E. Naugle, in the fall of 1958 the SSB moved to assume "a major operational role" in mission planning and experiment selection, with NASA relegated to supporting the experiments and executing the missions selected by the board.[10] There was precedent for this—the Vanguard program of 1955–1958 had been run in that way, with the National Academy of Sciences in charge of experiment selection and NRL directed to support them. Newell, in his position as Vanguard's science coordinator, had the unenviable job of matching up the academy's changing requirements with NRL's launch vehicle—among other things he had to deal with a radical change in configuration when the academy decided a spherical satellite would simplify atmospheric density calculations. Since the original satellite design was bullet-shaped and could be launched without an aerodynamic shroud this change forced a major engineering effort to alter the upper part of the rocket.[11] The convoluted administrative structure at NRL responsible for making these changes to the rocket in response to outside scientific requirements clearly frustrated Newell. In the immediate wake of Sputnik 1, he responded to a request for comment from a congressional aid with a scathing letter in which he complained about "a tremendous amount of time" wasted on briefings and reports and called for "a permanent, competent, and adequate staff of scientists" inside the responsible government agencies "to provide leadership in basic and applied research."[12] He was not about to see the Vanguard approach applied at NASA. Accordingly, he worked diligently to control both rocket and satellite configurations in-house.

Newell learned of the SSB plans for an operational role at one of their meetings in the fall of 1958 where he represented NASA. It was followed by a meeting attended by Newell; NASA's director of space flight development, Abraham Silverstein; Vanguard director John P. Hagen, for whom Newell had worked at NRL; and John W. Townsend, a former deputy in NRL's Rocket Sonde Research Branch, at which what Naugle calls "two significant decisions" were made: In addition to supporting basic research, Newell's Space Sciences Division was tasked to "prepare scientific experiments and payload systems for sounding rockets, and scientific experiments for earth satellites and space probes," while Hagen's Vanguard Division received the "responsibility for the integration of scientific experiments from the Space Science Division as well as from outside groups into payload systems for satellites and space probes." Everyone recognized that this would create friction with the SSB, as it would inevitably put NASA's in-house scientists in competition with those from academia and industry.[13] This began what Newell called a "love-hate" relationship between NASA and the SSB.[14]

In addition, as director of space sciences at NASA Newell quickly found himself in conflict with the Jet Propulsion Laboratory (JPL), which was transferred from army to NASA control shortly after the new agency opened its doors in the fall of 1958. Much of JPL's staff was committed to developing planetary missions and had evolved a program called Vega that would combine a modified Air Force Atlas intercontinental ballistic missile with JPL-developed upper stages to launch JPL-developed space probes to the Moon and planets. While the army might have been content to allow JPL latitude to develop everything in-house, NASA chose to limit the lab to developing the scientific probes and to fly them on an Air Force Atlas-Agena launcher developed for the SAMOS spy satellite program.[15]

During 1959, Newell made decisions that exacerbated NASA's relations with leaders both at JPL and at the SSB. First, at the suggestion of astrophysicist Robert Jastrow, Newell met with University of California chemist and Nobel laureate Harold Urey. In his book *The Planets*, Urey had observed that the Moon's distorted shape implied that it was geologically dead—something that could be proved by placing seismographs on the surface. Up to this point, while the Moon was a natural focus of attention, mainly in terms of Cold War competition between the United States and the Soviet Union, Newell thought it held little scientific interest. Based on Urey's input, Newell recommended an urgent project to soft-land instruments on the Moon, and directed Jastrow as lunar project officer to accomplish it. The decision to oversee the lunar science program from some place other than JPL irked the lab's director, William J. Pickering, who had worked aggressively to transform the lab into a science organization. Now, it appeared that Newell had forestalled JPL primacy in this arena. They eventually reached an accommodation that allowed for oversight from Newell's organization at NASA headquarters while the majority of the project management and engineering work was conducted at JPL.[16]

At the same time, Newell's efforts received criticism because satellite launches failed far too often. This allowed those who disagreed with the direction Newell wanted to take the space science program license to criticize his leadership of the program. Shortly after its creation, NASA inherited both NRL's troubled Vanguard Earth satellite program and a pair of competing lunar probes—both called Pioneer—that had been run by the army and air force. Launch vehicles at that time were extremely unreliable, and during the year six out of ten NASA launch attempts failed, all of which led to negative reactions by the public and investigations by Congress.[17] Newell later referred to a "double standard," noting that NASA's failure rate was no worse than that of defense department programs in the same era, but "early launch failures made the entire [NASA] program a symbol for failure in the public mind." Thus, "success had to be sought on the first try, and every reasonable effort bent toward achieving that outcome." From those early failures and the public reaction came NASA's intrusive management approach, with overlapping layers including headquarters *program* managers and center *project* managers all working to supervise and monitor both industrial contractors and scientific experimenters—to the annoyance of all concerned.[18]

Relations with the academic science community, particularly with regard to selecting experiments for flight on NASA spacecraft, became a significant issue during this same period. Prior to 1960 experiments had been selected on an ad hoc basis. Since 1946, selection of experiments for flight on suborbital sounding rockets had been the responsibility of an unofficial, but government-backed, committee with a succession of names, most recently the Rocket and Satellite Research Panel (RSRP). The Eisenhower-era Vanguard Earth satellite program had been run as a joint effort by NRL and the National Academy of Sciences with Newell as science coordinator responsible for ensuring that the satellites met the academy's requirements.[19] The sounding rocket and Vanguard experiment selection processes were essentially collegial—the key experimenters were members of (or their organizations were represented on) the various panels and committees involved. But in the wake of Sputnik 1 this broke down under pressure to launch. According to Naugle, "Personal acquaintance, experience with rockets, the ability to get clearance to work with classified launch vehicles, and proximity to the manufacturer of the spacecraft, as well as the scientific merit of the proposed experiment, began to influence the selection of space scientists."[20]

One example of how this ad-hoc approach to experiment selection could work to the advantage of some scientists was the instrumentation aboard the first successful US satellite, Explorer 1. James Van Allen, who chaired the RSRP and was a member of the Vanguard Panel on Internal Instrumentation, had been working on a cosmic ray detector intended to fly on a Vanguard satellite. Concerned by what he saw as a low probability that Vanguard would work, he decided before Sputnik to hedge his bets and made arrangements so that his detector could be made to work on a smaller satellite developed by JPL for launch on an Army Jupiter-C booster. When the Eisenhower administration authorized the Army to launch what became the Explorer 1 satellite in the wake of Sputnik 1 and the spectacular failure of the first attempt to launch a small Vanguard test satellite, Van Allen's cosmic ray instrument detected the radiation belts that came to bear his name.[21]

Selection of experiments for the army and air force originated Pioneer probes that NASA inherited in 1958 and attempted to launch in 1959 was managed on an even more ad-hoc basis than other projects. The probes were built by Space Technology Laboratories (STL), an industrial contractor, which selected in-house scientist Charles Sonett as principal investigator.[22] According to Newell, their performance was not well thought of by other scientists:

> During the discussion of the space probes, several members of the [SSB] referred to STL's performance on the recent Pioneer operation as very poor. [Princeton chemistry professor and future Presidential science advisor Donald] Hornig described the science work as shockingly careless in its approach. Van Allen was less severe in his criticism but concurred that STL's performance was poor—"greenhorn" as he called it. [University of Chicago physicist John] Simpson was strong in his feeling that the STL work on the science package was poor. Specific complaints were that the checkouts of equipment such as the ionization chamber were incomplete and inadequate; not enough care was given to calibrations; not enough care was given to total systems integration and testing.[23]

Early in 1960, according to Naugle, Newell "gave up trying to use ad hoc arrangements to solve his space science issues" and asked former NRL electrical engineer Jack Clark to write out a coherent set of instructions. Clark was assisted by former NACA electrical engineer Morton Stoller, and the resulting document "became the equivalent of the Ten Commandments" for selection of flight experiments at NASA.[24] Technical Management Instruction (TMI) 37–1–1 specified the procedure used to select experiments, assigning to Silverstein as director of space flight programs responsibility "for overall direction of the NASA space sciences program," while a new Space Sciences Steering Committee (SSSC) appointed by the director was "responsible for the review and approval for submission to the Director," of programs, "experiments, experimenters and contractors," and "space science assignments for the Goddard Space Flight Center and the Jet Propulsion Laboratory."[25] Clearly, the chair of the Steering Committee would be an extremely powerful position. Silverstein formally assigned this task to Newell, who by this time had been promoted to deputy director of what was now the Office of Space Flight Programs, in NASA Circular 73, a follow-up to TMI 37–1–1. It also defined steering subcommittees on aeronomy, ionospheric physics, particles and fields, astronomy and lunar science, and planetary and interplanetary science—the last of which Newell chaired personally.[26]

The fact that Newell chose to chair the planetary and interplanetary subcommittee personally reflected the importance he gave it—and headed off yet another fight with JPL director William Pickering, who had objected to Jastrow's chairmanship of the lunar science planning group, since he worked for the Goddard Space Flight Center (GSFC), which Pickering saw as a rival to JPL.[27] By chairing the committee himself, Newell eliminated any such charge, but it is also clear that he considered planetary research of special importance. Since 1959 he had been working with Jastrow and UCLA geophysicist Gordon MacDonald to create "a 'Home' for planetary sciences" at the AGU. In their initial paper on the subject, Newell and his coauthors wrote that "the techniques to be employed [by planetary probes] are those of the earth sciences, and not those of astronomy." As historian Joseph Tatarewicz has pointed out, this staked out a position independent of the preexisting field of planetary astronomy.[28] Doing so was natural for Newell, as his work prior to NASA had focused on the use of sounding rockets (and eventually artificial satellites) to measure properties of the terrestrial atmosphere.[29] Newell's efforts were rewarded with the presidency of the AGU's brand new planetary sciences section (and his election as an AGU fellow) in 1962.[30]

While Newell was thus instrumental in both establishing a reasonable process for incorporating academic science in NASA missions and gaining academic legitimacy for planetary research, he still faced a significant problem in relations between NASA's engineering-dominated in-house culture and the outside scientific community. For example, in 1961 J. Allen Hynek (director of Northwestern University's Dearborn Observatory) wrote a two-page letter to NASA administrator James Webb complaining about what he felt was a discourteously brief reply to a proposal from his department for a study of balloons as a method of landing instruments on Mars: "We at Northwestern received no word as to how our proposal had been received, no comment as to its merits or demerits, and no comment on whether the proposal was completely out of the ball park." He added that NASA was treating "communications made by scientists to scientists within your organization as mere commercial proposals, bid on the open market."[31] Webb referred this to Newell, who recommended: "That all notices of NASA's rejection of a research proposal contain a general statement of the basis for rejection and an invitation to contact the appropriate Program Chief for details."[32] When Hynek did so, Lunar and Planetary Program chief Oran

Nicks confessed that "[p]articipation with the scientific community to the extent required in the NASA Lunar and Planetary Programs is somewhat of a new experience to many of us at NASA."[33] While Newell had scientific training and experience working in both university and government research settings, many of his subordinates (including Nicks and Newell's deputy Ed Cortright) were engineers from the old NACA who had little experience dealing with the scientific community.

The communication problems were not all one way: Hynek replied to Nicks with an invitation to "favor us with a colloquium" on the planned Mars probe.[34] While gracious, that suggestion was hardly compatible with the workload at NASA Headquarters at that time, where staff worked almost around the clock. Newell told a writer the following year that "[t]he people of this office are working most of the time—evenings, Saturdays and Sundays—to keep the program moving at the pace which we feel is necessary." He personally took two briefcases of material home on weekends.[35] This did not allow time to organize and attend academic colloquia—something that was eventually explained to Hynek over the telephone.[36]

Problems in relations between NASA and outside scientists were not limited to headquarters. In 1964 Newell took four pages of notes from a meeting with physicists Herbert Brooks of MIT and John Simpson of the University of Chicago—both experimenters on the Mariner probe to Venus (figure 1.2)—regarding JPL and its arrogance in dealing with academic scientists. The disgruntled physicists recited a litany of problems in JPL's relations with the scientific community:

> Unclarity as to whom to approach for decisions... Basic document continuously being compromised because [it] can't be met in time available... Too many people in the loop... Can't sign off on subcontract because of JPL indecision... Bickering over costs of experiment... Can't get response from JPL on major items of correspondence... JPL acts so as to effect choice of experiments by delaying input from potential contestants... Delays in testing at JPL prevent feedback of compatibility with info on parts for use in flight units... Erosive comments... Wouldn't go through this again.[37]

In spite of these problems and others—particularly the failure of JPL's first six Ranger lunar probes, which led to the dismissal of the lab's lunar program manager—JPL remained the lead center for most NASA planetary missions. There were exceptions: In the wake of multiple Ranger failures and what Newell and his staff regarded as inadequate progress on the Surveyor lunar soft-lander, JPL lost the Lunar Orbiter photoreconnaissance and mapping mission to the Langley Research Center, and despite JPL resistance Newell tapped the Ames Research Center to follow its successful Pioneer series of solar-orbit probes with two probes to the outer planets, which became Pioneers 10 and 11.[38] Significantly, the Ames program was approved in part because Cortright was "looking for a way to give some competition to the Jet Propulsion Lab."[39] This broke the JPL monopoly on planetary missions

Figure 1.2 The Mariner 2 probe to Venus was a huge success. This Tournament of Roses Parade float on January 1, 1963, commemorated this mission. The parade's Grand Marshall in 1963 was Dr. William Pickering, director of the Jet Propulsion Laboratory (JPL). The float had a prominent position in the parade in recognition of the success of Mariner 2, the first spacecraft to encounter another planet, Venus, less than a month earlier in December 1962. (Credit: JPL.)

as NASA's centers, especially Langley, Ames, and Goddard, all got pieces of the planetary exploration program. During the 1970s this was reconsolidated under JPL but later redistributed in the 1990s to various other organizations such as the Johns Hopkins University's Applied Physics Laboratory with the NEAR spacecraft. More recently, the Goddard Space Flight Center built the Lunar Reconnaissance Orbiter and thereafter worked on OSIRIS-Rex.[40]

Newell's problems with the planetary science community were further exacerbated by the slow development of adequate launch vehicles for deep space missions. Through most of the 1960s and 1970s, these missions depended on a high-performance, cryogenic upper stage called Centaur that began its troubled life as an air force project, and then was transferred to NASA's Marshall Space Flight Center in Huntsville, Alabama, where it was seen as an undesirable distraction from the Saturn launch vehicles being developed for Project Apollo. Newell's papers are rife with Centaur-related problems: In 1962 an air force decision to reduce the rocket's payload forced him to delete several experiments from a planned Mariner Mars probe. He wrote to the affected experimenters explaining that reduced booster power would force NASA to limit the payload "to very light, simple instrumentation and direct transmission to Earth rather than by use of a capsule-bus telemetry system."[41] Only a few days later, he went before a Congressional Committee to answer questions when the first complete Atlas-Centaur exploded less than one minute in flight. Later that year, after an exchange of memoranda and telephone calls among Newell, Cortright, Marshall Space Flight Center director Wernher von Braun, and his deputy Eberhard Rees, Centaur management was transferred from MSFC to the Langley Research Center. Langley and prime contractor Convair (later General Dynamics) eventually turned Centaur into a successful upper stage that was used for a wide range of lunar and planetary probes, but its first completely successful mission did not come until the Surveyor 1 lunar soft landing in 1966. While NASA waited for Centaur, planetary missions were limited to what was possible with the less powerful Atlas-Agena launch vehicles, and thus carried much smaller payloads—as little as 11 kilograms—to the frustration of experimenters.[42]

It is worth noting that NASA's robotic lunar and planetary programs were closely linked during Newell's tenure—a single program office headed by Oran Nicks supervised all lunar and planetary programs for most of the 1960s and a single Lunar and Planetary Missions Board provided outside advice to NASA. The linkage between lunar and planetary missions extended beyond program management and outside advice to the actual hardware. Early Mariner planetary probes were based in large part on the Ranger lunar probe design, and the Atlas-Centaur launch vehicle for later Mariner missions was also used in the Surveyor program. While lunar programs had the highest priority among NASA's robotic programs the close connection between early planetary and lunar missions meant that both benefitted from the DX procurement priority assigned Project Apollo. Accordingly, the robotic precursor missions of Ranger, Surveyor, and Lunar Orbiter enjoyed an overall priority comparable to the human Moon landing program and had access to funding and other resources that would not have been available otherwise.[43]

After the Moon, Mars loomed as the most desirable target for a landing, in no small part because many scientists—including Newell—believed that it might well support some form of life. He told attendees at a 1962 conference that Mars was "the most likely candidate" planet for life beyond the Earth, based in part on infrared spectra taken from balloons.[44] His opinion received a second from many other elements of the space science community. A Space Science Board report that same year called the planet "our most important biological objective."[45] Thus, landing a set of biological experiments on Mars received a high priority in Newell's space science organization. He pursued a program to look for life on the red planet named Voyager (completely unrelated to the later probes with the same name to the outer planets), which he viewed as the flagship mission for the Space Science Office and if successful in finding evidence of life on Mars believed could well be the most important scientific discovery of the Space Age.[46] While never flown, Voyager-Mars turned out to have a surprising impact on planetary science, especially planetary astronomy, in which Newell played a very direct role.

In 1963 Newell became aware of new Martian atmospheric density measurements more than ten times lower than previous estimates, which if correct would require a drastic redesign of Voyager's landing approach. In essence, the question revolved around whether or not the spacecraft could use a parachute to brake for a soft-landing in the thin atmosphere of Mars or whether it would require a landing rocket. A parachute would be less expensive, less sophisticated from an engineering perspective, and more foolproof but if the density of the Martian atmosphere was too thin a parachute landing system would not work. To nail down a precise value for the planet's atmosphere, Newell called a rare colloquium at NASA headquarters that fall, which failed to clarify the situation since participants offered a wide range of values. Worse, it became clear that the astronomical community on whom NASA depended for such information regarded the Martian atmosphere as a low priority for research. Frustrated, Newell told Nicks that he was prepared to fund a dedicated ground-based planetary observatory to solve the problem—taking funds from a planetary probe if necessary. This represented a radical turnabout in NASA's astronomy program, which until Newell's intervention concerned itself mainly with space-based instruments.[47] At Newell's instigation NASA funded planetary observatories at the University of Texas McDonald Observatory, and the University of Hawaii's observatory at Mauna Kea through Sustaining University Program facility grants to see what they could learn about Mars.[48]

None of this planning for a major NASA mission to land a spacecraft on Mars came to anything at this time, in no small part because Newell could not manage competing interests and divergent ideas among the planetary science community. In that instance, based on recommendations from planetary scientists NASA's Office of Space Science had formulated a $2 billion program in 1960s dollars to search for life on Mars. While some scientists supported the Voyager mission, many thought it too risky and expensive. A public dispute spilled into the Capitol before the general public. In the summer of 1967, because of conflicting testimony from scientists and a general shortage of funds due to the cost of the Vietnam War and the needs of the Great Society, infighting among space scientists prompted presidential and congressional questioning and eventually forced NASA to cancel the Voyager project.[49]

In the fall of 1967, frustrated by the congressional action and irritated by this strife, NASA administrator James E. Webb stopped all work on new planetary missions until the scientists could agree on a planetary program. Thereafter, the scientific community went to work hammering out a mutually acceptable planetary program for the 1970s. Retrenched and restructured, a program emerged that led to a succession of stunning missions throughout the 1970s, even as budgetary pressures and reduced political support remained. Newell and the space science community learned a hard lesson in practical politics from the Voyager fiasco. Most important, they learned to resolve differences through internal discussions, rather than public complaints to the media or in testimony before Congress. They also learned that while strong scientific support could not necessarily guarantee political support for a mission, lack of agreement among the space science community would certainly ensure a program's demise.[50]

Newell claimed credit for one significant personal accomplishment indirectly related to planetary science: In 1966, after his office had been expanded to cover both space science and applications satellites, he backed University of Wisconsin meteorologist Verner Soumi's proposal to fly a spin-scan sensor aboard the first advanced technology satellite in geosynchronous Earth orbit. As with funding for ground-based planetary astronomy, he did this over the objections of his own staff. Up to this point, NASA had used television for meteorology satellite imaging systems, but standard television at the time consumed excessive power and was limited to only 525 lines of resolution. The spin scan technique, by contrast, put a simple sensor on the side of a rotating spacecraft, and tilted the sensor to scan a separate swath of the Earth on each rotation. Over a period of 20 minutes, an extremely high-resolution image could be built up. Spin-scan was an immediate success and formed the original basis for the World Meteorological Organization's worldwide weather watch; but in the author's opinion it has significance far beyond that: It broke the hold of television and opened civil space platforms for the use of charged couple devices (CCDs) and other imaging sensors.[51] Instruments exploiting spin-scan technology were carried by the Pioneer X and XI probes, which provided the first close-up images of Jupiter and Saturn in 1973 and 1979, respectively.[52]

By then Newell was no longer in charge of NASA's robotic space program. In the wake of the 1967 Apollo 1 tragedy, in which three astronauts burned to death during what should have been a routine test, Newell was promoted to become the agency's associate administrator. This put him in a position where his responsibilities extended beyond robotic spacecraft and space science to include crewed space flight and aeronautics.[53] In his new role, Newell was far more focused on internal NASA matters than relations with the external scientific community, and this led to a near rebellion by academic scientists over planning for a crewed planetary mission in 1970.

A manned Mars mission had been the subject of numerous studies at NASA centers over the years, and at one point had been identified as "an extraordinary scientific opportunity" by the SSB,[54] but no such mission had been authorized during the 1960s. At a press conference on the eve of the Apollo-XI launch, however, Vice President Spiro T. Agnew—who chaired a presidential advisory group considering what NASA might do after Apollo—expressed "my individual feeling that we should articulate a simple, ambitious, optimistic goal of a manned flight to Mars."[55] This came as a surprise to NASA: Newell apparently learned of Agnew's comments from a television broadcast, and was unclear at first what time frame he had given for a Mars landing (it turned out to be the end of the century).[56] In the ensuing discussions, Newell attempted to play his usual role as NASA's ambassador to the space science community—among other things, seeking input from advisory committees. At this point opposition appeared: After a joint meeting of the Lunar and Planetary and Astronomy Mission Boards, Newell wrote that the chairmen of both boards "like the substance and content" but "worry about [the] realism and size" recommended for NASA's post-Apollo program, and that both chairmen "favor a pluralistic goal" rather than focusing on any single large-scale mission. Concerns were raised that manned space projects could "swallow up" funding for the "fragile" space science program. He added that both boards expressed "major concern over single goal" preferring a "triple theme: science—applications—human adventure."[57]

After hearing this, Newell should certainly have been aware that a major push for a single goal like Agnew's human Mars mission would not be greeted with cheers from the scientific community—yet in a *Science* article a few months later, he noted "remarkable unanimity" of opinion in support of a draft post-Apollo plan, with one notable exception: the timing (and by implication, priority) of human planetary missions. The same article quoted two scientists, University of Rochester biologist Rolf Vishniac and Lawrence Radiation Laboratory physicist Luis Alvarez, in support of a manned Mars landing, but also quoted University of California geophysicist Gordon MacDonald, a Lunar and Planetary Missions Board (LPMB) member, in opposition: "My intuitive feeling is that [human space flight] is going to be very limited."[58]

When the resulting plan was publically announced, with three of four options featuring a manned Mars mission as NASA's top post-Apollo priority, relations between the agency and the scientific community nearly disintegrated. Newell accepted some of the blame for this. In *Beyond the Atmosphere*, he wrote that as the plan was put together, "only hours were available for making hasty revisions." This created a problem with the LPMB: "In the course of one of these quick changes, part of the planetary program was modified. NASA people supposed that the change was in keeping with the desires of the [LPMB]—but it wasn't." That resulted in "talk of the members resigning in mass," not only because of the failure to notify them of a policy change affecting their part of the program, but also because the board had never been briefed on the overall package. The chairman told Newell "that if the [LPMB] had known of all the program possibilities that were being considered…some of the board's recommendations would have been quite different."[59]

What made this situation especially unfortunate is that there was no chance whatever of NASA receiving presidential authorization for a manned planetary mission: Agnew did not have President Nixon's support for the concept and was told as much before the plan was published.[60] It took several months for this to be realized by Newell and other top NASA leaders.[61] Based on his papers and correspondence, Newell spent his remaining years as NASA's associate administrator struggling to preserve as much of the space program as possible in a time of declining budgets, which forced the agency to make painful choices. Early in 1970, Newell sent a letter to the National Academy of

Sciences Physics Survey Committee, advising them that space physics would have to be "deliberately de-emphasized" due to budget constraints, resulting in fewer opportunities to fly even "small inexpensive" instruments.[62] He also alerted NASA's top management to the problem in relations with the scientific community, sending a "a Review of Present Ferment and Foment in the Scientific Community" to NASA administrator Thomas Paine.[63] Only a few months later, Newell learned of Paine's resignation while attending a meeting of the SSB, during which he noted, "A general background of concern over any large scale projects for science—must be highly productive" and "preference for smaller projects."[64]

By this time, despite functioning as acting NASA deputy administrator while the Nixon administration dithered over a permanent replacement for Paine, Newell may have been better connected with the scientific community than he was with his immediate superior, acting NASA administrator George Low. At the end of the year, when a controversy erupted between the SSB and NASA over the priority assigned to a "grand tour" of the outer planets, which the board initially supported but eventually opposed on grounds of excessive cost, Newell admitted in his notebook that he was "surprised" to learn the Grand Tour had been given top priority among NASA's robotic missions; he "had not been here during budget discussions—could have 'straightened Low out' on that."[65]

While Newell no longer had direct responsibility for the robotic space program, he remained interested in planetary missions. His notes included a briefing on the Viking Mars probes (which replaced the ill-starred Voyager) in 1971, and at least some participation in the discussions that led to installation of the famous plaque designed to commemorate our species that was installed on the Pioneer X and XI probes (the first human-made artifacts to leave the solar system). At a conference in Boston, Newell told *Christian Science Monitor* journalist Eric Burgess—who had originally suggested the probes carry such a message—"that the plaque was on the spacecraft, but there would be no publicity until the launch."[66]

Newell spent most of his last two years at NASA focusing on the space shuttle and Earth orbit missions, though he continued to attend scientific meetings—including a JPL colloquium on Mars.[67] At the end of 1973 he took early retirement, but continued to serve on advisory boards and wrote an excellent history of early US space science efforts as a contractor to the NASA history office. He died in 1983.

What are we to make of Newell's role in fostering the emerging discipline of planetary science? We might begin with the telescope on Mauna Kea that Newell insisted on funding to provide engineering data for future Mars probes—it was the first telescope to detect an object in the Kuiper Belt beyond the orbit of Pluto, which is representative of the revolution in planetary astronomy of the late twentieth century.[68] Newell deserves a large measure of credit for that, according to Smithsonian National Air and Space Museum historian David DeVorkin. At a 2008 NASA history conference, he showed a graph illustrating a stunning increase in the number of working astronomers that began in 1960, which he credited to NASA funding for space-based (and to a lesser extent ground-based) astronomy projects, and to Newell personally. According to DeVorkin, NASA provided well over half of all federal funding for astronomy projects between 1966 and 1972, as a direct result of announcements of opportunity for flight experiments and university grants issued over Newell's signature. This not only increased the number of astronomers but changed the character of the entire field: Space-based projects of interest to NASA tended to be physics-based. The traditional model of astronomy as an observational science based purely on images was replaced by one in which quantitative measurements were performed by interdisciplinary teams, and astronomy was only one of many fields affected by NASA funding.[69]

There was a cost to this: Beyond the discomfort of established astronomers (and other scientists) finding themselves almost lost in fields that changed beyond recognition in a matter of only a few years, the rapid peak and decay of NASA funding in the 1960s meant that more than a few young scientists completed their education only to find that positions were not available. DeVorkin quoted a 1972 report suggesting that an unemployment rate of 600 percent was possible among astronomers,[70] and Newell himself mentioned the problem in a 1971 interview.[71]

Despite the drop in funding after the 1960s, a new discipline clearly had emerged. Writing in 1990, Tatarewicz noted that the AGU's planetary science section had been joined by the larger American Astronomical Society Division of Planetary Sciences (DPS) in 1968. Tatarewicz identified ten universities granting PhDs in planetary science, and estimated the total number of researchers at several hundred by the mid-1970s.[72]

By establishing the space science steering committee, Newell created the procedures by which NASA evaluated proposals to choose experiments for flight aboard spacecraft, still used in 2009, according to NASA science directorate chief scientist Paul Hertz.[73] The instruments that measured the radiation flux around Jupiter, and the temperature of the acid clouds in the atmosphere of Venus, provided evidence for the presence of water on the Moon and Mars, and images going to the very edge of the observable solar system have, almost without exception, been selected after review by the Space Science Steering Committee.[74] And while no longer the only academic home for planetary science, the AGU's planetary sciences section still publishes papers and provides peer review for those working in the field.[75]

While most of NASA's budget (and public attention) were focused on the manned lunar landings, Newell and his staff managed an extremely active series of planetary missions in the 1960s that landed instruments on the moon, flew past Mars and Venus, measured interplanetary fields and particles, and laid the groundwork for landings on Mars and exploration of the outer planets in subsequent decades.[76]

The one blot on an otherwise tremendous record of accomplishment came after Newell's elevation to associate administrator in 1967. Up to that point Newell has served effectively as an ambassador between NASA and the academic science community. By 1969, though, he had become so focused on internal matters (particularly strategic planning) that he seemed to have lost touch with the scientific community. This had disastrous results, first when the LPMB rebelled over plans for a manned Mars mission, and later when the SSB turned out to oppose plans for an unmanned grand tour of the outer planets. Given the limited budgets available in the 1970s, it was probably inevitable, as Newell wrote, that some scientists were going to complain at having their pet programs cut—but the complete breakdown in communication at the level of the LPMB and SSB went far beyond that. Given that Newell was too busy with other matters to keep in touch with the scientific community, he should have delegated responsibility for that task to someone else at NASA. Unfortunately, he did not.[77]

That particular error must be weighed against an otherwise highly successful career, which set the stage for the phenomenally successful robotic space science missions of the late twentieth and early twenty-first centuries. In the three decades since the Moon landings, no human being has gone any further than Earth orbit, and no clearly defined application has been demonstrated that requires humans in space, other than for the purpose of studying human reactions to space flight (a circular argument).[78] In the same period, the early deep space probes launched under Newell's supervision have been succeeded by several generations of robots that have now visited every planet in our solar system except Pluto, and many moons.[79] That would please Newell. It would not surprise him.

Newell's personal journey in science began with looking up into the skies at the stars, and down at the rocks in streams and quarries of his native New England.[80] We can only wonder what that precocious but practical child would have thought if he had been told that he would spend his life helping to raise instruments above the atmosphere to get a better look at those stars and planets, and helping train the astronauts who brought home rocks from the Moon.

Notes

1. Most of the material in this chapter is from John D. Ruley, "The Professor on the Sixth Floor: Homer E. Newell, Jr. and the Development of U.S. Space Science," MS thesis, University of North Dakota, 2006.
2. Homer E. Newell, Jr., "Resume," April 1974, Record Group (RG) 255.3, National Archives and Records Administration (NARA), Archives II, College Park, MD.
3. "National Aeronautics and Space Act of 1958," July 29, 1958, in John M. Logsdon, gen. ed., *Exploring the Unknown: Selected Documents in the History of the U.S. Civil Space Program* (Washington, DC: NASA SP-4407, 1995), pp. 334–35.
4. Homer E. Newell, Jr. *Beyond the Atmosphere: Early Years of Space Science* (Washington, DC: NASA SP-4211, 1980), pp. 11–14.
5. Joseph N. Tatarewicz, *Space Technology and Planetary Astronomy* (Bloomington: Indiana University Press, 1990), pp. 116–34.
6. This is the story told in John E. Naugle, *First Among Equals: The Selection of NASA Space Science Experiments* (Washington, DC: NASA SP–4215, 1991).

7. Homer E. Newell Jr., “Staff Meetings Record,” December 2, 1958, RG 255.3, NARA.
8. Charles M. Atkins, NASA *and the Space Science Board of the National Academy of Sciences* (unpublished draft in the files of the NASA History Office), as cited in Naugle, *First Among Equals.*
9. G. K. Megerian, “Panel Report 1959–1,” January 29, 1959, RG 255.3, NARA, p. 4.
10. Naugle, *First Among Equals.*
11. Constance McLaughlin Green and Milton Lomask, *Vanguard—A History*, http://history.nasa.gov/SP-4202/chapter5.html and http://history.nasa.gov/SP-4202/chapter6.html (accessed May 14, 2011).
12. Homer E. Newell, Jr., Letter to Edwin L. Weisl, December 9, 1957, RG 255.3, NARA.
13. John E. Naugle, http://history.nasa.gov/SP-4215/ch4–2.html (accessed May 16, 2011).
14. Newell, *Beyond the Atmosphere*, p. 205.
15. Ibid., pp. 258–68; R. Cargill Hall, *Lunar Impact: A History of Project Ranger* (Washington, DC: NASA SP-4210, 1977), http://history.nasa.gov/SP-4210/pages/Ch_1.htm (accessed May 16, 2011).
16. Robert Jastrow, *Journey to the Stars: Space Exploration—Tomorrow and Beyond* (New York: Bantam Books, 1989), pp. 11–14.
17. Naugle, http://history.nasa.gov/SP-4215/ch5–1.html#5.1.2 (accessed May 16, 2009).
18. Newell, *Beyond the Atmosphere*, pp. 159–60.
19. Ibid., pp. 15–46.
20. Naugle, http://history.nasa.gov/SP-4215/ch1–3.html#1.3.3 (accessed May 16, 2011).
21. David H. DeVorkin and Allan Needell, “Van Allen, James” (Oral History Interview), February 18–August 8, 1981, National Air and Space Museum, Washington, DC, pp. 218–20.
22. Gideon Marcus, “Pioneering Space,” *Quest: The History of Spaceflight Quarterly* 14, no. 2 (2007): 52–59.
23. Homer E. Newell, “Conference Report, October 24–25,” quoted in Naugle, http://history.nasa.gov/SP-4215/ch4–2.html (accessed May 16, 2011).
24. Naugle, http://history.nasa.gov/SP-4215/ch6–2.html#6.2.1 (accessed May 16, 2011).
25. NASA TMI 37–1–1, “Establishment and Conduct of Space Sciences Program—Selection of Scientific Experiments,” April 15, 1960, RG 255.3, NARA.
26. NASA Circular 73, “Membership of Space Sciences Steering Committee and Subcommittees,” May 27, 1960, RG 255.3, NARA. Naugle, http://history.nasa.gov/SP-4215/ch6–3.html (accessed May 16, 2011).
27. W. H. Pickering, Letter to Dr. A. Silverstein, December 17, 1959, RG 255.3, NARA.
28. Tatarewicz, *Space Technology and Planetary Astronomy*, p. 119.
29. Newell, *Beyond the Atmosphere*, pp. 58–84.
30. “Past President Biographies: 1961–1980,” Washington, DC: American Geophysical Union, undated, http://www.agu.org/inside/pastpres_bios_1961–1980.html.
31. J. Allen Hyneck to James E. Webb, October 11, 1961, JPL Archives, Pasadena, CA.

32. Homer E. Newell, "Administrator's Briefing Memorandum: Letter of Complaint from Dr. Hynek, Dearborn Observatory, dated October 11, 1961," November 7, 1961, JPL Archives.
33. Oran W. Nicks to J. Allen Hynek, November 24, 1961, JPL Archives.
34. J. Allen Hynek to Oran W. Nicks, November 27, 1961, JPL Archives.
35. Shirley Thomas, *Men of Space: Profiles of the Leaders in Space Research, Development and Exploration, Vol. 5* (Philadelphia: Chilton Books, 1962), p. 131.
36. Roger to Oran Nicks, December 19, 1961, JPL Archives.
37. Homer E. Newell, "Notes: 30 Aug 62 to 13 Apr 64," RG 255.3, NARA.
38. William F. Burrows, *Exploring Space: Voyages in the Solar System and Beyond* (New York: Random House, 1990), p. 169.
39. Mark Wolverton, *The Depths of Space: This Story of the Pioneer Planetary Probes* (Washington, DC: Joseph Henry Press, 2004), p. 15.
40. Jim Bell and Jaceuline Mitton, eds., *Asteroid Rendezvous: NEAR Shoemaker's Adventures at Eros* (New York: Cambridge University Press, 2002); Howard E. McCurdy, *Low-Cost Innovation in Spaceflight: The Near Earth Asteroid Rendezvous (NEAR) Shoemaker Mission* (Washington, DC: NASA SP-2005 –4536, Monographs in Aerospace History, No. 36, 2005); R. R. Vondrak, J. W. Keller, and C. T. Russell, eds., *Lunar Reconnaissance Orbiter Mission* (Chichester, UK: Springer Praxis, 2010).
41. Homer E. Newell, memorandum to Robert E. Bottrdeau, May 4, 1962, quoted in Edward Clinton Ezell and Linda Neuman Ezell, *On Mars: Exploration of the Red Planet, 1958–1978* (Washington, DC: NASA SP-4212, 1984), http://www.hq.nasa.gov/office/pao/History/SP-4212/ch2.html (accessed May 16, 2011).
42. Eberhard Rees, "Centaur Programming Guidelines," July 3, 1962, JPL Archives; Edgar Cortright, telephone interview with John D. Ruley, June 13, 2007; Virginia P. Dawson and Mark D. Bowles, *Taming Liquid Hydrogen: The Centaur Upper Stage Rocket* (Washington, DC: NASA SP-2004–4230), pp. 59–90; Ezell and Ezell, http://www.hq.nasa.gov/office/pao/History /SP-4212/ch2.html (accessed May 16, 2011).
43. Newell, *Beyond the Atmosphere*, pp. 217–18; Ezell and Ezell, http://www .hq.nasa.gov/office/pao/History/SP-4212/ch2.html (accessed May 16, 2011); Hall, *Lunar Impact*, http://history.nasa.gov/SP-4210/pages/ Ch_13.htm (accessed May 16, 2011); Eugene M. Emme, comp., *Aeronautics and Astronautics: An American Chronology of Science and Technology in the Exploration of Space, 1915–1960* (Washington, DC: National Aeronautics and Space Administration, 1961), pp. 94–105.
44. Homer E. Newell, "What We Have Learned and Hope to Learn From Space Exploration," in *Proceedings of the NASA-University Conference on the Science and Technology of Space Exploration, Volume 1* (Washington, DC: NASA SP-11, 1962), pp. 59–78.
45. Space Science Board, *A Review of Space Research* (Washington, DC: National Academy of Sciences, 1962), pp. 1–11.
46. Ezell and Ezell, http://www.hq.nasa.gov/office/pao/History/SP-4212 /ch4.html (accessed May 16, 2011).
47. Tatarewicz, *Space Technology and Planetary Astronomy*, pp. 58–63; Ezell and Ezell, http://www.hq.nasa.gov/office/pao/History/SP-4212/ch4.html (accessed May 16, 2011).

48. Tatarewicz, *Space Technology and Planetary Astronomy*, pp. 63–85.
49. Michael A. G. Michaud, *Reaching for the High Frontier: The American Pro-Space Movement 1972–84* (Westport, CT: Praeger Publishers, 1986), http://www.nss.org/resources/library/spacemovement/chapter10.htm#n19 (accessed September 16, 2011).
50. Roger D. Launius, "Guest Blog Human Spaceflight on the Brink of Extinction What Might We Learn from the 1967 Planetary Science Crisis," *Space News*, July 21, 2010.
51. Gary Davis, "History of the NOAA Satellite Program," *Journal of Applied Remote Sensing* 1, no. 12504 (January 25, 2007): 9.
52. Eric Burgess, *By Jupiter: Odysseys to a Giant* (New York: Columbia University Press, 1982), pp. 38–47, 131.
53. Arnold S. Levine, *Managing NASA in the Apollo Era* (Washington, DC: NASA SP-4102, 1982), pp. 51–65.
54. Space Science Board, *A Review of Space Research*, pp. 1–9.
55. Science and Technology Division, Library of Congress, *Astronautics and Aeronautics, 1969: Chronology on Science, Technology, and Policy* (Washington, DC: NASA SP-4014, 1970), http://history.nasa.gov/AAchronologies/1969.pdf (accessed May 16, 2011).
56. Homer E. Newell, "AA Log of Phone Calls 28 May 1969—31 Dec. 1970 (Book II)," RG 255.3, NARA.
57. Homer E. Newell, "Notes: Aug. 67 to July 69," RG 255.3, NARA.
58. Luther J. Carter, "Post-Apollo: NASA Seeks a Mars Flight Plan," *Science* 165, no. 3897 (September 5, 1969): 987–91.
59. Newell, *Beyond the Atmosphere*, p. 219.
60. John Ehrlichman, *Witness to Power: The Nixon Years* (New York: Simon and Schuster 1982), pp. 144–45, as quoted in Jules Witcover, *Very Strange Bedfellows, The Short and Unhappy Marriage of Richard Nixon and Spiro Agnew* (New York: Public Affairs, 2007), pp. 66–67.
61. Homer E. Newell, "Notes: July 69 to May 70," RG 255.3, NARA.
62. Physics Survey Committee, *National Research Council, Physics in Perspective, Volume 2, Part 3* (Washington, DC: National Academy of Sciences, 1972–73), p. 976.
63. Robert W. Smith, with contributions from Paul A. Hanle, Robert Kargon, and Joseph N. Tatarewicz, *The Space Telescope: A Study of NASA, Science, Technology and Politics* (Cambridge, UK: Cambridge University Press, 1989), p. 195.
64. Homer E. Newell, "Notes: May 70 to Jun 72," RG 255.3, NARA.
65. Newell, "AA Log of Phone Calls 28 May 1969—31 Dec. 1970 (Book II)."
66. Burgess, *By Jupiter*, pp. 145–46.
67. Homer E. Newell, "Notes: Dec 73 to March 74," RG 255.3, NARA.
68. Barry Parker, *Stairway to the Stars: The Story of the World's Largest Observatory* (Cambridge, MA: Perseus Publishing, 1994), pp. 1–16, 39–62, 309.
69. David DeVorkin, "How Did Astronomy Change in Response to Sputnik?" Presentation at NASA 50th Anniversary History Conference, October 28–29, 2008, Washington, DC.
70. Ibid.
71. Robert Sherrod, "An Interview with Dr. Newell," July 22, 1971, Johnson Space Center Archives, Rice University Library, Houston, TX.
72. Tatarewicz, *Space Technology and Planetary Astronomy*, pp. 122–32.

73. Dr. Paul Hertz, chief scientist, Space Science Directorate, NASA HQ, email to author, April 20, 2009.
74. Naugle, http://history.nasa.gov/SP-4215/ch8–1.htm.
75. "Exploring All Things Planetary," Washington, DC: American Geophysical Union, undated, http://www.agu.org/sections/planets/ (accessed May 14, 2011).
76. Newell, *Beyond the Atmosphere*, pp. 327–369.
77. Ruley, "The Professor on the Sixth Floor," pp. 328–77.
78. William Langewiesche, "Columbia's Last Flight," in Dava Sobel, *The Best American Science Writing 2004* (New York: Penguin, 2004), p. 237.
79. Fernand Verger, Isabelle Sourbes-Verger, and Raymond Ghiradi, *The Cambridge Encyclopedia of Space: Missions, Applications and Exploration* (Cambridge, UK: Cambridge University Press, 2003), pp. 191–200, 225–73, 279–314.
80. Homer E. Newell, "Man and the Universe—Matter or Mind?" in Marcy Babitt, *Living Christian Science: Fourteen Lives* (Englewood Cliffs, NJ: Prentice-Hall, Inc., 1975), pp. 239–55; Homer E. Newell, "Rockhounding in the Space Age," Part I in *The Lapidary Journal* 29 no. 8, p. 1478.

Chapter 2

The Survival Crisis of the US Solar System Exploration Program in the 1980s*

John M. Logsdon

After almost two decades of spectacular successes in the US program of solar system exploration, 1981 was a year in which the program's survival was literally very much in question. Initial Reagan administration budget cuts, the cancellation of a previously approved planetary mission, and the unsuccessful attempt to gain White House support for a US mission to Halley's Comet eventually threatened the program with almost total termination.[1]

The National Aeronautics and Space Administration, under severe White House pressure to reduce its budget, identified the planetary exploration program as its lowest priority scientific activity, and said that it would drop the program entirely if forced to accept the budget reductions being proposed by the Reagan administration. Only in December 1981, and only on the basis of interventions having much more to do with institutional and political interests than with the scientific or societal merits of planetary exploration, was the program saved from termination. Although no new missions were approved at that time, this reprieve provided an opportunity for the solar system exploration community to rethink its program strategy and to gain support for the program's continuance, albeit with reduced expectations and at reduced budget levels. Those adjustments provided the foundation for the solar system exploration program of the 1980s and 1990s; however, it also meant a long pause in the program. The United States launched no missions beyond Earth orbit between 1977 and 1989; the first mission reflecting the revised exploration strategy was Magellan, launched in May 1989. (It should be noted that almost three years of this "mission gap" resulted from the grounding of the Space Shuttle after the 1986 Challenger accident.)

This essay traces the events surrounding this survival crisis of the US planetary exploration program. In the years since 1981, there have been both periods of relative stability and periods of uncertainty with respect to that program; many of the factors at play in 1981 have reappeared in subsequent years.[2] The solar system exploration program has continued to exhibit the many issues associated with the politics of program termination, retrenchment, and continuity, and with the process of setting priorities among different areas of big science.

Background to the 1981 Crisis

According to one account, in the period immediately following World War II, "only a handful of astronomers in the world were giving much attention to the local problems of the Solar System."[3] For most, the interesting scientific questions lay far beyond the solar system: in the stars, interstellar matter, other galaxies, and the large-scale structure of the universe.

That situation changed a decade later, when the progress of space technology and its links to high-priority political, military, and scientific goals in the United States and the Soviet Union made possible a large and ambitious program of space exploration. The initial impetus was provided by the Soviet launch of Sputnik 1 in October 1957 and the US reaction to that launch, including the creation of a new civilian space agency, the National Aeronautics and Space Administration (NASA). During 1958 and 1959, NASA planners were designing an American space program for the next decade that included exploration of the solar system by spacecraft traveling to the Earth's Moon and to at least the more accessible of the other planets, Venus and Mars.[4]

But those early mission planners discovered a void in contemporary knowledge of the solar system due to a half-century or more hiatus in ground-based study of the planets. This void made the design of a scientifically valid program of space-based lunar and planetary exploration very difficult. Faced with this situation, in the early 1960s NASA "responded with a multifaceted program that transformed the field. [Ground-based] observatories were constructed, instruments acquired, astronomers trained, research programs funded, and other activities supported."[5] In essence, NASA created anew the field of planetary astronomy by supporting a vigorous ground-based research effort; by luring other scientists, particularly geologists, into the field with generous research grants; by supporting the training of new planetary astronomers; and most fundamentally by offering scientists the opportunity to place their instruments on space missions traveling to the planets.

Accompanying the creation of a scientific community interested in solar system research was the involvement with NASA of a premier engineering organization to plan and carry out planetary missions. As part of the government reorganization that created a civilian space agency, in 1958 the Army's Jet Propulsion Laboratory (JPL), an institution that had been founded by Theodore von Karman and his associates, was transferred to NASA sponsorship. The laboratory had emerged during and immediately after World War II as a center of US competence in rocketry.[6] The JPL was (and is) formally a part of the elite California Institute of Technology (Caltech), and its employees are Caltech employees. Its status was that of a Federally-Funded Research and Development Center, a unique form of organization created so that top-level scientists and engineers could carry out government-funded work while not becoming government employees. The US Army was JPL's sponsor in its formative years; in late 1957 and early 1958, JPL engineers teamed with Wernher von Braun's rocket team at the Army's Redstone Arsenal and with University of Iowa scientist James Van Allen to develop and launch America's first satellite, Explorer I. After President Eisenhower transferred sponsorship of JPL from the army to NASA later in 1958, the laboratory quickly identified lunar and planetary exploration as that portion of the emerging civilian space effort most likely to provide the engineering and operational challenges it sought.

Although other NASA research centers occasionally became involved in solar system missions in the 1960s and 1970s, JPL and the associated scientific community it helped to nurture remained at the center of the US planetary effort for the next 20 years. The accomplishments of the US program of robotic solar system exploration became a hallmark of the US effort in space, second in public visibility only to the manned Apollo lunar landing program. From the initial Mariner spacecraft flyby of Venus in 1962, through missions that studied the Moon, Venus, and Mars, to the Viking landings on Mars in 1976 and the Pioneer and Voyager flybys of Jupiter and Saturn in the late 1970s and the 1980s, there was a constant flow of new data and spectacular images.[7]

Underneath this surface appearance of great success, however, was almost constant uncertainty about the future of the solar system exploration program. In 1967, Congress canceled a very ambitious and expensive mission to launch two automated spacecraft to land on Mars aboard a giant Saturn V Moon rocket; in response, NASA administrator James Webb ordered a complete rethinking of NASA's planetary exploration program. Out of that planning effort came many of the successful missions launched during the 1970s.[8]

Key to the long-term vitality of any area of space science is the flow of new data required to address outstanding scientific questions.[9] These new data come from a continuing series of missions, which, if the relevant scientific community has its way, are carried out on a schedule and in a sequence keyed to its priorities. "New starts," that is, approval to begin development of a new mission, are thus the lifeblood of a vigorous area of space science. Approval for new starts for solar system missions proved difficult to obtain during the 1970s, as the civilian space program's overall national priority and budget were reduced, as other areas of space science developed ambitious plans, and as a major new development program, the Space Shuttle, took an increasingly large share of NASA's available resources. Noel Hinners, head of the space science office in NASA's Washington headquarters, told Congress in February 1976 that the planetary program was on a "going out of business trend."[10] Indeed, between the 1975 peak of funding for the Viking project to land two spacecraft on Mars (a much less ambitious mission than the one canceled in 1967), and 1977, funding for the planetary program fell by a factor of four.

The planetary community did get one major new start early in 1977, when the outgoing Ford administration approved two ambitious space science missions, an Earth-orbiting large space telescope that eventually became known as the Hubble Space Telescope, and a complex spacecraft to orbit Jupiter and its satellites and to send a probe into the Jovian atmosphere. This latter mission, later named Galileo, would have a troubled history, largely due to its links with the Space Shuttle and other launch systems; four years later, in 1981, its proposed cancellation became the issue on which the fate of the planetary exploration program hinged.

There were no planetary new starts in the president's budget proposals in 1978, 1979, or 1980; by then, the journal *Science* reported "planetary science [is] on the brink again" and pointed out that planetary scientists faced "a difficult uphill battle in the next decade of selling less glamorous but scientifically vital missions with ever-increasing price tags." The Viking mission had cost over $1 billion, Galileo was estimated to cost at least $500 million, and the next mission waiting for approval, a spacecraft orbiting Venus to carry out a mapping mission using a powerful radar, also had a cost estimate of over $500 million. *Science* estimated that there were only "six hundred or so" planetary scientists in the United States; this was a dramatic increase from the few scientists of 20 years earlier, but still a relatively small group in the overall context of US science, and even of US space science.[11]

Adding to the uncertainty regarding the future of the planetary program at this point was an emerging conflict within the interested community over the appropriate strategy for gaining approval for future missions. On one side was the leadership of the overall space science community as well as of the planetary science community. The Committee on Planetary and Lunar Exploration (COMPLEX) of the National Academy of Sciences Space Science Board had developed a strategy for planetary and lunar exploration based almost solely on scientific merit.[12] Most planetary scientists believed that this science-based strategy should be their primary guide in assigning priority to and advocating for particular missions and should be followed by NASA in determining which missions to propose for funding.

On the other side were individuals like Bruce Murray, director of JPL beginning in 1976, and Carl Sagan, increasingly a public figure as well as a working scientist. Murray had argued from the time he took over JPL that to gain public and political support for future, expensive missions, they must combine both scientific and technical merit and "pizzazz"—that is, public interest. Top priority, argued Murray, should be given to missions that combined elements of *exploration*—the discovery of new places—with their scientific objectives. He argued that the cost of space science missions was so high compared to the costs in other areas of terrestrial science that there had to be broader justifications beyond scientific merit alone for undertaking such missions. This put him in opposition to the leaders of the space science community as represented by the Space Science Board, who argued that scientific merit should be the controlling factor in setting mission priorities. In the 1977–1981 period Murray became an unceasing advocate for a US mission to Halley's Comet during its 1985–1986 appearance, on the grounds that such a mission would capture public imagination as well as yield important scientific results.[13]

Many in the scientific community were skeptical of the realism of Murray's strategy, pointing out that NASA's string of glamorous firsts in the solar system could not go on indefinitely, but that a tremendous amount of good but less spectacular science remained to be done. These scientists preferred a course of action that counted on NASA, the White House, and Congress to provide the funds required to carry out that science, and provided the scientific community the authority to assign mission priorities primarily on the basis of scientific merit.

In the fall of 1980, there were signs that the approach of the scientific community might bear fruit. While Murray, after not getting approval for an ambitious mission to rendezvous with Comet Halley, also had been unsuccessful during the year in convincing NASA to insert a hastily conceived Halley flyby mission in its plans, NASA had put forth as its top scientific priority a "new start" on the Venus Orbiting Imaging Radar (VOIR) mission. This was the mission that had been on the top of the planetary scientists' wish list for the past three years; it was based on the use of a large and powerful synthetic aperture radar instrument to penetrate the clouds constantly shrouding the Venusian surface, so that the planet could be mapped with resolution of better than 100 meters. The Carter administration announced a few days before the presidential election (in an apparent attempt to win a few votes in California) that it intended to include VOIR in its fiscal year 1982 budget, which would go to the Congress in January 1981. Even after Carter's defeat, the mission stayed in the final Carter budget proposal to Congress, and for a few months at least most of the planetary community thought its future less uncertain.

New Administration, New Priorities

A budget submitted by a defeated administration has only limited significance until its contents and underlying philosophy are validated by the incoming president and his associates. In 1981, this validation did not occur. Ronald Reagan had won in a landslide by promising, among other things, to reduce federal spending and to redirect government priorities. To help implement that goal, one of president-elect Reagan's early decisions was to designate as director of the Office of Management and Budget Representative David Stockman, a tough-minded and extremely competent 34-year-old Congressman from Michigan.

Stockman was a fiscal conservative, and with vigor began identifying areas of the federal budget for reduction. Rumors of a Stockman "black book" that contained a draconian list of proposed budget cuts quickly began circulating in Washington, and when the Reagan revision of the FY 1982 budget was sent to Congress on February 17, those rumors were confirmed. Overall, $41.4 billion in budget cuts were proposed, with areas such as social and urban programs bearing the brunt of the reductions.[14]

In this context, the NASA budget fared fairly well. The proposed Carter FY 1982 $6.7 billion budget for NASA was reduced by $604 million, but this amount still represented an 11 percent increase over the FY 1981 budget. Most of the additional NASA funds went to the troubled Space Shuttle program. However, the Office of Management and Budget required NASA to cancel one of its three approved space science missions, the Hubble Space Telescope, the Galileo mission to Jupiter, or the US spacecraft that was part of International Solar Polar Mission, a joint US/European project to send two spacecraft over the poles of the sun. NASA chose to cancel its part of the Solar Polar project, greatly angering its European partners.[15] This meant that the planetary program had avoided the immediate prospect of Galileo, its only approved mission, being canceled. But the new start for VOIR was rescinded, and there was no commitment from the White House to any vision of the future of the country's efforts in space.

Much worse was to come as the year progressed. It was some time before key administration positions relevant to the nation's space program were filled. The Reagan administration did not announce its choices for NASA administrator and deputy administrator until April 23, and a presidential science adviser was not named until May 19. These were among the last major administration appointments to be announced. While the planetary science establishment and its allies within NASA were trying to develop a strategy for convincing the new administration to provide adequate support for the future missions they favored, Bruce Murray continued his personal campaign to gain White House approval for a flyby mission to Halley's Comet.[16] To make his argument, Murray went outside NASA channels to lobby Congress, the White House, and the media, in the process alienating both NASA management and those who believed that the integrity of the space science program, as certified by the leaders of the scientific community, was its primary political asset. At a time when the planetary science community needed to be unified in order to withstand threats to its continued support, Murray's activities as the head of the NASA facility charged with managing the planetary program and the associated campaign for the Halley mission mounted by The Planetary Society (the public membership group founded by Murray and Carl Sagan) were causing significant divisions among those interested in solar system exploration.

Threats to the program were more likely to be acted on than anyone except the top management circles of NASA and those handling space budgets within the Office of Management and Budget recognized. The new NASA administrator was James Beggs, who had been a senior manager at NASA in the late 1960s, then an executive in the aerospace industry; he had extensive Washington experience. He and especially his wife Mary were also well-connected politically to the upper levels of the Reagan administration. The new NASA head was disturbed by the FY 1983 OMB budget target for NASA of $6.5 billion, which had been given the agency in March, and on August 17 he wrote Presidential Counselor Edwin Meese's deputy, retired admiral Robert Garrick, that "I have come to the conclusion that some fundamental policy decisions need to be made before we can formulate the FY 1983 budget." Beggs pointed out that NASA was committed to doing three things: (1) completing the Shuttle program; (2) maintaining a space science and exploration program; and (3) maintaining an aeronautical research program. In his view, "given the current budget numbers,...we cannot continue to do all these things simultaneously." Beggs indicated his preference was "to cut out one of these activities and for this we need policy guidance."[17]

Beggs was to repeat this request frequently in the next several months, but the only vehicle through which policy guidance was provided was the budget process. Ignoring the early OMB target of $6.5 billion, on September 15 NASA submitted to OMB a FY 1983 budget request of $7.572 billion in new budget authority and $7.186 billion in budget outlays. Beggs identified the cuts that could bring the new budget authority down to $7.1 billion, but argued that a reduction below that level would require major cuts in the Shuttle program (which he knew were not acceptable to the White House) or "dropping out of one or more major program areas, such as planetary exploration." Beggs took an aggressive position (figure 2.1), refusing to give OMB a budget at a level less than $7.1 billion without first getting the policy guidance he had requested.[18]

The NASA budget request fell on unsympathetic ears. In fact, President Reagan told a nationwide television audience on September 24 that unless additional budget cutting measures were taken immediately, the Federal

Figure 2.1 President Ronald Reagan and NASA administrator James Beggs at the twenty-fifth anniversary celebration of NASA's formation in 1983. (Credit: NASA.)

deficit would increase to unacceptable levels. As one step, the president announced an additional 12 percent across-the-board cut in the FY 1982 budgets for nondefense government programs; the fiscal year was due to start in less than a week. This was not a propitious environment in which to argue for substantial budget increases for an agency such as NASA; NASA funding was in the part of the Federal budget that Reagan wanted to cut. In addition, it had become clear over the summer that additional funds would be required to keep the Shuttle on its planned schedule to achieve operational capability as soon as possible. That same day, David Stockman provided NASA its official budget target for FY 1983. Rather than the preliminary $6.5 billion target that NASA had already ignored, the ceiling was to be $6.041 billion in FY 1983 outlays, with an additional cut to $5.687 billion in outlays to come in FY 1984.[19]

NASA Sets its Priorities

Beggs's reaction to these low-budget targets was quick and sharp. He told Stockman on September 29 that meeting the OMB guidelines while maintaining "viable programs in some areas" would mean closing down "other major programs that NASA has operated since its inception."[20] The planetary exploration program was at the top of the list of efforts that NASA was "willing" to give up, if forced to accept major budget cuts. Beggs offered the following rationale:

> The planetary exploration program is one of the most successful and viable NASA programs. However, it is our judgment that in terms of scientific priority it ranks below space astronomy and astrophysics. Planetary exploration is much more highly dependent on launch vehicles, and it is our opinion that the most important missions that can reasonably be done within the current launch vehicle capability have, more or less, been done. The next step in planetary exploration is to do such things as landing missions and sample return missions, and these require full development of the Shuttle and the ability to assemble elements in earth orbit before sending the assembled spacecraft on its way.
>
> In our judgment, it is ultimately better for future planetary exploration to concentrate on developing the Shuttle capabilities rather than to attempt to run a "subcritical" planetary program given the current financial restrictions we face. Of course, elimination of the planetary exploration program will make the Jet Propulsion Laboratory in California surplus to our needs.[21]

This statement embodied the worst fears of the planetary community. The scientific payoffs from their work were assigned secondary priority, and their program's fate was tied to the Space Shuttle, rather than to the expendable boosters that had launched all planetary missions to date. If NASA's ties to JPL were severed, the engineering and operations teams required to carry out the complex missions of the future would be broken up or assigned to other, non-NASA work.

Influences on NASA Priorities

A variety of factors led NASA's leaders to single out the planetary program for potential termination. One was the fact that the planetary science community was relatively small compared to scientists working in space physics or space astronomy and astrophysics, and had not developed a position of influence within the space community. Reinforcing the higher status of the space astronomy and astrophysics community was the completion of the National Academy of Science's survey report *Astronomy and Astrophyics for the 1980's*, generally known as the Field Report after its primary author, Harvard astronomer George Field. This report gave highest priority within the overall area of astronomy to a series of Shuttle-launched, Earth-orbiting facilities known as the "Great Observatories," among which were the already approved Hubble Space Telescope and Gamma Ray Observatory. Another factor was that most future planetary missions then being proposed would indeed be very costly and likely to return less dramatic data and images than their predecessors. The divisions within the planetary community itself on future priorities and on scientific and political strategies limited its ability to maintain its funding priority within the NASA space science program.

Another important influence was NASA's strong commitment to the Space Shuttle as its only means for launching future space science missions. Projects in areas of space science such as astronomy and astrophysics seemed well-matched to the shuttle's capability to put heavy payloads into low Earth orbit, and such missions could be launched at almost any time. By contrast, planetary missions required a shuttle-launched upper stage to propel them from the shuttle's orbit to a deep space trajectory, and they had to be launched at widely spaced times called "launch windows" determined by the alignment of the planets. The mismatch between the requirements of planetary exploration and the capabilities of the Space Shuttle certainly contributed to a NASA preference for missions in other areas of space science.

The two top NASA officials, Administrator James Beggs and Deputy Administrator Hans Mark, seemed to have been following different approaches to priority setting at this time. At his confirmation hearings in June, Beggs had said that "the potential for stopping our planetary exploration program or putting large gaps in it is very disturbing to me. I think planetary exploration is a hallmark of the agency. It would be a disaster if we gave it up."[22] By threatening to terminate whole areas of activities if NASA were forced to take large budget cuts and by putting the planetary program on the top of the termination list, Beggs was playing budgetary hardball. The Shuttle program was sacrosanct due both to its association with the public appeal of humans in space and to its links to national security. The planetary program was NASA's only other widely known activity. In addition, it had its roots in Southern California, the home base of the president and many of his top advisers. Beggs's calculation was that shutting down the planetary program would not be an acceptable option to the White House, and thus that NASA would get a budget allocation adequate to keep going both the planetary program and other activities to which NASA had assigned higher priority.[23]

By contrast, the situation in September 1981, and his position in NASA's front office, gave Hans Mark the opportunity to put into practice some long-held views. Mark, who has his doctorate in nuclear physics, is an individual with wide-ranging interests beyond the technical arena and a relish for being provocative in ideas and actions. Mark and Beggs were not previously close associates and temperamentally were very different individuals. Mark had come to NASA from his position as undersecretary and then secretary of the air force and director of the National Reconnaissance Office during the Carter administration. In that role, he had been the chief defender of the Space Shuttle program within the Department of Defense at a time when Carter was considering canceling the effort. From 1969 to 1977, Mark had been director of NASA's Ames Research Center, and so was quite familiar with the agency's programs.[24] Mark had for some years been skeptical of the value of the NASA space science program. In 1975 he had written:

> In the last decade, the United States has spent on the average a half a billion dollars on space science. This budget is roughly equal to that of the National Science Foundation and I, personally, find it difficult to believe that we have a cultural or intellectual justification for continuing our space science effort at the same level for the indefinite future...
>
> My concern stems from the fact that unfortunately the results of space science to date have not been of major significance. While there have been a number of valuable findings, it is fair to say that no *fundamental* or *unexpected* discovery has been uncovered in the course of our exploration of the planets and the regions surrounding the Earth...

Mark did find one field of science that might meet his criteria for scientific excellence and to which observations from spacecraft might make important contributions. This was astrophysics, the study of stars and galaxies beyond the solar system.[25]

In addition to his views on space science, Hans Mark had long held a strong opinion on the appropriate future of the JPL. As Caltech had begun a search for a new JPL director in 1974, Mark had been asked if he wanted to be considered for the job. His reply was negative; he noted that "the basic problem faced by the laboratory is that it's [*sic*] purely NASA business [i.e., planetary exploration] will probably decline...It is absolutely essential for the health of the laboratory to seek new business opportunities in the most aggressive manner possible...The major opportunities for new business lie in the Department of Defense."[26] Mark doubted that such a redirection would be acceptable to Caltech and thus judged that he should not be a candidate for the JPL job.

These two themes—that planetary science was not of the highest priority and that JPL ought to apply its skills to defense and intelligence-related work—were interwoven in Mark's activities as NASA struggled with the need to cut its budget. In August 1981, Mark and his engineering assistant Milton

Silveira produced a document titled "Notes on Long Range Planning." In it they argued that making the Space Shuttle operational should "have the highest programmatic priority in NASA for the coming years to realize a return for this large investment" and that a space station "should become the major new goal of NASA." With regard to space science, "[I]n the coming decade, scientific investigations conducted in Earth orbit will be the most important because they will take advantage of the unique properties of the Shuttle." Finally, they concluded that "planetary exploration will be de-emphasized somewhat until we have a Space Station that can serve as a base for the launching of a new generation of planetary exploration spacecraft."[27]

As director of JPL, Bruce Murray, particularly in light of his lack of success in gaining approval for the Halley mission that he thought was essential to the future viability of JPL and the solar system exploration program, had also come to the conclusion that JPL had to seek other sources of support if it were to maintain its vitality as a premier technological organization (figure 2.2). He found in Mark a very receptive accomplice. On August 16, Mark wrote Murray that he wished "to encourage and support in every way possible your present efforts to expand JPL activity in Department of

Figure 2.2 Carl Sagan, Bruce Murray, and Louis Friedman, the founders of the Planetary Society at the time its organization. These advocates helped to lead the effort for an aggressive planetary science mission. The fourth person is Harry Ashmore, an advisor, a Pulitzer Prize–winning journalist and leader in the Civil Rights movement in the 1960s and 1970s. (Credit: NASA/JPL.)

Defense (DOD) space program activities, with the objective of sustaining JPL's unique capabilities by taking on work that is related to the strengths of the institution."[28] Two months later, in a handwritten note to Murray, Mark made his combined themes very clear:

> Where you and I have differed over the years is in our judgment of whether the popular support enjoyed by the planetary exploration program can be translated into the necessary long-term political support to assure a stable level of funding large enough to carry out what the planetary community thinks of as an adequate program. I have never believed that this could be achieved and I still do not believe that it can be done. It is for this reason that I have urged—and that I continue to urge—that the leadership at JPL must take immediate and aggressive steps to get a strong and stable defense-related program going at JPL. After having watched "big science" closely in the United States for almost three decades, there is no doubt in my mind at all that national defense is the only truly *stable* source of large research and development funds.[29]

Though for different reasons, Beggs and Mark clearly put the NASA planetary program in jeopardy by assigning it the lowest priority of all of NASA's major activities. In the ensuing several months, OMB was quite happy to accept NASA's ranking and to propose the planetary program's cancellation as a way of controlling NASA's budget, not only in FY 1983 but in subsequent years.

Disagreements in the Budget Process

Throughout October and November, James Beggs continued, unsuccessfully, to push for a meeting with OMB Director Stockman, Presidential Counselor Meese, and White House Chief of Staff James Baker (who collectively had been designated in July as the top-level Budget Review Committee). NASA resisted submitting a formal budget FY 1983 request until mid-October. Finally, it was agreed to use the September 15 NASA budget submission as the basis for OMB review. As mentioned earlier, that submission had requested $7.572 billion in FY 1983 budget authority. Included in the request was a proposed $276 million budget for planetary exploration; this amount provided $87 million for Galileo and $20 million for restoring the new start for VOIR.

NASA received its tentative budget allowance from OMB late in November. The overall budget had been reduced by $1.313 billion from the NASA request, to $6.259 billion. The planetary budget had been reduced to $118 million, and included no funds for either Galileo or VOIR.[30]

Beggs appealed the OMB allocations to Stockman on November 30, 1981. He told the budget director that "as someone who has devoted his entire professional career to working for American pre-eminence in space and aeronautics, I cannot accept the proposition that national economic imperatives compel the draconian funding reductions you have proposed on programs which have had such an extraordinary history of success." Beggs pointed out that he had "repeatedly asked for meetings with senior policy level officials in the Administration to resolve these policy questions." In his appeal, Beggs asked for restoration of full funding for the planetary program.[31]

Beggs's appeal set the stage for the final decisions on the fate of the planetary exploration program. The focal point for those decisions was the Budget Review Board, which scheduled a meeting on the NASA appeal on December 9.

In preparation for that meeting, NASA and interested Executive Office of the President agencies summarized their conflicting views in brief position papers. NASA argued that the scientific return of the planetary exploration program had been "extraordinary, and the implications for the future are boundless. Americans have taken enormous pride in the nation's planetary exploration endeavors which have been a true reflection of the greatness and vigor of the United States." The NASA appeal also pointed out that

> the precipitous reduction of activity at the Jet Propulsion Laboratory risks loss of a major national asset. It is our understanding that DOD is planning to increase their reliance on the Jet Propulsion Laboratory for assistance in development of advanced sensor systems for national security applications. An unstructured phase-down of JPL would result in the loss of the most talented members of JPL staff to the detriment of planned DOD activities.[32]

The OMB staff justification for the proposed budget cuts noted that "given the urgent need for fiscal restraint and noting particularly the high out-year cost implications, OMB staff believe that lower priority programs such as planetary exploration must be curtailed—even if they have been successful in the past." The OMB paper also noted "that the context in which NASA [in Beggs' 29 September letter] earlier provided an unsolicited statement that planetary exploration is of relatively lower priority than astrophysics and space-based astronomy has not changed." Canceling Galileo and not starting VOIR, estimated OMB, "could save about $1.2 billion."[33]

The advocates of canceling or deferring indefinitely the US planetary exploration program had gained an ally during the fall as Presidential Science Adviser George A. "Jay" Keyworth was put in charge of an overall review of US space policy and programs. Keyworth was a physicist who had been a mid-level manager at Los Alamos National Laboratory before coming to Washington. He was not well-known to members of the academic science community other than those who had worked on nuclear weapons or laser programs. Although both Keyworth and Hans Mark had close ties to Edward Teller, and although their positions on planetary exploration came to resemble one another, they were not personal friends, nor did they consult .with each other on their approach to space policy.

Keyworth initially had seemed sympathetic to arguments supporting both the scientific and political payoffs from planetary exploration,[34] and had taken an active and somewhat supportive role in the final stages of attempts to mount a US mission to Halley's Comet. But the *Washington Post* reported on December 2 that Keyworth "has recommended halting all new planetary space missions for at least the next decade—an idea he said the White House seems to be buying."[35] Keyworth's position, while resembling that of Hans Mark with its emphasis on using the Space Shuttle to support Earth orbiting astronomical and astrophysical facilities such as the Hubble Space Telescope and other "Great Observatories," also apparently recognized the possibility of a redefined, less ambitious, and thus less expensive planetary exploration program. He told *Aviation Week* that

> for years, some scientists who have been visionary enough to have seen this [budget limits] coming have been asking what type of planetary exploration could be done that is somewhat less expensive. What I wish to do is very much encourage the scientific community to start evaluating what can be done, so we can have a program that is balanced across planetary, astronomy and astrophysics, and solar and terrestrial. ...There is something special about planetary. It's more than science: it's exploration...It's a symbol of US leadership in science and technology. From that sense, I think keeping a healthy planetary program alive is important beyond just the bounds of science.[36]

What Keyworth most objected to was the high cost of the planetary missions that NASA was proposing, compared to their likely scientific returns. He noted that the United States had already done initial exploratory missions to most planets, and that a project like VOIR was just a "higher resolution experiment."[37]

Whatever his public stance, in his arguments to the Budget Review Board, Keyworth indicated that "I totally concur" with OMB's decision to cancel Galileo and VOIR, because those missions would "revisit the planets at much higher cost without commensurate additional scientific payoffs." He suggested that "the shuttle offers us a new capability to expand our horizons through...new astrophysical initiatives," and that "NASA is not in principle opposed to this philosophy. Their basic concern is over continued stability at the Jet Propulsion Laboratory." In summary, Keyworth indicated that "*the cut in planetary exploration represents an example of good management.* If 'business as usual' were to continue in planetary exploration, an unjustifiable increase in the overall space program would result."[38]

As the Budget Review Board meeting approached (it apparently was postponed to December 15), it appeared that NASA had few allies in the inner circles of the White House who could block the proposed budget cuts, and with them the end of a significant US program of planetary exploration. If help was to arrive, it would have to come from outside, or from the president himself.

Trying to Save the Program

Potential sources of support for the planetary program included the planetary science community, those in the public with a particular interest in solar system exploration, those in potentially influential positions in and out of government who had become Bruce Murray's allies as he tried to gain approval for a US mission to Halley's Comet, and similarly influential individuals whose primary interest was in the health of the California Institute of Technology. The planetary program had become identified with Caltech's JPL, and had brought worldwide attention and prestige to the university. In addition, the annual fee paid by NASA to Caltech for managing JPL had become an important component of the overall Caltech budget.

It did not take long for news of NASA's September 29 response to the OMB budget guidelines to reach the various elements of the planetary community. The *Washington Post* on October 6 reported that "NASA Weighs Abandoning Voyager,"[39] by now on its way to a 1986 flyby of Uranus after its August 1981 encounter with Saturn. *Aviation Week* in its October 12 issue reported that "termination of U.S. planetary spaceflight and closure of the Jet Propulsion Laboratory would be considered. This would include cancellation of the Galileo Jupiter-orbiter/probe...Shutdown of the NASA deep space tracking network, thus preventing data acquisition from Voyager on its 1986 Uranus and 1989 Neptune flybys, has been suggested."[40]

Coincidentally, the Division of Planetary Sciences of the American Astronomical Society was meeting during the week of October 12. This meant that the scientists who would be most affected by the termination of the planetary program were gathered in one place. Not surprisingly, their response was outrage. Eugene Levy, chairman of COMPLEX, the top scientific advisory body for solar system exploration, was particularly vocal. "At this moment," he commented, "not one of us knows whether, a year from now, the U.S. will have a program of solar system exploration." Levy continued,

> We are *not* faced with an invigorating, open-minded appraisal of where we are in our scientific investigations of the solar system. We are *not* seeing an administration eager to assess national scientific programs, and committed to moving forward vigorously with those that have particular intellectual, cultural and national importance. Instead, highly placed government officials *assert* that most of the important things in planetary exploration have already been done! They *announce* that "the era of planetary investigations is over!" Decisions are being made without serious study of the issues, without significant consultation with individuals and institutions that grasp the scientific questions, and with reliance instead on personal preconceptions. We may see important policy-level decisions, affecting major scientific activities of the United States, formulated at the whim of a few randomly placed people in the administration—people who are neither informed on these issues, nor sensitive to the importance of science and technology for our society in the large.[41]

At the meeting there was significant controversy over how to respond to the threat of program termination. While some thought it appropriate to be active advocates in favor of their area of science, others believed that the integrity of the scientific community would be compromised by such open advocacy.[42] All agreed that a letter reflecting the community's concerns should be sent to the most senior White House official identified as having policy responsibility for space, Presidential Counselor Edwin Meese. Accordingly, David Morrison, outgoing chairman of the Division of Planetary Science, and Carl Sagan, in his role as president of The Planetary Society, on October 14 wrote Meese "to ask your support to ensure the survival of planetary exploration in the United States." They argued that "a thousand years from now our age will be remembered because this is the moment we first set sail for the planets," and told Meese that "we and millions of Americans will appreciate any help you give to the enterprise of the planets."[43]

Meese's response was speedy and seemingly positive. On October 22, he wrote Morrison and Sagan, saying: "Your points are extremely well taken and will be definitely taken into consideration within budgetary limitations. Please know that this Administration is dedicated to the exploration of space, as [*sic*] have been the history of our nation. I have shared your concern with the President."[44]

Six weeks later, when the news that OMB had indeed recommended terminating the planetary program and that Science Adviser Keyworth was supporting the OMB position reached the science community, there was an attempt to organize a letter and telephone campaign to members of Congress, but no concerted approach to the executive branch by the planetary community. Pioneer space scientist James Van Allen attempted to have the Space Science Board of the National Academy of Sciences take the lead in protesting the proposed cuts,[45] but this would have meant that the board would be supporting a particular area of space science, something it had always been hesitant to do. Indeed, National Academy president Frank Press privately rebuked COMPLEX chairman Eugene Levy for appearing to speak for the academy in calling Science Adviser Keyworth "intellectually naive" regarding the scientific arguments underpinning the planetary program.[46]

Similarly, there was no organized campaign of public protest over the potential termination of the planetary program mounted in the October–December 1981 period. The vehicle for mobilizing public protest would have been The Planetary Society. The dramatic images from the Voyager flybys of Jupiter and Saturn, the high public profile of Sagan and his public television series "Cosmos," and an effective direct mail membership campaign had led to the society's membership mushrooming from twenty-five thousand to seventy thousand within a little more than a year. The Planetary Society membership had been mobilized in August 1981 for a letter-writing campaign in support of a US mission to Comet Halley. The White House received some ten thousand letters from society members; they were routed to NASA unopened, and never answered. The idea of another mobilization of The Planetary Society membership was considered in early December; one proposal was to send a mailgram calling for immediate protests to the White House to one-quarter of the society's members and a letter with the same message to all members. But the combination of the lack of payoff from the earlier campaign and the difficulty and costs of gathering enough support to influence executive branch decisions in the short run led to the abandoning of the idea for such a campaign. The Planetary Society did remain active in the behind-the-scenes efforts to rescue the program.[47]

Finding Powerful Allies

The final recognition that there would be no US mission to Halley's Comet had left Bruce Murray deeply concerned about the future of the JPL. He pursued two major lines of action with respect to ameliorating JPL's prospects. One was to gain Caltech faculty approval for a significant increase in Department of Defense support, including classified projects, for JPL. Murray, with Hans Mark's support, had been marketing JPL's capabilities to the Air Force Space Division and to the Central Intelligence Agency, with particular attention to satellite surveillance activities.[48] One obstacle to this campaign was the press reports that NASA was considering cutting its ties to JPL. Murray wrote Beggs in mid-October, telling him that "we have encountered people in DOD who are very concerned about continuing discussion of new DOD tasks with us because they surmise we are going to be declared surplus by NASA. They don't want to be involved in any action which somehow might lead them to become institutionally responsible for JPL." Murray asked Beggs for a public statement of NASA's intent to retain its ties with JPL, whatever budget cuts were made.[49] Murray received Caltech faculty approval, with little controversy, for increasing the DOD share of JPL's workload up to 25–33 percent at a October 20 meeting of the faculty.

Recognizing the uncertain future of JPL, the Trustees of Caltech in January 1981 had created a "Trustees Committee on JPL." That committee had a number of members of national reputation and influence; it was chaired by Mary Scranton, wife of former Pennsylvania governor and Republican presidential aspirant William Scranton, and herself an individual with high-level political connections. The Caltech Trustees Committee on JPL met for the first time on October 23, and approved Murray's plan to make JPL into an institution that maintained its primary affiliation with NASA while taking on significant DOD work. Key to that plan, of course, and thus to JPL's stability as an institution, was maintaining a significant planetary mission workload; if that objective were not achieved, noted Murray, "JPL could become an unintended casualty in the rearrangement of federal priorities."[50]

Concern over JPL's future had already been brought to White House attention by Arnold Beckman, chairman of Beckman Instruments and a Caltech Trustee. Beckman had written Edwin Meese on October 5, saying that the NASA response to administration FY 1982 and FY 1983 budget cuts "threatens to create total chaos and a rapid disintegration of a 5,000 person, $400 million Southern California enterprise...There are obvious implications to the support of the President and to his Party should the Administration permit such a catastrophe to take place."[51]

As reports of OMB's budget recommendations surfaced in early December, Beckman (at Bruce Murray's urging) once again wrote Meese, saying that he could not "emphasize strongly enough the gravity of such a decision [to cancel Galileo] and its negative effect on JPL and the California Institute of Technology." Beckman urged Meese "not to allow the emasculation of the technical and scientific capabilities of the Jet Propulsion Laboratory."[52]

Similar letters of support for JPL and the planetary program were sent to Meese by conservative California Representative John Rousselot (the letter had been drafted by JPL) and by Thomas Pownall, president of Martin Marietta. Pownall argued that his company's work on solar system missions had convinced him "that we and our planetary program associates and competitors have enhanced significantly our ability to satisfy the critical needs of our primary Aerospace customer, the Department of Defense, because of the extraordinary challenges we have met and managed and the disciplines we have developed in the process."[53]

Most active of JPL's politically connected supporters at this point was Mary Scranton. She reported to Bruce Murray that Keyworth's early December public statements on canceling the planetary program had provided "a rallying point around which to arouse interest and sympathy." In response, Scranton contacted Senators Charles Percy, Charles Mathias, and Mark Hatfield and Vice President George H. W. Bush. She also spoke with Fred Bernthal, top assistant to Senate Majority Leader Howard Baker. Scranton reported that the vice president had already been briefed on the JPL situation by prominent California Republican Robert Finch, and that she had only asked Bush to "look at the political problem that cancellation of such program might bring to the Republican party in the future."[54]

Caltech president Marvin (Murph) Goldberger made an early December trip to Washington in support of JPL. He met, among others, with a group of senators interested in the planetary program and other Caltech activities. In particular, Goldberger urged Howard Baker to express his support for a continued program of planetary exploration.[55] Goldberger was a Democrat, and although he had good connections with the liberal Republicans in the Senate, he had limited ability to influence the conservative Californians in the Reagan inner circle.

The various approaches to Senate Majority Leader Baker bore fruit. On December 9, he wrote President Ronald Reagan in support of Galileo saying, "I urgently request that $270 million be restored to the NASA budget for FY 1983 to continue the Galileo mission as originally planned."[56]

Although Baker may have had his budget figures wrong (the proposed FY 1983 budget for Galileo was actually $87 million), the political impact of his intervention was decisive. Baker originally intended to hand his letter directly to the president on December 9, but did not do so. So he called the White House on both December 9 and December 10 to make sure that President Reagan had indeed seen the letter and to "underscore his interest." Baker stressed that the letter was not "a pro forma request nor a matter of parochial Tennessee interest." Rather, Baker indicated that "he personally feels strongly about this issue." The Baker letter was routed to David Stockman for action; it is not clear whether in fact it ever reached the president.[57]

The Budget Review Committee met on December 15. Science Adviser Keyworth took the lead in suggesting a compromise in which $80–90 million would be added to NASA's planetary exploration budget in order to avoid the cancellation of the Galileo mission. This alternative, noted Keyworth, "would permit the stability and excellence of the Jet Propulsion Laboratory to be continued." The Budget Review Board asked NASA "to consider this alternative and report back immediately." It also hoped that OMB and NASA could settle the issue and that "an appeal to the President . . . be avoided."[58]

As a result of Baker's intervention, the immediate possibility of the demise of the US program of solar system exploration had passed. But the program had hung on by its fingertips; for the fifth year in a row, no new planetary mission was approved, for no funds for VOIR were restored to the NASA budget. What was gained was a year's breathing space, and the opportunity for NASA and the planetary community to come forward with a program that could gain the support of the Reagan administration.

Redesigning the Planetary Exploration Program

The planetary exploration community, both within and outside of NASA, was prepared to take advantage of its reprieve from summary termination. In late 1980, then NASA administrator Robert Frosch had approved the creation of a Solar System Exploration Committee (SSEC) as an ad hoc subcommittee of the NASA Advisory Council. The SSEC was to have a two-year lifetime (November 1, 1980–October 31, 1982), was to include as members representatives from all space science and technical disciplines interested in planetary exploration, and was to develop a strategy to encompass solar system missions proposed for initiation in the 1985–2000 time period.[59] It was within the SSEC framework that the planetary program was restructured to become politically and financially acceptable to the Reagan administration.

The idea for a committee to rethink the planetary program came from John Naugle, who had been in charge of the NASA space science effort from 1967 to 1974. Naugle had retired from NASA in 1974, only to be called back to service by Administrator Frosch as the agency's chief scientist in 1977. Naugle was no stranger to the need for planetary program planning. As mentioned earlier, in 1967 NASA Administrator James Webb had canceled all of NASA's future planetary activities in a pique after Congress had rejected plans for landing two automated spacecraft on Mars using a Saturn V booster. Naugle's first assignment in 1967 as space science chief was to propose a new planetary program; to do that, Naugle worked with a Lunar and Planetary Mission Board composed of concerned non-NASA scientists and chaired by astronomer John Findlay of the National Radio Astronomy Observatory.

The results of the Lunar and Planetary Mission Board's activities from 1967 to 1971 formed the basis for the extremely successful planetary program of the 1970s, including the Pioneer missions to Jupiter and Saturn, the Mariner 9 mission to orbit Mars, the Viking Mars landers, and the Voyager spacecraft to fly by Jupiter, Saturn, and eventually Uranus and Neptune. As Naugle assessed NASA's situation in 1979 and 1980, he recognized that the plans developed by the Mission Board had been carried out and that, because there was no accepted NASA long-range approach to planetary exploration in the 1980s and beyond, each planetary new start proposal was being assessed on an ad hoc basis, and therefore was vulnerable to shifting political winds because it was not seen as part of an integrated strategy.[60]

Naugle convinced Tim Mutch, NASA associate administrator for Space Science, and Angelo (Gus) Guastaferro, head of the planetary program in NASA's headquarters, of the value of a planning process similar to that carried out a decade earlier. Mutch asked Naugle to chair the SSEC, which met for the first time on November 10–11, 1980. The committee was originally comprised of 13 members (other members were added during 1981 and 1982); Guastaferro served as its executive secretary.

A starting assumption for the SSEC was that the scientific strategy for solar system exploration just completed by COMPLEX would serve as the starting point for SSEC deliberations. This strategy assigned priorities to unanswered scientific questions regarding the solar system; its goal was making major steps in understanding the process by which the planets formed from the solar nebula, how they evolved with time, and how the appearance of life in the solar system was related to the chemical history of the system. The COMPLEX strategy did not translate top-priority scientific objectives into a particular set of planetary missions and then develop a strategy for their implementation; this was to be the purpose of the SSEC.

After an initial meeting in late 1980, the SSEC began its work in earnest during 1981; it was clear that it was operating in a very different environment than had the Lunar and Planetary Mission Board a decade earlier. The differences were emphasized in a June 1981 presentation to the SSEC by Don Hearth, who had been director of the NASA Headquarters Planetary Office at the time of the Mission Board activity. Hearth noted that, while both in 1967 and now in 1981 the planetary program was in a "going-out-of-business" situation and there was a lack of consensus on program content, compared to the situation a decade earlier:

- there was much greater competition for resources within NASA;
- the planetary program was no longer the dominant activity within the Office of Space Sciences;
- in 1967 very little of the solar system had been explored, whereas in 1981 there was a substantial record of achievement;
- in the late 1960s, there had been widespread recognition that the nation must have a planetary program;
- in the late 1960s, the position of the Soviet Union in planetary exploration had been much more challenging; and
- in the late 1960s, five NASA centers were participating in planetary activities—now only JPL was active.

Hearth noted that all of these factors, in addition to the hostile attitude of the Reagan administration, would make it difficult to gain approval for anything but low-cost planetary efforts.[61]

The overall approach of the SSEC to its assignment emerged relatively quickly. The committee met frequently (November 1980, and January, February, April, and June 1981) leading to a week-long "summer study" in August 1981. The work of the SSEC was supported by intensive studies carried out by NASA centers and contractors, particularly the JPL and NASA's Ames Research Center. The committee also "took testimony" from a variety of interested individuals such as physicist and visionary Gerard O'Neill, astronomer Carl Sagan, and *New York Times* science writer Walter Sullivan.

Two major conceptual issues were central to the SSEC discussion. One was whether to recommend a broadly conceived, "balanced" (a code word for giving roughly equivalent attention as objects of study to the inner planets, to the outer planets, and to comets and asteroids) approach to solar system exploration along the lines recommended by COMPLEX, or to pursue an approach that focused on a particular scientific issue or a specific solar system body such as Mars. The other was whether it was in fact possible to develop a scientifically valid strategy for solar system exploration that could be carried out at significantly lower cost than had been the case during the 1970s.

The question of a broad versus a focused approach arose at the first SSEC meeting; there was strong support for both approaches voiced by different members of the SSEC. In subsequent meetings, the committee considered various candidates for a focus, including:

- providing a basis for better understanding the Earth through the comparative study of the planets;
- providing a scientific basis for the future exploitation of near-Earth resources; and
- providing precursor information required to undertake subsequent manned exploration of Mars.[62]

At the January 1981 SSEC meeting, committee member Charles Barth of the University of Colorado raised the possibility that there existed productive planetary missions, each with limited objectives, that might cost approximately $100 million each; the SSEC asked Barth to develop his idea in more detail.[63] At its February meeting, the SSEC heard James Pollack of Ames argue for embedding such small missions in an "Explorer-type line" in the planetary exploration budget. In the Explorer program of lower-cost Earth-orbiting missions, each project was not treated as a "new start" requiring separate budget approval, but rather was funded out of an annual budget provided to the Explorer program overall.[64] Many SSEC members expressed "skepticism that any planetary mission of value can be undertaken" for less than $250–300 million and questioned "the receptiveness of the OMB and Congress to new level-of-effort line items such as Explorer." The committee did recognize, however, "the importance of thinking through the potentiality of relatively low cost specialized planetary missions."[65] (This idea foreshadowed by a decade the "faster, better, cheaper" approach to space science of the 1990s.)

At its June meeting, the SSEC heard a presentation by JPL's Don Rea, a committee member, on the lab's study of what were called "Mariner Mark II" missions. These had originally been identified as "targeted missions" but the name had been changed to associate the effort with the earlier Mariner program, since, "with its distinguished lineage," this name would provide "the connotation of modest cost with excellent return."[66] The goal of the study was to develop a capability for outer planet missions "characterized by reduced mission costs, cost-effective advanced technologies, high inheritance over 4–5 missions, and a requirement for chemical propulsion only."[67] The concept of a basic spacecraft for use in a variety of outer planet missions fit well into the overall SSEC approach (as well as providing for JPL a steady program of work) and was quickly adopted by the committee.

By its June meeting, the committee was able to reach consensus on a statement of a basic rationale for and approach to solar system exploration. The elements of that consensus were:

1. The fundamental motivation for the planetary program remains the broadly based exploration of our solar system that has produced a multitude of major discoveries during the last two decades. Beyond intrinsic exploratory rewards, this program continues to produce a rich harvest of scientific information…The exciting exploratory phase is far from complete.
2. We advocate a mix of missions varying in complexity and cost to pursue this program of exploration. Some of the objectives can be met by means of smaller focused or dedicated spacecraft. Others will require larger systems capable of returning samples to Earth. Within each of these categories, we are seeking…major cost savings by maximizing inheritance and minimizing the development of new systems for a given mission.
3. The SSEC should identify the relationship between solar system exploration and NASA's human activities in space. This orientation includes an interest in assessing the potential of mineral and volatile resources in the near-Earth environment.[68]

Gus Guastaferro left NASA Headquarters in April 1981; his position as executive secretary of the SSEC was assumed by Geoffrey Briggs, deputy director of the NASA Headquarters Planetary Office. At the June SSEC meeting, Briggs suggested to the committee an approach to dividing the missions required to achieve the goals of the planetary program for the rest of the century into three categories based on mission complexity and cost. These classes he described as Pioneer-class (least expensive); Mariner-class (of moderate expense); and Viking-class (expensive). Viking-class missions, suggested Briggs, should only be proposed if they could be tied to a "key date, such as the 500th anniversary of Columbus' discovery of the New World." The committee's reaction to Briggs's plan was "guarded."[69]

By the time of the 1981 SSEC summer study, which took place in La Jolla, California, from August 10–14, Noel Hinners had replaced John Naugle as SSEC chairman. Naugle was leaving NASA to work in industry and believed that he had been successful in getting the SSEC study started in a productive fashion; in addition Naugle had become ill over the summer and could not attend the La Jolla meeting. Hinners, like Naugle, was a former head of NASA's Office of Space Science and in 1981 was the director of the Smithsonian Institution's National Air and Space Museum.

The summer study resulted in an interim SSEC report. Developing that report meant assigning tentative mission priorities and adopting an overall programmatic approach. Two approaches were considered. One would identify a minimally viable core planetary program plus options that could be added to the core if resources were available; the other was the three-tier approach that had been suggested by Geoffrey Briggs, with more details provided by John Niehoff of SAIC, a NASA support contractor.

During its summer study the SSEC adopted the three-tier approach to classifying future missions and developed a core plan that included only Pioneer- and Mariner-class missions. Pioneer-class missions were estimated to cost between $100 and $150 million each; Mariner-class missions, from $300 to $500 million. The cost of most Viking-class missions was estimated at a billion dollars or more, and the SSEC recognized that there was no short-term chance of gaining political acceptance for such missions, however scientifically attractive they might be, in the budget climate of the early 1980s. As candidates as the initial Pioneer-class missions, the SSEC identified a Mars orbiter to locate water on the planet, a Mars geochemical orbiter, and a lunar geochemical orbiter; as the initial Mariner Mark II missions, a rendezvous with comet Tempel II, with an asteroid flyby en route, and a Saturn orbiter were proposed.

Even though Viking-class missions were considered too expensive to include in the core plan, SSEC members were unwilling to accept the idea that no such ambitious missions would be approved in the future. They wanted potential large, expensive missions to be identified and studied in enough detail to understand their scientific and exploratory payoffs, their technological requirements, and their likely costs.[70]

The results of the SSEC summer study were presented to the NASA administrator, to the NASA Advisory Council, and to the Division of Planetary Science of the American Astronomical Society. In addition, a brief summary of SSEC activities was published in *Science*.[71] In effect, then, the interim conclusions of the committee were widely known within the concerned community as the policy and budgetary conflicts over the future of the planetary program heated up in the October–December 1981 period.

Following its summer study, and now that the main outlines of its findings were in place, the SSEC planned to spend the year remaining in its charter refining its conclusions and involving a broader segment of the scientific community in its activities. To those ends, both JPL and Ames embarked on more detailed studies of missions that had been identified by the SSEC

(Ames studied only Pioneer-class missions; JPL both Pioneer- and Mariner Mark II-class missions), and four science working groups were established, on Outer Planets, Terrestrial Planets (Solid Body), Terrestrial Planets (Atmospheres), and Small Bodies.

The 1981 work of the SSEC thus created a basis for a new approach to planning and advocating planetary missions. The December White House decision not to cancel Galileo and thus future planetary exploration presented an opportunity to put that approach into practice.

A Future for Planetary Exploration

By the time of the early February 1982 meeting of the SSEC, acting NASA administrator for space science and applications Andrew Stofan could report that "policy makers now all seem to agree that NASA would stay in the planetary exploration business, where before they had favored taking NASA out of the business." Negotiations between OMB and NASA had resulted in a FY 1983 planetary budget of $154.6 million, an increase of only $36.6 million over the original OMB allocation. The Galileo budget was $92.6 million; no funds directly related to VOIR were included. To keep the overall planetary budget as low as possible while still funding the Galileo mission, funds for mission operations and data analysis (for ongoing missions) and for research and analysis (using data from completed missions) were significantly reduced.

Thus, while one approved planetary mission remained in NASA's future plans, the White House provided only minimal support for the planetary science community overall. The SSEC members were therefore not comforted by Stofan's message, arguing that what was being requested was still a "get out of the business" budget except for Galileo.[72]

Nevertheless, a corner had been turned. Science Adviser Keyworth as early as mid-December (in his interview with *Aviation Week*) had indicated that his real goal was to bring the costs of future planetary missions into line with other elements of the NASA space science program. On May 13, 1982, in a speech to a group of planetary scientists, Hans Mark modified his position, saying:

> You all know that I have raised questions about the relative priority and the value of planetary exploration when compared to other scientific missions in space. I still believe that such questioning is important and that we should, periodically, go through the exercise of looking critically at the relative, as well as the absolute scientific value of the work we do. You should also know, however, that as a public official responsible for assisting in the formulation of our space exploration program, I am thoroughly committed to continuing planetary exploration, not only for scientific reasons but also because of the fact that, in the long run, we are learning things that will eventually allow us to exploit the resources of the solar system. I believe that it is possible to structure a program of planetary exploration based on these justifications, and you have my personal commitment that I will work very hard in that direction.[73]

Given the willingness of the White House to continue the planetary program, it remained for NASA to decide which mission to propose for a new start in the next budget year, FY 1984. The SSEC was asked at its February 1982 meeting to recommend that mission, with the following guidelines: cost between $150 and $300 million; industry involvement in the project; and international cooperation if feasible. Candidates for a new start, thought NASA management, might be a Venus Mapping Mission, less expensive than VOIR, a lunar polar orbiter, or a Mars mission of some character.[74] Immediately following this discussion, the SSEC had its initial exposure to JPL's thoughts on a less expensive Venus Mapping Mission. John Gerpheide of JPL described a mission using spare hardware and a less costly approach to mission development and operations that could accomplish most of the mapping objectives that had been established by COMPLEX.[75]

This approach was very attractive to the SSEC, because it provided a way of achieving what had been for four years now the top priority scientific objective of the solar system program, mapping the surface of Venus at high resolution, but doing so at the relatively lower cost that the SSEC was arguing was possible. The committee asked that the Venus Mapping Mission be studied in more detail.[76] When those studies confirmed that the mission, now called Venus Radar Mapper (VRM), could indeed be carried out for less than $300 million, the SSEC at its June meeting endorsed the mission as the first new start in the restructured NASA program of planetary exploration.[77] NASA had already decided to put forward the mission as its top candidate for a FY 1984 new start in space science.

There was little controversy over the inclusion of VRM (the mission was later renamed Magellan) as the NASA budget underwent OMB review in October and November 1982. When President Reagan sent his FY 1984 budget to Congress in January 1983, the mission was NASA's only new start in space science.[78] *The New York Times* headlined its story "Plans to Explore Planet Revived," noting that "after years of steady decline, the nation's planetary exploration program appears to have been rescued by the Reagan Administration." *The Times* added that "the tactic of winning approval for new missions by designing lower-cost vehicles . . . may soon become a basic strategy," although "space agency officials emphasized that the Administration has made no commitment" to missions beyond VRM. However, "only a program based on low-cost missions, they said, stood much chance for the foreseeable future."[79]

Conclusion

There has been much discussion over the years about the need to set priorities among areas of science and among various proposed "big science" projects. Common to these discussions is the search for some framework or process within which to make the difficult choices among competing uses for scarce resources on some sort of objective basis.

The events described in this essay suggest that a different approach to priority-setting, one much more political in character than is preferred by leaders of the scientific community, actually operates. Government-funded activities create vested interests in their continuation, including both the interests of individuals and of institutions. In an environment of resource scarcity, these activities also give rise to alternative claimants who argue for a revision of the status quo and a redistribution of benefits. All interests attempt to persuade those with the power to allocate resources to favor their point of view. This is nothing more than a description of the American political process in operation.

What happened during the "survival crisis" of 1981 was a political struggle over the future of the US space science program and of the institutions through which it was carried out. The element of the overall program, which had been in ascendancy in the 1960–1975 period, the solar system exploration effort, put on a last ditch struggle to maintain that position. The planetary community had seen its share of the space science budget shrink during the second half of the 1970s, as scientifically attractive mission proposals were put forth by the astronomy, astrophysics, and solar-terrestrial physics elements of the NASA science program. An attempt spearheaded by JPL director Bruce Murray and scientist-author Carl Sagan to gain support for the planetary program on the basis of its exploratory character rather than solely on the basis of scientific merit failed when no US mission to Halley's Comet was approved. A different approach, putting forth a scientific strategy to underpin particular planetary mission proposals, also was unsuccessful in arguing that the scientific payoffs were worth the high costs of achieving them.

When two individuals, Hans Mark and George Keyworth, whose views on space science priorities meshed (although they were seldom in agreement on other issues), ended up in key positions in the Reagan administration, the battle over space science priorities was joined in earnest. The desire to reduce the NASA budget and the continuing high budget demands of the Space Shuttle program provided the background for arguments that other areas of space science should be given priority for the time being. The American political process—even the inner workings of the White House and the Executive Office—is open to scrutiny and engagement by those strongly concerned with particular policy, institutional, and budget decisions. Thus the stakes in the December 1981 budget appeal process were known to all parties, and those who stood to lose from the likely outcome mobilized to protect their interests. In doing so, they attempted to forge useful alliances with all possible sources of influence on the decision process.

In 1981 the key intervention was made by a powerful Congressional leader. What his involvement demonstrated was that an action potentially justified on other grounds must also be politically acceptable, if it is to gain both White House and Congressional approval. The White House became convinced that eliminating or indefinitely postponing the planetary exploration program, even if it made sense in scientific or programmatic terms, was not going to be accepted by key actors in Congress; in addition, other concerned actors in industry, academia, and other relevant communities made known their unhappiness with respect to such a course of action.

The decision process then turned to finding an approach that was politically acceptable and still achieved the key objectives of adjusting priorities and controlling budget growth. It was the good fortune of the planetary community that it could quickly bring forward a responsive alternative, in the form of the interim conclusions of the SSEC. The SSEC not only had developed what was thought at the time to be a lower-cost approach to planetary exploration, but also had done so through the involvement of key leaders of the concerned scientific community. Thus the SSEC approach had both substantive and political utility; it also, however, meant that the kind of large-scale exploratory missions advocated by Murray, Sagan, and The Planetary Society would in the future face an uphill battle to gain NASA and White House approval.

None of this was neat, and it resulted from particular individuals occupying particular positions at particular times. But the end product clearly was a shifting of priorities for the 1980s among areas of space science, away from solar system exploration toward astronomy and astrophysics. That the "Great Observatory" missions being proposed by the astronomy and astrophysics community were a better fit to the capabilities of the Space Shuttle, NASA's top priority effort, reinforced this shift.

Even with this shift, the NASA program of solar system exploration of course continued. The SSEC published its initial report, *Planetary Exploration Through the Year 2000: A Core Program*, in 1983. That report identified, in order of priority, the following initial missions in a recommended core program:

1. Venus Radar Mapper
2. Mars Geoscience/Climatology Orbiter
3. Comet Rendezvous/Asteroid Flyby
4. Titan Probe/Radar Mapper

The four missions that the SSEC recommended in 1983 had different fates, but in combination they served as the basis for NASA's solar system exploration program until that program was once again "reinvented" in the "faster, better, cheaper" period beginning in 1992.[80] The Venus Radar Mapper, which became Magellan, was a total success. The Mars Geoscience/ Climatology Orbiter, which became known as Mars Observer, failed as it arrived at Mars in 1993. The Comet Rendezvous/Asteroid Flyby mission was first combined with the Titan Probe/Radar Mapper mission in an attempt to argue that the two missions could be flown for the 150 percent of the cost of a single mission. This argument was not successful, and the Comet Rendezvous/Asteroid Flyby mission, known as CRAF, which was to be the first of the Mariner Mark II missions, was canceled. This allowed additional resources to be allocated to the Titan Probe/Radar Mapper, which became the Cassini-Huygens mission to explore the total Saturnian system, not just Titan. That mission was finally launched in October 1997 and as of this writing continues to send back data about Saturn and its moons.

Key to the SSEC recommendations was the argument that solar system exploration should be treated as a coherent program, not as a series of separate missions. Such a program, estimated the SSEC, could be sustained at a total budget level of about $300 million per year.[81] This recommendation was never put into practice, and in the subsequent decades the planetary program has continued to struggle for resources both within and external to NASA. The factors at play in 1981 in shaping the program have remained influential, although there has never again been a threat to totally terminate the effort. The reality that political considerations will always be an important factor in the approval of publicly funded activities such as space science is still resisted by the scientific community, but nonetheless persists.

Notes

*This chapter is a revision of an account originally prepared for the NASA History Office in 1989, but never before published. I want to thank James Beggs, Hans Mark, Bruce Murray, Louis Freidman, and Sylvia Fries for their comments on the original version of the chapter.

1. For an account of attempts to gain approval for a mission to Comet Halley, see John M. Logsdon, "Missing Halley's Comet: The Politics of Big Science," *ISIS* 90 (June 1989): 254–80. Many of the conflicts described in this chapter were first manifested in the attempts to gain support for the Halley mission.
2. See, as an example of later program turbulence, Scott Hubbard, *Exploring Mars: Chronicles from a Decade of Discovery* (Tucson, AZ: University of Arizona Press, 2012).
3. Fred Whipple, "Discovering the Nature of Comets," *Mercury*, January–February 1986, p. 5.
4. For accounts of early planning for lunar and planetary missions, see Edward Clinton Ezell and Linda Neumann Ezell, *On Mars: Exploration of the Red Planet 1958–1978* (Washington, DC: NASA SP-4212, 1984).
5. Joseph N. Tatarewicz, "'A Strange Plea'—The Campaign for Planetary Astronomy in Support of Solar System Exploration, 1959–1962," in National Air and Space Museum, *Research Report. 1985*, p. 92.
6. For a discussion of the history of the Jet Propulsion Laboratory (JPL) through the mid-1970s, see Clayton Koppes, *JPL and the American Space Program* (New Haven: Yale University Press, 1982). For JPL history from the time that Bruce Murray became its director in 1976, see Peter J. Westwick, *Into the Black: JPL and the American Space Program, 1976–2004* (New Haven: Yale University Press, 2006).
7. The Pioneer mission was managed by NASA's Ames Research Laboratory, not JPL.
8. For an account of this rethinking, see Amy Paige Snyder, "NASA and Planetary Exploration," in John M. Logsdon, gen. ed., with Amy Paige

Snyder, Roger D. Launius, Stephen J. Garber, and Regan Ann Newport, *Exploring the Unknown: Selected Documents in the History of the U.S. Civil Space Program, Vol. V: Exploring the Cosmos* (Washington, DC: NASA SP-2001–4407, 2001), pp. 280–85.

9. See the 1986 report of the NASA Advisory Council, *The Crisis in Earth and Space Sciences*, for a discussion of this issue.
10. US Congress, Senate Committee on Aeronautical and Space Sciences, *NASA Authorization for Fiscal Year 1977*, Hearings, Part 2, p. 1138.
11. Richard Kerr, "Planetary Science on the Brink Again," *Science*, December 14, 1979, pp. 1288–89.
12. The COMPLEX strategy was spelled out in three separate reports: Committee on Planetary and Lunar Exploration, Space Science Board, *Strategy for the Exploration on the Inner Planets. 1977–1987* (Washington, DC: National Academy of Sciences, 1978); *Strategy for Exploration of the Outer Planets, 1978–1988* (Washington, DC: National Academy of Sciences, 1979); and *Strategy for the Exploration of Primitive Solar System Bodies——Asteroids. Comets. and Meteorites, 1980–1990* (Washington, DC: National Academy of Sciences, 1980).
13. For a personal account of the planetary program's rise and fall, see Bruce Murray, *Journey into Space: The First -Three Decades of Space Exploration* (New York: W.W. Norton, 1989).
14. *The New York Times*, February 18, 1981, p. Al.
15. See Joan Johnson-Freese, "Canceling the U.S. Solar-Polar Spacecraft," *Space Policy*, February 1987, for a discussion of this decision.
16. For Murray's account of his attempts to gain support for a mission to Halley's Comet, see Murray, *Journey into Space*, part V.
17. Letter from James Beggs to Rear Admiral Robert M. Garrick, August 17, 1981. Copies of this and other documents cited in this chapter are on file in the NASA Historical Reference Collection at the NASA History Office, NASA Headquarters, Washington, DC. Such documents are identified in these references by the acronym NHRC.
18. Letter from James Beggs to David Stockman transmitting NASA's FY 1983 budget recommendations, September 15, 1981 (NHRC).
19. Letter from David Stockman to James Beggs, September 24, 1981 (NHRC).
20. Letter from James Beggs to David Stockman, September 29, 1981 (NHRC).
21. Ibid.
22. *Aviation Week and Space Technology*, June 24, 1981, p. 56.
23. Interview with James Beggs, February 2, 1989. Other interviews carried out in connection with this research include: Hans Mark, December 10, 1988; George A. Keyworth II, April 4, 1989; Bruce Murray, May 17, 1988; Geoffrey Briggs, August 4, 1988; John Naugle, June 24, 1988; Eugene Levy, June 3, 1988; Angelo Guastaferro, May 16, 1988; and Louis Friedman, executive director of The Planetary Society, May 17, 1988.
24. For Mark's own account of his career, see Hans Mark, *Space Station: A Personal Journey* (Durham: Duke University Press, 1987).
25. Hans Mark, "New Enterprises in Space," *Bulletin of the American Academy of Arts and Sciences* XXVIII, no. 4 (January 1975), 19. In an April 2012 communication to the author, Mark noted that he had been correct in his

judgment of scientific significance, pointing out that two Nobel Prizes had been awarded based on findings from NASA astrophysics missions.
26. Letter from Hans Mark to Robert Sharp, April 9, 1974 (NHRC).
27. The memorandum is reprinted as Appendix 4 to Mark, *The Space Station*. The quoted passages are on p. 239.
28. Letter from Hans Mark to Bruce Murray, August 16, 1981 (NHRC).
29. Note from Hans Mark to Bruce Murray, October 14, 1981 (NHRC).
30. These figures are drawn from the material prepared by NASA to appeal the OMB allocations and transmitted to the White House by a letter from NASA comptroller C. Thomas Newman to Craig Fuller, director of cabinet administration, December 5, 1981 (NHRC).
31. Letter from James Beggs to David Stockman, November 30, 1981 (NHRC).
32. FY 1983 Budget Appeal, National Aeronautics and Space Administration, Planetary Exploration, December 5, 1981 (NHRC).
33. Office of Management and Budget, "Summary of OMB/NASA Positions: Space Science and Related Programs (Including Planetary Exploration)," undated (NHRC).
34. Interview with Eugene Levy, who had met with Keyworth at the latter's home in Santa Fe in June, before Keyworth officially assumed his position in Washington.
35. Philip J. Hilts, "Science Board to Advise President Proposal," *Washington Post*, December 9, 1981.
36. Alton K. Marsh, "Adviser Urges Shuttle Emphasis," *Aviation Week and Space Technology*, December 14, 1981, pp. 16–17.
37. M. Mitchell Waldrop, "Planetary Science in *extremis*," *Science*, December 18, 1981, p. 1322. Interview with George Keyworth.
38. Paper prepared for Budget Review Board, "Selected White House Views. Department: NASA. Issue: Planetary Exploration," December 8, 1981 (NHRC); emphasis in the original.
39. Thomas O'Toole, "NASA Weighs Abandoning Voyager," *Washington Post*, October 6, 1981.
40. Craig Couvalt, "NASA Assesses Impact of Budget Cut Proposal," *Aviation Week and Space Technology*, October 12, 1981.
41. *Science News*, October 24, 1981, p. 260; emphases in the original.
42. One result of the debate over how to respond to threats to continued government funding of space science activities, including planetary exploration, was the creation of a Space Sciences Working Group under the auspices of the American Association of Universities, the Washington-based organization representing the interests of major US research universities. This group began its operations in early 1982, and attempted to influence Congress on those portions of the NASA space science budget of most interest to academic researchers.
43. Letter from David Morrison and Carl Sagan to Edwin Meese, October 14, 1981 (NHRC).
44. Letter from Edwin Meese to Carl Sagan, October 22, 1981. An identical letter was sent to Morrison (NHRC).
45. *Aviation Week and Space Technology*, December 7, 1981, p. 17.
46. Interview with Eugene Levy.

47. Interview with Louis Friedman. For a history of the early years of The Planetary Society, see *The Planetary Report*, January/February 1986, pp. 3–11.
48. Memorandum from E. D. Hinkley and C. B. Farmer to Bruce Murray, "Technical Meeting at CIA Headquarters," October 20, 1981 (NHRC).
49. Letter from Bruce Murray to James Beggs, October 16, 1981 (NHRC).
50. Letter and enclosures from Bruce Murray to Mrs. William Scranton, October 16, 1981; memorandum from Bruce Murray to JPL senior staff, "JPL Directions for the Future," October 30, 1981; letter from Bruce Murray to Hans Mark, November 3, 1981 (NHRC).
51. Letter from A. O. Beckman to Edwin Meese, October 5, 1981 (NHRC).
52. Draft of letter from A. O. Beckman to Edwin Meese, December 10, 1981. Draft was prepared by JPL and transmitted to Beckman by a December 10 letter from Bruce Murray.
53. Letter from John Rousselot to Edwin Meese, December 10, 1981, and letter from Thomas Pownall to Edwin Meese, December 10, 1981 (NHRC).
54. Letter from Mrs. William W. Scranton to Marvin Goldberger, December 6, 1981 (NHRC).
55. Letter from Bruce Murray to Arnold Beckman, December 10, 1981 (NHRC).
56. Letter from Howard Baker to the president, December 9, 1981 (NHRC).
57. Memorandum from Powell Moore to the president, December 10, 1981 (NHRC). Interviews with George Keyworth and James Beggs.
58. Budget Review Board Decisions, National Aeronautics and Space Administration, December 11, 1981. The date on this document places the timing of the Budget Review Board meeting in question. Originally scheduled for December 9, most evidence suggests it was postponed until December 15. This means either that the date on this document is incorrect or that, after Baker's intervention with the president, the Budget Review Board met on the NASA appeal on December 11.
59. "Purpose of Solar System Exploration Committee," November 10, 1980 (NHRC).
60. Interview with John Naugle.
61. Summary Minutes of the Solar System Exploration Committee, June 1–2, 1981, pp. 4–5 (NHRC).
62. Solar System Exploration Committee, NASA Advisory Council, *Planetary Exploration through Year 2000: A Core Program* (Washington: Government Printing Office, 1983), p. 65.
63. Summary Minutes of the Solar System Exploration Committee, January 26–28, 1981 (NHRC).
64. Explorer-class missions are low-cost science and technology missions involving suborbital or Earth-orbiting spacecraft. Explorer missions usually study the Earth and its immediate environment.
65. Summary Minutes of the Solar System Exploration Committee, February 23–24, 1981, p. 3 (NHRC).
66. Memorandum from D. G. Rea to Bruce Murray, "Title of 'Targeted Missions Study,'" May 19, 1981.
67. Minutes of SSEC Meeting of June 1–2, 1981, p. 8.
68. Attachment 5 to Minutes of SSEC Meeting of June 1–2, 1981.
69. Ibid., pp. 6–7.

70. Summary Minutes of the Solar System Exploration Committee, October 26–27, 1981, p. 3 (NHRC).
71. M. Mitchell Waldrop, "To the Planets, Cheaply," *Science*, September 18, 1981, p. 1350.
72. Summary Minutes of the Solar System Exploration Committee, February 8–9, 1982 (NHRC).
73. Speech for the "Saturn" Meeting, University of Arizona, May 13, 1982, p. 5 (NHRC).
74. Minutes of the SSEC Meeting of February 8–9, 1982.
75. Ibid.
76. Ibid.
77. Summary Minutes of the Solar System Exploration Committee, June 3–4, 1982.
78. The Venus Radar Mapping mission was renamed Magellan and was launched to Venus on May 4, 1989. At the time of the SSEC discussions in 1982, the launch target for the mission had been 1988.
79. John Noble Wilford, "Plans to Explore Planets Revived," *New York Times*, February 20, 1983, p. A33.
80. For a discussion of this approach, see Howard McCurdy, *Faster, Better, Cheaper: Low Cost Innovation in the U.S. Space Program* (Baltimore, MD: The Johns Hopkins University Press, 2001).
81. NASA Advisory Council, *Planetary Exploration through the Year 2000*, pp. 15–16.

Chapter 3

Faster, Better, Cheaper: A Sociotechnical Perspective on Programmatic Choice, Success, and Failure in NASA's Solar System Exploration Program

*Amy Paige Kaminski**

In the 1990s the National Aeronautics and Space Administration (NASA) reformulated its program of robotic solar system exploration missions.[1] "Flagship" spacecraft like Viking, Voyager, and Galileo had dominated the program in the previous decade. Although wondrous and prolific missions, each took many years and a billion or more dollars to develop, allowing the agency to launch just a few of them. The 1990s instead found NASA deploying much smaller spacecraft to a variety of destinations within the solar system, including the renowned Mars Pathfinder, the first remotely controlled rover to reach another planet's surface, and for a fraction of the cost of its predecessors. The agency's plan was to concentrate on spacecraft with focused objectives and to use lean management techniques to reduce the cost of each mission, freeing resources to develop and launch more spacecraft more often to generate a steadier flow of data than infrequent, large missions could allow. But while NASA launched and achieved its goals for several missions developed under this "faster, better, cheaper" philosophy, five space science probes produced in this way failed before the decade's end. NASA soon thereafter backed away from this mode of planetary mission acquisition.[2]

More than a decade later, this ephemeral strategy receives mixed reviews. Many have pointed to the mishaps in 1999 of multiple missions developed under the strategy to refute the validity of the philosophy's claims, noting that a mission might be developed "faster" and "cheaper" but will not necessarily also be "better" in any material terms.[3] Some have criticized NASA management's use of the approach, arguing that its success rate was disappointing, and downright embarrassing, for an accomplished spacefaring nation such as the United States and that to continue taking such risks would border on irresponsible.[4] Others look back on the approach's successes as reflecting an exciting and momentous era in NASA's history, one that, while imperfectly executed, evoked a return to the agency's cultural roots as an organization committed to pushing back the frontiers of technology. Defenders of the approach insist that failure and risk-taking are both parts of technoscientific advancement and have aimed to put into perspective the nation's gains and losses from it.[5] Many believe NASA's general penchant for engineering conservatism for the past several decades has not served space exploration well in the long run.[6]

These opposed viewpoints indicate a lack of agreement about the goals and purposes the nation's planetary exploration efforts ought to serve. They also raise an interesting question: why did NASA choose to move away from "faster, better, cheaper" when some thought the agency's innovative, risk-taking approach was prudent? Immediately following the 1999 mishaps, NASA organized independent investigation boards to uncover what went wrong. Such inquiries are standard following technological accidents and disasters in many, if not most, domains. But in the pages of major newspapers and in the halls of Congress, the sentiments were that NASA had failed in its use of the "faster, better, cheaper" approach. The claim of "failure," though, implies a value judgment. What constitutes failure? What constitutes success? And for whom? Just two years earlier, jubilant crowds within NASA, the White House, and the general public had rejoiced at Pathfinder's extraordinary bounce-landing on Mars. Several other missions developed using the lean strategy also had worked as planned. One cannot understand NASA's move away from "faster, better, cheaper" strictly in terms of technical failure, the challenges of building reliable, low-cost spacecraft, or ineffective management techniques. In this essay I offer an analytical perspective on why NASA's use of this engineering, management, and science strategy was so fleeting, doing so by examining both why NASA began to employ it in the early 1990s and why the agency abandoned it before the decade was out. The full explanation, I argue, lies in the fact that NASA's planetary program is shaped through conflict, cohesion, and feedback among missions' technical performance; national economic conditions; and a broad set of actors' goals, values, expectations, and definitions of success for space exploration. Like colliding ocean waves, the process is unpredictable and is always in the state of becoming.

Adopting this explanation requires two critical recognitions. First, one must appreciate that solar system exploration and the advancement of planetary science, like virtually all scientific disciplines, is highly dependent on the creation and availability of particular technological artifacts and processes. In the case of planetary studies, robotic spacecraft capable of traveling through and enduring in the harsh environment of space have become the technological centerpieces of modern solar system investigations, capable of surveying worlds beyond our own to a degree ground-based and even Earth-orbiting telescopes simply cannot achieve. The interactions of these machines; the ground systems that communicate with them; the bits, bytes, photons, and pixels they return from space; and the scientists and engineers who develop, launch, control, and extract data from them are essential to the production of human knowledge about the solar system.

This description of what it takes to sustain robotic exploration of the planets, however, is only partially complete. Robotic solar system exploration is not exclusively the product of the spacecraft and the mission managers, scientists, and engineers who create them. Since NASA's foray into the solar system with robotic probes began in the early 1960s, many other actors have also contributed to shaping the program's development. National policymakers set federal spending policies and priorities and hold the strings to the purse that makes planetary exploration possible. These actors, and NASA administrators, determine the extent to which the agency will emphasize robotic planetary projects relative to its other programs at any given time. NASA leadership also sets the tone for the agency's operating culture. The broad community of planetary scientists, mostly resident in academic institutions, thinks up potential investigations and recommends which of them NASA should prioritize for funding. Private corporations team with NASA to build spacecraft systems. Career journalists and bloggers report and offer views on the agency's work, thereby helping to shape public knowledge and views while providing a form of feedback to NASA about its performance. As this essay will show, each of these players approaches and assesses the program's merit and achievements according to its own, unique set of standards for program success and tolerances for risk.

Using concepts from the sociology and history of technology, one can conceive of the practice of robotic solar system exploration as a network of human agents and organizations, technological artifacts, and economic and natural phenomena whose interactions determine the course, stability, and performance of the robotic planetary exploration program that NASA executes. I draw here on the "actor-network theory" of Michel Callon, Bruno Latour, and John Law, which maintains that the form a technological system takes is based on how human actors and material objects assimilate. As the involved entities struggle to "enroll" each other in serving or sharing their respective interests, the technological system is subject to being constantly remade. This idea thus is useful for examining how such networks form and change.[7] Law has extended it in his notion of "heterogeneous engineering," which notes that factors such as economics and natural phenomena (usually ignored by scholars as background conditions) also can influence a network's development.[8] Applying these concepts to the issue at hand, one can regard NASA's approach to robotic planetary exploration as the product of how the agency integrates the various elements that support or make their presence felt in regard to this endeavor: human actors, spacecraft, the space environment, and economic conditions alike.[9] That spacecraft may thrive or malfunction; budget availability wax or wane; and the involved human actors mesh or differ in their goals and values, or come and go from their positions (especially with election cycles), invites constant, evolving conflict and pressures on NASA to resolve differences and harmonize the network that sustains the robotic planetary program. NASA's move toward the "faster, better, cheaper" regime of planetary spacecraft development and then out of it so quickly is, I argue, a reflection of this continual struggle.

Herein I maintain that NASA embraced and subsequently abandoned "faster, better, cheaper" planetary missions when spacecraft performance, economic conditions, and particular actors' goals, values, expectations, and definitions of success for the program failed to synergize at the beginning and at the end of the 1990s, respectively. I begin by showing that during the 1980s the interactions among planetary scientists, NASA planetary mission managers, NASA leadership, Reagan-era and earlier policymakers led NASA into a pattern of developing high-cost, high-risk planetary missions.[10] The next section of the essay demonstrates how the responses of the planetary science community, journalists, Bush and Clinton administration officials, and a new NASA administrator to this situation and to a clampdown on federal spending government-wide pointed the way for NASA to adopt in the early 1990s a new strategy of smaller, more frequently launched planetary spacecraft developed with less agency oversight. Following that, I show that while this strategy had some skeptics it generally satisfied the involved actors for several years as NASA was able to develop and operate many small spacecraft. As the essay's penultimate section reveals, by the end of the decade, when five "faster, better, cheaper" spacecraft failed in a single year and several actors found this result unacceptable, NASA responded by once again revising its approach to planetary mission development. I conclude by emphasizing the instrumentality of each network element to NASA's choice of approach to robotic solar system exploration while illuminating the disparate definitions of success and collective high standards that actors hold for NASA's planetary program and space exploration efforts more generally.

Pathway toward Peril

The network for robotic solar system exploration was born in the years following Sputnik's launch, emerging not only as a technical possibility but also as a scientific and political desirement. As the prospect of reaching worlds beyond Earth with rockets and artificial satellites materialized, astronomers who had lost interest in planetary studies in favor of extragalactic objects during the first half of the twentieth century rekindled their interest in solar system research.[11] Policymakers, too, saw value in planetary exploration, albeit of a different sort: they recognized that demonstrating a US capability to reach the Moon and to travel beyond could support the nation's strategy to outdo Soviet achievements in space. In directing federal dollars to the purpose of planetary (including lunar) exploration and charging the fledgling NASA to "expand knowledge of phenomena in the atmosphere and space," President Eisenhower and national lawmakers set the stage for a community of planetary scientists to come into its own and for NASA to become the institutional home for this activity.[12] NASA rapidly developed the capability to achieve such feats, launching 18 robotic probes during the

1960s to orbit and land on the Moon ahead of the Apollo human landing missions and also to fly by Venus and Mars.

These probes delivered fine technical performances and yielded unprecedented looks at new worlds, giving scientists vast amounts of data to mull through. The robotic planetary program, however, never was nor ever has become a priority for NASA management or the nation's political leadership: the nation's space technology showpiece has, for these actors, clearly been its human space flight systems. As the nation's political leadership turned its attention to Vietnam and domestic issues, NASA's overall budget, including planetary mission funding, dropped following the height of spending on Project Apollo in the late 1960s; what political attention was paid to NASA centered on the development of the nation's next human space transportation system: the Space Shuttle. A few years prior, the Congress had refused to fund an ambitious and costly set of robotic orbiters and landers to explore Mars. Nonetheless, President Johnson and Congress accepted a revised proposal from NASA, based on science community recommendations, for a more moderate set of planetary missions, making clear that the nation would continue the agency's legacy of missions to solar system destinations.[13] Over the next few years, with the understanding that these missions could be accommodated without impacting priority funds for the Space Shuttle, the Congress appropriated funding for planetary projects including the Mars Viking missions, the Pioneer 10 and 11 and Voyager spacecraft bound for Jupiter and the solar system's outer reaches, and the Pioneer Venus Orbiter and Multi-Probe.

All of these spacecraft went on to meet scientists' expectations, but an 11-year gap would separate the Venus missions' launch from that of the next planetary mission. Planners for the Voyager and Viking spacecraft had grossly underestimated the costs of these missions and their sophisticated instrument suites. NASA, still ensconced in a high-spending mentality following the Apollo days, covered the increased costs at the expense of future planetary missions rather than de-scoping or terminating them. When the Reagan administration took office in the early 1980s, the Office of Management and Budget attempted to tighten funding for NASA while preserving funds for the Space Shuttle, in part by eliminating at least one new solar system exploration mission.[14] Planetary scientists remained a minority within the space science community and, unlike the larger body of astrophysicists, had not established a consensus on mission priorities, thus rendering their projects more susceptible to cuts. The administration ultimately spared the targeted mission—the Galileo orbiter and probe to Jupiter—when supporters played upon the importance of the mission to the viability of the Jet Propulsion Laboratory (JPL) in Pasadena, California.[15] The experience, however, clearly demonstrated the disparity that loomed between the goals of planetary scientists and the new political regime.

In 1983 NASA's Solar System Exploration Committee, an advisory group of external scientists that the agency had formed three years earlier to identify ways to reduce mission costs and to provide strategic planning for planetary missions, responded to the extant budget and political situation with a proposal for a program called Planetary Observers, which would support missions costing less than $60 million (FY 83$) by relying on modified Earth-orbital platforms and mature instrumentation.[16] NASA received a new funding start for a first Planetary Observer, the Mars Geoscience/Climatology Orbiter (later renamed Mars Observer), in fiscal year 1984. The mission to map the Martian surface and atmosphere in exceptional detail, however, quickly lost any identity it had as a small mission when mission planners, cognizant of the scarcity of planetary flight opportunities, selected a very elaborate and expensive payload. The cost also escalated because the spacecraft required more modification than originally expected and then NASA had to reconfigure it for flight on a Titan III booster rather than as planned for the Shuttle following the *Challenger* accident. Meanwhile, Galileo mission planners had begun to add capability to their spacecraft when they realized it would be the only launch to Jupiter for the foreseeable future. The mission price, not including launch, topped $1.3 billion (FY 89$) by the time it launched in 1989.

The discrepancy between political support, scientific interest in planetary exploration, and the inherent costliness and challenge of building probes that could achieve their missions ultimately set the program into a spiral. Having limited resources for planetary exploration, scientists and engineers took steps to maximize the science return and chances of technical success for their particular missions. They increased missions' ambitiousness and took measures to increase mission reliability such as building redundant systems and back-up spacecraft, avoiding unproven technologies, and holding elaborate mission reviews to lower the chances of technical failures. These strategies, however, had the effect of elevating missions' complexity, costs, and development schedules and drawing down funds available for future mission opportunities. During the 1980s NASA launched just two planetary missions: Galileo and the Magellan mission to Venus. Moreover, the extreme efforts to bolster missions did not necessarily assure their success. In Galileo's case, a failure to relubricate the spacecraft's high-gain antenna after a long wait on the ground before launch led the antenna to not deploy fully and forced science teams to rely on the spacecraft's small antenna, which received data at lower rates. Even more significantly, Mars Observer became NASA's most costly robotic mission failure when in 1993 mission operators lost and never regained contact with the spacecraft while preparing to insert it into orbit around Mars.

The Pendulum Swings

Viking, Voyager, and Galileo were, in terms of the massive amounts of scientific data they delivered and the thousands of images of exotic worlds they provided to the global public were wildly productive. But the strategy of fortifying planetary missions at tremendous costs had proven unsatisfactory to many of the actors involved in the robotic solar system exploration network.

With few missions launched and operating successfully during the 1980s and early 1990s, many planetary scientists' research interests remained unfulfilled. NASA's bloated planetary missions had not earned admiration within the George H. W. Bush administration either. Staff members of the White House's National Space Council wondered why NASA was still building behemoth spacecraft when other US government agencies had demonstrated that space flight activities could be conducted on reduced budgets and shortened schedules. Meanwhile, beginning in the early 1990s, pressure mounted on the entire federal discretionary budget, within which NASA is funded, due to the Budget Enforcement Act of 1990 and the eventual Clinton administration and Congressional commitments to lowering federal spending. No parties were entertaining the idea of abandoning the solar system exploration program, as the Reagan administration had contemplated, but what solution would effectively fulfill the collective yet incongruent goals and values of the actors enrolled in the network that sustained the endeavor? The prospect of a program of low-cost planetary missions emerged as a solution that could satisfy the disparate actors.[17]

The planetary science community had not given up on the idea of a small-missions program since proposing one in the early 1980s. NASA's space science advisory committee had expressed concern in 1986 that solely pursuing flagship missions was not in space scientists' best interest.[18] Three years later, NASA organized a workshop in which participating scientists proposed a new, low-cost planetary initiative modeled upon the agency's long-standing Explorer program of small spacecraft for astrophysics and solar physics missions. In 1990 the solar system exploration subcommittee of NASA's space science advisory committee recommended that the agency again consider establishing a budget line for small planetary missions so as to increase the stability, breadth, and diversity of planetary science research and avoid problems experienced by missions such as Galileo.[19] The planetary science community, however, had failed to discipline itself to adhere to such programs in the past, and NASA had been complicit in allowing mission costs to soar. Commitment to the concept would come with the backing of political actors who shared a belief that the time for low-cost missions had arrived. Critical to the transformation was the Bush administration's emplacement of a NASA administrator who would impose the discipline to make small missions become reality.

The National Space Council's staff had grown frustrated with the seeming unwillingness of NASA's management to move away from expensive space systems, whether for robotic or human space flight. Space Council staff members were enamored with the fact that the Strategic Defense Initiative Organization in the mid-1980s had demonstrated a space-based ballistic missile defense system relatively inexpensively, quickly, and using a lean management style. They believed NASA would similarly need to commit to lower-cost ways of developing space projects for the nation to be able to afford the Bush administration's Space Exploration Initiative—a colossally ambitious program that would entail establishing a human settlement on the Moon and then mounting a human expedition to Mars—as well as other space flight programs. Bush administration officials also saw merit in the Defense Department's custom of setting mission requirements and relying on external entities to propose how best to execute them through contract competitions and thought NASA could realize savings by following suit.[20] Concluding that replacing NASA's top leadership was imperative if the agency was to change, they identified TRW's Daniel S. Goldin (figure 3.1) as the right individual to lead NASA's transformation and, in the words of Howard McCurdy, to "deconstruct the old culture of high spending."[21] Having built spy and communications satellites for 25 years, the outspoken Goldin saw all of NASA's programs, including the robotic planetary program, as in drastic need of refocusing and restructuring in order to ensure the agency's future as a premier research and development organization. Highly critical of the cost overruns that had evolved within the agency's planetary program, he firmly believed that the space science and engineering communities within NASA, academia, and industry needed to dislodge the mindset of cramming a spacecraft with scientific capability and taking expensive measures to avoid failure.[22] He posited that flying less complex spacecraft and instruments and using smaller launch vehicles and advanced technologies would translate into less costly missions with shorter development times, allowing NASA to pursue more frequent launch opportunities.

The Senate confirmed Goldin as NASA's ninth administrator in April 1992, by which point NASA's space science program managers as well as political actors on both ends of Pennsylvania Avenue were converging to

Figure 3.1 NASA administrator Daniel S. Goldin was the longest serving of anyone in the position, 1992–2001, and profoundly shaped the agency, especially with his "faster, better, cheaper" emphasis. Here he speaks with US secretary of state Madeleine Albright at the launch of STS-88, the first US launch for the International Space Station, on December 3, 1998. (Credit: NASA photo no. KSC-98PC-1769.)

pursue a course of small planetary projects. Aware of NASA and the planetary science community's attention to establishing a low-cost planetary missions program, the Senate supplemented NASA's FY 1992 appropriations bill with direction to "prepare a plan to stimulate and develop small planetary or other space science projects, emphasizing those which could be accomplished by academic or research communities."[23] During the month in which Goldin arrived at NASA, the agency responded to the Senate with the statement that small planetary missions would be "the centerpiece of NASA's new programs for the 1990s."[24] The incoming Clinton administration retained Goldin at the helm of NASA. While the administration's major challenge regarding NASA would be to rein in costs for Space Station Freedom (to become the International Space Station in 1993) and to get the Congress to agree to a spending plan and to bring the Russians in as partners, Clinton's science adviser, physicist and former Office of Technology Assessment director Jack Gibbons, maintained that NASA should sustain a healthy space science program alongside its human space flight endeavors and endorsed the small-missions approach.[25] Office of Management and Budget director Leon Panetta agreed.[26] The administration requested—and received—Congressional approval to begin a small planetary missions program called Discovery in fiscal year 1994. Mars Observer's disappearance that same year—and the media's ensuing questions about how NASA could have let such a debacle happen—only served to affirm for NASA and the nation's political leadership the prudence of a program of small probes that could distribute mission risks.[27] With the Republican Revolution of 1994 taking a no-holds-barred approach to reducing the federal deficit and the Clinton administration also tightening government spending, NASA's overall budget steadily declined throughout the 1990s, so the small-missions approach seemed to make good economic sense as well.

Proving "Faster, Better, Cheaper"

Having secured funding for new small solar system exploration missions programs, NASA set to work to demonstrate the viability of conducting productive planetary missions at low costs. NASA imposed strict guidelines on the Discovery program. Approximately every eighteen months, the agency would solicit proposals for complete missions to any solar system destination that would fit under a particular cost cap (the original was $150 million [FY 92$]) and take no longer than three years to develop.[28] Participation was open to teams led by a single individual—a "principal investigator"—and comprised of representatives from universities, aerospace industry firms, NASA field centers, and other government agencies. NASA would evaluate submitted proposals' scientific value, cost, and technical feasibility and select one or two of the most meritorious concepts to fund and develop. Selected missions whose teams failed to remain within given technical and cost constraints would be subject to termination. The agency also managed a second program of low-cost, small spacecraft approved by Congress for a fiscal year 1995 new start. Called Mars Surveyor, the program was intended to recover Mars Observer's science objectives and to replace an earlier concept for a system of spacecraft that would land on the red planet and monitor its environment. The program would investigate Mars's climate and resources, and seek signs of water and life (figure 3.2) by sending an orbiter and a lander to the planet every 26 months over 10 years within an annual program budget of $156 million (FY 96$).[29] Unlike Discovery, Mars Surveyor would not solicit proposals for complete mission packages but was managed by JPL, which would seek the community's input on instrument payloads and experiments consistent with the science objectives NASA established.

Goldin pressed scientists and engineers at NASA, in universities, and in the aerospace industry to understand that he was serious about emplacing a completely new philosophy for how NASA planetary missions were developed. He stressed that mission teams should accept higher levels of technical risk to minimize project costs and to shorten development cycles, touting that an occasional failure of one or two among several small missions was preferable to losing a flagship mission. Within such a "portfolio approach," Goldin conceded, individual missions might not be as capable as their predecessors, but they would distribute benefits and risks across the program; in doing so, he believed they would better serve scientists' interests while inspiring students to pursue science and engineering careers, and fostering

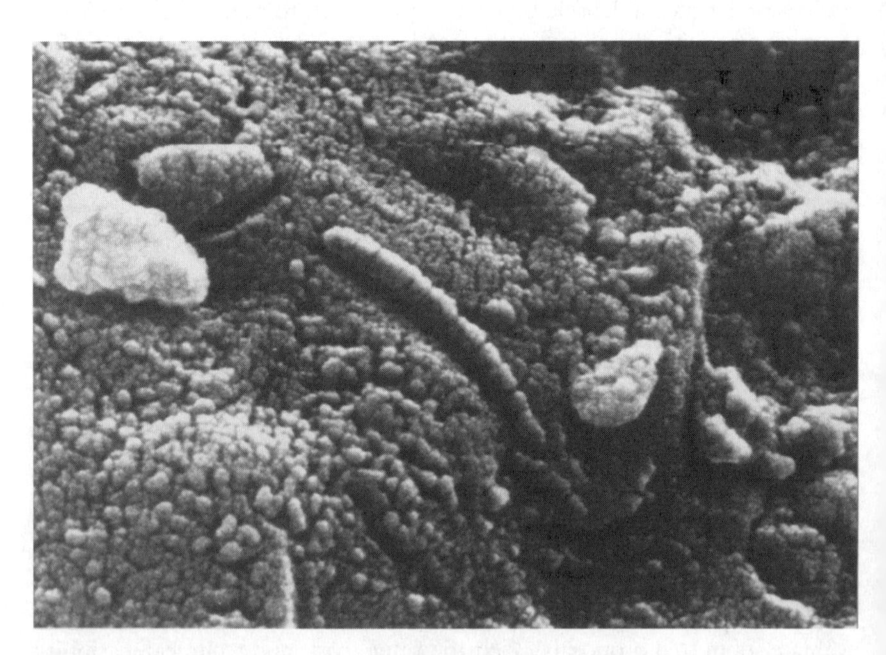

Figure 3.2 This meteorite from Mars landed on Earth and was discovered in 1984. This could have been the result of an impact of a celestial body on Mars, sending material from Mars into space. The suggestion that it may have evidence of biological material in it sent the scientific community into an excited spin in the mid-1990s and sparked intense interest in further exploration of the red planet. (Credit: NASA photo no. ALH84001-EM1.)

public interest in American space efforts. "We need to stretch ourselves. Be bold—take risks," he told JPL employees. "A project that's 20 for 20 isn't successful. It's proof that we're playing it too safe. If the gain is great, the risk is warranted. Failure is OK, as long as it's on a project that's pushing the frontiers of technology."[30] Without a doubt, Goldin promoted a radically different approach than the previous, "science-at-all-costs" mode of planetary mission development. For Goldin, planetary missions were not just about near-term science objectives: he also believed these spacecraft could and should be sites for trying new things, learning, and investing in NASA's future space technology capabilities and its long-term ability to remain exciting and worthy of the nation's support. He wanted to return NASA to assuming the "can-do" attitude that Diane Vaughan argues the agency lost as its bureaucracy expanded during the Apollo era.[31]

NASA's earliest experiences with the Discovery and Mars Surveyor programs suggested that, through Goldin's "faster, better, cheaper" approach, the agency perhaps had found a strategy to synchronize and stabilize the planetary exploration network. The approach offered much to please the planetary science community, policy officials, the media and the public, and those who believed NASA should be pushing the boundaries of technology. As mission teams were finding creative ways to bring down the costs of missions, NASA was able to achieve an unprecedented 13 new project starts between Discovery and Mars Surveyor during the 1990s—even as the NASA science budget's purchasing power began to decline in the mid-1990s.[32] Following the strategy Goldin advocated, these missions achieved unheard-of low costs by adhering to focused science objectives, simple space system designs, creative engineering, and streamlined management techniques. Launched in February 1996, NASA's first Discovery mission, the Near Earth Asteroid Rendezvous (NEAR), was developed in just two years and at a cost of $138 million (FY 98$) due to the fact that its builder, the Johns Hopkins University's Applied Physics Laboratory, possessed instruments developed for previous space missions that, while not designed for asteroid studies, could be used for the mission with little modification.[33] Similarly, simplicity of design and hardware proved key to Lunar Prospector's low cost and selection in the first Discovery competition: mission planners at NASA's Ames Research Center and Lockheed Martin selected a basic spacecraft bus, engines, batteries, transponder, and four instruments from existing "off-the-shelf" hardware. Realizing its objectives for $63 million (FY 98$), the mission had been selected by NASA over more sophisticated, yet more expensive, proposals to make the point that missions could be conducted at costs even well below the Discovery cap. The Genesis mission, designed to gather and return to Earth samples of the "solar wind" of charged particles flowing from the Sun, achieved a $216-million (FY 97$) price tag due to the mission's simple operations scheme. The mission saved on fuel and instrument development costs because, by solar system standards, it had only a short distance to travel: a "mere" one million miles to the destination at which it would collect samples and then back to Earth.

But part of Goldin's challenge to mission planners was to accept the risk of incorporating novel technologies into space flight missions in the quest to reduce costs and improve capabilities for these and future missions. Several did and reaped such dividends. The Lunar Prospector team's ability to keep costs low was helped by the decision to use an unproven Lockheed Martin Athena II solid rocket rather than a tried, and more costly, vehicle. Mars Global Surveyor became the first spacecraft to employ aerobraking—a technique of lowering and circularizing a spacecraft's orbit using atmospheric drag each time it passes closest to a planet—as its primary means of achieving its planned orbit.[34] The technique, which allowed the mission to save on the costs of the fuel otherwise required to slow and coax the spacecraft into orbit, would be used on subsequent Mars-orbiting spacecraft. Mars Pathfinder (figure 3.3) gambled on using a shroud of protective airbags to avoid the costs of developing a propulsion system to land the spacecraft and rover on the Martian surface and, in doing so, proved a technology that the agency would use again to land the Spirit and Opportunity rovers on Mars in 2004. The "faster, better, cheaper" strategy also sought through the New Millennium program initiated in 1996 to test new technologies in the space environment as dedicated projects separate from missions intended to meet scientific goals. Flown in 1998, the first New Millennium mission, Deep Space 1, tested 12 technologies, including an ion propulsion engine, which was incorporated in the Dawn asteroid flyby Discovery mission launched in 2007.

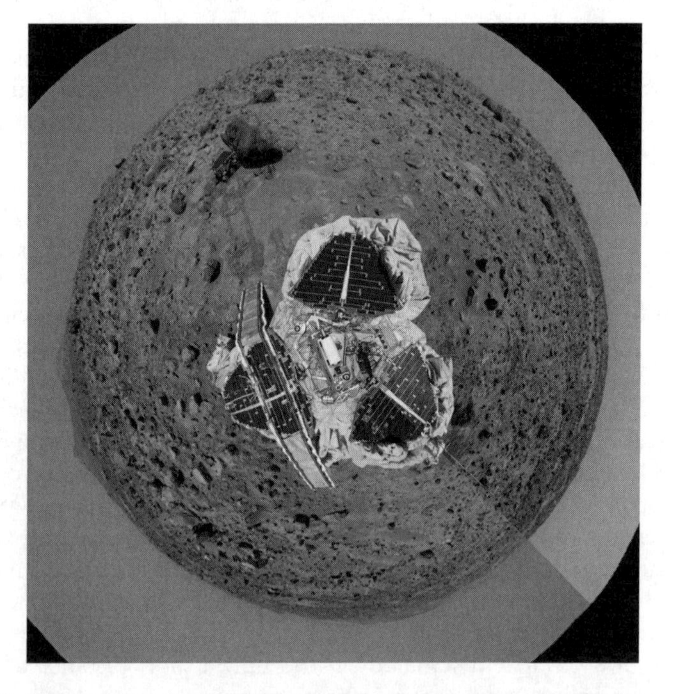

Figure 3.3 This image was created from several other images to give the impression of an overhead photograph of the Mars Pathfinder with the Sojourner rover at a nearby rock. (Credit: NASA/JPL photo no. PIA02652.)

Also to credit for yielding missions that met the tighter cost and schedule parameters was Goldin's empowerment of mission teams. Mars Pathfinder's planners at JPL, for example, controlled costs by using "skunk works" project management methods, keeping working teams small and holding them responsible for managing their work rather than subjecting them to formal reviews led by NASA Headquarters personnel less connected to the project.[35] Brian Muirhead, Mars Pathfinder's flight system manager, has explained:

> On Pathfinder, because the core of the team of engineers was multidisciplinary...we were able to function with a much smaller staff than had been used on similar projects in the past. For the entire management team, in particular, I insisted on generalists and let them know they were expected to think and act that way. A community of generalists allows faster decision-making, makes the transitions between different phases of the project less inefficient and more fluid, and makes it easier to cross-train workers.[36]

Through the competitively selected Discovery program missions, too, NASA and university scientists were motivated to think creatively and given newfound amounts of autonomy to manage projects.[37] Tasking each mission's principal investigator with ensuring mission cost, schedule, and technical performance inspired mission scientists and engineers to innovate to complete a project within the program's specified parameters. It also spurred among mission teams a degree of accountability for performance that did not exist in the era of flagship missions and top-down management.[38]

All of these cost-saving and schedule-reducing measures would be for naught, of course, if the spacecraft did not perform at their intended destinations as the involved network of actors expected. But their performance delighted many. The missions not only demonstrated technologies that would be useful for future space missions but also had begun to prove the possibility of conducting low-cost, focused scientific investigations within the inner solar system. They could take credit for a host of discoveries. For example, seven papers reporting the results of Mars Pathfinder's investigations of the Martian surface, atmosphere, and weather systems appeared in the journal *Science*.[39] Mars Global Surveyor detected the presence of a magnetic field around Mars, provided new insights into Mars's mineral composition and deeply layered terrain, and became the first spacecraft to record the full evolution of a Martian dust storm.[40] Lunar Prospector became the first mission to create an operational lunar gravity map and to directly measure water ice at the Moon's poles.[41] Broadly speaking, the "faster, better, cheaper" approach had begun to deliver on scientists' desire to obtain a steady stream of data from a variety of destinations while giving a wider range of scientists more opportunities to exercise their imaginations and to propose new investigations.[42]

Beyond what science came out of the missions, however, was what the "faster, better, cheaper" approach did to fire the public imagination. In the days of flagship missions, people connected to NASA planetary exploration through the magnificent images sent back to Earth from spacecraft that did not themselves receive much attention. The new brigade of small spacecraft drew public attention not just to the results of exploration but to the adventures and antics of the mechanized explorers themselves. Eventually, the world would witness NEAR entering orbit around and then landing gently on an asteroid—an incredible feat never before achieved—while Deep Impact would hurl a penetrator into the nucleus of a comet and then fly through the dusty wake. But first, Mars Pathfinder captured worldwide attention when it bounce-landed its way across the Martian surface on July 4, 1997, and proceeded to return its first image: a self-portrait of its tiny rover resting on a rusty-red planet five million miles from Earth. Within four days following its landing, the mission's website and mirror sites registered 150 million hits, a staggering amount of Internet traffic for the mid-1990s. Mars Pathfinder became the talk of television network morning shows and dominated the front pages of newspapers with early images and headlines heralding: "Pathfinder a Picture of Success," "Mars: Giant Leap for U.S.," and "Hallelujah! Back on Mars!"[43] Americans had not reached Mars successfully in two decades, and ordinary people were excited to follow the rover's journey on the legendary planet.[44]

To be sure, "faster, better, cheaper" skeptics remained. The early Discovery and Mars Surveyor missions were, on a per-spacecraft basis, clearly not in the same scientific league as their flagship predecessors: Viking's two orbiters and two landers boasted a total of 34 instruments and endured for six years, while Mars Pathfinder's rover carried 3 instruments and lasted for three months. The planetary science community had not abandoned interest in more ambitious missions, having posited in 1994 that a mix of small and larger missions would best serve science.[45] Some journalists questioned how the scientific return of missions produced under the new regime compared with that of their larger brethren—a matter the Senate in 1998 requested that the National Research Council examine.[46] There were also sentiments in NASA as well as in policy circles that perhaps NASA was trying to do too much with too little budget.[47] At the same time, that Pathfinder had generated so much national pride and excitement while meeting its mission objectives by and large validated the "faster, better, cheaper" approach for many trade press and mass media outlets, NASA, and the Clinton administration, the lattermost noting that Pathfinder marked "the beginning of a new era in the Nation's space exploration program" and "our new way of doing business" at NASA.[48] With missions in the works that would ultimately fly in front of a comet's nucleus (Stardust), capture samples of the solar wind (Genesis), and achieve still other scientific and technological firsts, the space agency seemingly had found a formula by which to conduct many of the investigations the planetary science community sought to accomplish, satisfy political stakeholders, and to restore its public image as an organization capable of wondrous discoveries and technological advancements.

Turning Point

The year 1999 started well for NASA's low-cost planetary missions. Although the NEAR spacecraft missed its planned January 1999 encounter with the asteroid Eros due to a flawed main engine burn 18 months prior, mission planners had figured out a way for the spacecraft to reach its target the following year. Stardust, the second competitively selected Discovery mission, launched successfully in February. Come March, however, NASA's Wide-Field Infrared Explorer, a small space-based telescope developed under the agency's Explorer program, failed to meet its main mission objective when an electronics design error prompted the telescope cover to open earlier than planned and depleted the cryogens necessary to conduct its sky-surveying mission.[49] Then trouble came for the planetary program when in September and December, respectively, the Mars Climate Orbiter and the Mars Polar Lander—plus two New Millennium probes that rode aboard the latter—seemingly vanished as they approached the red planet. NASA had lost five space science spacecraft in a single year.[50] That more missions were flown under the "faster, better, cheaper" regime and that each was accepting greater risk given leaner development processes meant there were more chances for failures. But a 37-percent failure rate was nowhere near the two-in-ten loss rate Goldin had had in mind as a "target" across the agency's planetary, astrophysics, solar physics, and Earth science low-cost probes.[51]

An investigating panel pinpointed the root cause of the Mars Climate Orbiter failure as the use of English instead of metric units in a flight trajectory software program, which had prohibited the spacecraft from being inserted properly into orbit around Mars.[52] While investigators could not say with certainty what happened to the New Millennium probes, they concluded that the Mars Polar Lander most likely plummeted to the Martian surface following premature shut-off of the main descent engine. In addition to these technical points of failure, however, they identified contributing factors stemming from how NASA, JPL, and their contractors had executed the "faster, better, cheaper" approach. Led by retired Lockheed Martin executive Tom Young, the Mars Program Independent Review Team (MPIAT) traced the fatal flaws of the Mars-bound spacecraft to NASA's attempt to manage the orbiter-lander pair as a single project with a development budget comparable to that of Mars Pathfinder. Given this budget, which the MPIAT estimated to be short by 30 percent, mission planners skipped key tests and did not prepare sufficiently for the missions' operations phases.[53] In the case of Mars Polar Lander, the team had failed to perform a full system test after changing out a major subsystem and had attempted to land on Mars with a sophisticated propulsion system for about half the cost of Pathfinder's airbag system. Insufficient communications between JPL and the contractor had led to poor systems integration and inadequate preparations for Mars Climate Orbiter's operations phase.[54]

The fates of these Mars missions substantiate the conclusions of analysts who have studied the performance of space missions: the balance between spacecraft cost, schedule, complexity, and reliability is extremely delicate. Indeed, experienced spacecraft engineers and managers are painfully aware that the line between technical success and failure for a space mission is razor-thin; thousands of things can go wrong with the launch or operation of a spacecraft, and a single oversight or misstep can doom a mission, rendering for naught the countless hours of planning, building, reviewing, and testing a team undertakes to ensure a functioning mission. The low-cost concept worked well for simple missions like Lunar Prospector. If, however, mission managers attempted to cut costs too low, abridge schedule too much, or load too many new instruments or technologies onto a spacecraft, risks would skyrocket, and projects would become highly prone to failure.[55] NASA had, in essence, asked JPL to build two less-than-simplistic Mars probes for the price of one.[56] Schedule and budget had held sway, prohibiting teams from raising issues that might have saved the missions.[57] Citing these very points, the MPIAT findings suggested that NASA management had pushed "faster, better, cheaper" too far. But the reviewers concomitantly noted that, with some adjustments, NASA could—and should—continue to use the approach as a program management technique.[58]

Goldin's "faster, better, cheaper" was also being judged by another set of actors and standards. The news media were the first to opine: "Mars Probe Lost Due to Simple Math Error," read one *Los Angeles Times* story headline.[59] Another depicted the Mars Climate Orbiter loss as a "blow" to the US space program.[60] *USA Today* reported bluntly: "NASA's Approach to Exploration Not Working."[61] Then, in March 2000, Jim Oberg of *UPI* published a story contending that NASA knew before the Mars Polar Lander left Earth that the spacecraft's propulsion system was flawed, yet they proceeded to launch to maintain schedule.[62] The story appeared one day before Goldin was scheduled to testify before NASA's Senate authorization committee on the performance of NASA programs. The senators in attendance pressed Goldin about the Oberg article's claims and about NASA's track record with "faster, better, cheaper." In doing so, they made clear that NASA's use of the approach was falling short of what they wanted and expected from NASA and its space exploration programs. "NASA is right to be ambitious. People expect that," Kay Bailey Hutchison of Texas averred.[63] But at the same time, the senators maintained that they, and the American people, expected NASA to deliver with the missions that tax dollars funded. "For $14 billion a year," Bill Frist said, "the American taxpayers deserve better," noting that the hearing was intended to discuss "how to get NASA back on track."[64] John Breaux of Louisiana pointed out that space exploration is an "area where the United States is so clearly the best in the world" and that when NASA experiences a mission loss, "it's a big one."[65] At a hearing on NASA's fiscal year 2001 budget request one month later, Kit Bond, the chairman of NASA's Senate appropriations subcommittee, called the losses "very embarrassing."[66] All of these reactions are perhaps summed up in House Science Committee chairman James Sensenbrenner's two-word response to the unit conversion error responsible for the first Mars failure: "I'm speechless."[67] While Goldin had envisioned that the approach would prove worthy by ensuring that NASA had a healthy portfolio of space science missions, members of Congress thought differently: each and every spacecraft's success mattered to them because even small spacecraft cost big money and America was a nation to do nothing but excel. And here, NASA had just lost five probes in a very short period of time.

Furthering the media's and Congress's scrutiny of NASA was a litany of concerns about human space flight safety, most recently that in July of 1999 the STS-93 Space Shuttle mission had brushed closely with hazard. Aborted on the pad just seconds before launch, the mission went on to fulfill its objectives after launching a few days later, but the Columbia orbiter experienced an electrical short and premature shutdown of one of its main engines during its ascent. In February 2000 the Aerospace Safety Advisory Panel released its annual report, accusing NASA of having jeopardized shuttle safety by reducing funding for the program's workforce sharply in the previous several years.[68] In light of these many criticisms of his efforts to cut costs across the agency, Goldin revised his stance on failure and risk-taking.[69] With respect to robotic planetary missions, Goldin took full responsibility for the mishaps, admitting: "I, Dan Goldin, pushed too hard and in doing so stretched the system too thin."[70] Consistent with the MPIAT's recommendations, he took measures to improve communications among mission teams and agency management, implemented requirements for more oversight and formal reviews of missions, and committed to more training for young engineers and project managers. Goldin made plain that he was not going to walk away from the general principles of "faster, better, cheaper," but he began to emphasize mission reliability and safety as priorities. While he had testified before Congress that additional funding was not necessarily the solution to NASA's challenges, the fact that the nation had moved into an era of budget surpluses meant that the Clinton administration, in its final year, and the Congress were inclined to augment NASA's budget for science and workforce purposes. Such news was encouraging to the planetary science community and other space scientists. The National Research Council had recently released the assessment of "faster, better, cheaper" missions' scientific merit, which the Senate had requested nearly two years prior, and stated in the report that the approach's stand on risk and its restriction on larger missions had compromised scientific objectives and outcomes.[71] Thanks to the economic windfall, combined with how the MPIAT, the news media, the Congress, and scientists had reacted to the performances of those five spacecraft, and the shuttle, Goldin resolved that going forward NASA would work to ensure that each and every planetary mission delivered on its objectives.

Even with the budgetary breathing room, the new planetary exploration program drew on the "faster, better, cheaper" approach's cost-consciousness and lessons learned from NASA's experience with flagship missions. For well over a decade, robotic planetary missions did not return to anywhere near the size of the missions of Viking or Galileo.[72] The agency's redesigned ten-year strategy for robotic Mars exploration, worked in close partnership with White House budget staffers and rolled out in October 2000, would look for signs of past and present water on the red planet using a mix of small and moderately sized orbiting, landing, and roving spacecraft.[73] As advocated by the planetary community, a new line of small, competed "Mars Scout" missions followed in the mold of the Discovery program, which would continue as well. An additional program, New Frontiers, gave scientists the ability to propose missions more complex than Discovery missions with larger, albeit still constrained, budgets. New Horizons, launched in 2006, became the first mission in this series, designed to fly by Pluto and into the Kuiper Belt beyond. Exhibiting that cost discipline still governed within the agency, NASA science program managers selected this mission after JPL's estimate to develop a similar mission had grown to over $1 billion. Save for meeting scientists' interests in a major mission to return samples from Mars or to probe the icy moons of Jupiter, the new planetary program closely reflected their priorities.[74]

But beyond retaining the cost caps, the new program's philosophy was quite different. Because the cost targets had been raised, every robotic planetary mission became strategic in value. If failures of small missions had been problematic in several circles, the new pursuit of costlier missions meant that the stakes became much higher to guard against losses. NASA consequently reverted to a very conservative posture in terms of how it built and managed the development of space systems. For example, NASA elected to send not one but two rovers to Mars in 2003, each launched separately, to hedge against coming up empty-handed during this attempt to explore the planet.[75] The agency's renewed conservatism also permeated how it evaluated proposals in its competitive mission programs. Proposals that attempted to infuse spacecraft with new technologies that could offer new capabilities and push the boundaries of science as well as provide the seed corn for future planetary exploration were downgraded for the potentially adverse impacts they could have on meeting a mission's objectives. Over the next several years, NASA's space science budget reached record levels; however, precious few resources were made available to invest in the enterprise's future—a situation the planetary science community and many NASA science managers would come to lament.[76] Risk had, once again, become an anathema to the program. Although the MPIAT had not explicitly advocated that the agency move so far in this direction, the overt pressures of the network of actors who sustained and shaped robotic solar system exploration had pressed NASA to embrace a new approach to this endeavor.

Conclusion

In this essay I have attempted to go beyond the conventional understanding that NASA retrenched from the "faster, better, cheaper" strategy of robotic planetary spacecraft development because it endured five spacecraft losses in a single year using the approach. By conceiving of planetary exploration as a network of linked actors, spacecraft, and economic and environmental factors and looking across the rise and fall of the approach's use, one can begin to see that the explanation is far more complex, grounded also in the goals and expectations several groups have for the planetary exploration enterprise and NASA and the American space exploration program more generally.

"Faster, better, cheaper" was one phase of the evolution of the American planetary exploration program. The planetary program conjoins a set of disparate people having different goals, technological systems, and economic and environmental conditions, all of which shape and sustain this effort. Each of these actors and factors is inherently powerful in that each has the capability to push on or influence the others to move toward a particular outcome; without any single entity, the program's direction may have been quite different. Moreover, no single entity is responsible for the program's direction. "A shrinking budget for NASA," for example, does not in and of itself logically explain the turn to "faster, better, cheaper" in the early 1990s, just as "a better budget outlook" doesn't elucidate the turn away from it some years later. NASA might well have proceeded with a single flagship mission or cancelled planetary exploration in response to the economic conditions in the early 1990s, just as the agency could have proposed in its budget requests during the plumper years of the 2000s to invest in a few flagships or in numerous technologies to advance the planetary program in the long run, or even to press on with more very small missions. NASA didn't embrace these alternatives because multiple actors and factors collectively drove the agency in particular directions. The "faster, better, cheaper" initiative emerged when the previous, flagship approach to planetary spacecraft development no longer seemed to satisfy the respective goals and values of the scientists and policymakers, or the coming budget conditions, that sustained the program. Members of these groups consequently engaged in discourse about bringing down mission costs, and Goldin was the forceful catalyst to see the proposition through. When, however, the strategy of cost-cutting seemed to *threaten* the goals and expectations of many scientists, members of Congress, and the media, Goldin—despite continuing to believe in the philosophy and despite the MPIAT's conviction that the strategy did not need to be abandoned but implemented differently–backed away from it in favor of an approach that shunned risk.

But why did these actors react so strongly to NASA's 1999 mission mishaps? And how exactly did Goldin come to be pressured by these players to retreat from "faster, better, cheaper?" The answer, I believe, lies in examining how the planetary exploration network's various actors relate to spacecraft. Scientists, of course, depend on spacecraft that operate as planned for their livelihoods; a single spacecraft failure is devastating for a scientist involved in that mission and is certainly a reminder to all others just how precarious the technological lifeblood of their profession is. A spacecraft success for the media, if they pay attention at all, is a symbol of human achievement, a pioneering spirit, and adventure; a loss, however, grabs headlines and is often a reason to doubt either NASA's competence or the cost value of planetary exploration. For elected officials in Congress, a spacecraft success, if it does get noticed, is a symbol of American exceptionalism, pride, and determination to explore; a failure, however, translates as an irresponsible use of taxpayer dollars, a lamentable disappointment, an embarrassment, and a reason to scrutinize NASA's management aptitude. A string of failures only magnifies these players' concerns.

Despite having particular goals and values vis-à-vis the robotic planetary program, these actors share an understanding of what makes missions "better": the near-term results and successes of spacecraft. Spacecraft failures challenged each one's definition of success. Goldin's "faster, better, cheaper" approach, however, was premised on a belief that only over the longer term, over a portfolio of missions, would NASA be able to realize and prove the strategy's effectiveness; the "better" in the approach's moniker, according to him, would become evident over time as risks were distributed across missions and failures taught NASA how to do better in the future. Goldin's definition of success and those of scientists, the media, and political officials did not mesh, as none among the latter is culturally attuned to think about space exploration success over the long term of a portfolio. For them, multiple spacecraft mishaps were not steps along the road to eventual success; they were catastrophes (or, in the case of the shuttle's management, near-misses). That this discourse came to prevail over talk of cost-cutting among a plurality of network actors integral to sustaining robotic planetary exploration provided the impetus for Goldin to put "faster, better, cheaper" to rest and to adopt a new philosophy for the program.

The actors in the network sustaining planetary exploration collectively expect the missions NASA flies to maximize scientific knowledge of the solar system, create excitement and adventure, and perform flawlessly, all while spending taxpayer dollars as effectively and efficiently as possible. It is a tall order when combined with the reality that not only is the cost of space flight still staggering, even for missions that were developed in the "faster, better, cheaper" style, but also, even after five decades of space flight, the technical challenges of reaching and operating in space remain monumental. Space flight is therefore still hardly routine, and the rarity of opportunities to conduct space flights contributes to making human actors so exacting. Accordingly, those enmeshed in the robotic planetary network will almost certainly continue to debate how best to explore our little corner of the universe.

Notes

* I wish to thank Kelley Boyer, Donald Clark, Charles Elachi, Scott Hubbard, Mallory James, Greg Jolley, Roger Launius, Sterling Mullis, Michael New, Sonja Schmid, Alan Stern, Ed Weiler, and David Winyard for their helpful comments on this manuscript.

1. I use the terms "planetary" and "solar system" interchangeably in this essay. My use of the term "exploration" encompasses any activities, including scientific investigations, that robots may perform at destinations throughout the solar system.
2. Although the origins of the term "faster, better, cheaper" are unclear, it was popularized by NASA administrator Daniel S. Goldin to describe his vision for NASA' turn to smaller space science missions. While others prefer the term "smaller, faster, cheaper" to eliminate the subjective "better," I use the former term consistent with its popular historical usage.

3. Howard E. McCurdy, *Faster, Better, Cheaper: Low-Cost Innovation in the U.S. Space Program* (Baltimore: Johns Hopkins University Press, 2001), 9; Liam Sarsfield, *Cosmos on a Shoestring* (Santa Monica, CA: RAND, 1998); David A. Bearden, "A Complexity-Based Risk Assessment of Low-Cost Planetary Missions: When Is a Mission Too Fast and Too Cheap?" paper delivered at the Fourth IAA International Conference on Low-Cost Planetary Missions, JHU/APL, Laurel, MD, May 2–5, 2000.
4. Senate Committee on Commerce, Science, and Transportation, Subcommittee on Science, Technology, and Space, Hearing on NASA's Performance, March 22, 2000; Senate Committee on Appropriations, Subcommittee on Veterans Affairs, HUD, and Independent Agencies, Hearing on NASA FY 2001 Budget, April 13, 2000; "Slow Down, Be More Careful," *Sun-Sentinel* (April 3, 2000): 14A.
5. See, e.g., Neil deGrasse Tyson, "Mathematically Challenged Americans Suffer from…Fear of Numbers," *Natural History* 110 (December 2001 /January 2002): 30.
6. Orlando Figueroa, interview with the author, January 8, 2010; G. Scott Hubbard, interview with the author, January 5, 2010.
7. Michel Callon, "Some Elements of a Sociology of Translation: Domestication of the Scallops and the Fishermen of St. Brieuc Bay," in *Power, Action, and Belief: A New Sociology of Knowledge?*, ed. John Law (London: Routledge and Kegan Paul), 1986), 196–233; Bruno Latour, *Science in Action* (Cambridge, MA: Harvard University Press 1987); John Law, "Technology and Heterogeneous Engineering: The Case of Portuguese Expansion," in *The Social Construction of Technological Systems: New Directions in the Sociology and History of Technology*, eds. Wiebe E. Bijker, Thomas P. Hughes, and Trevor Pinch (Cambridge, MA: MIT Press, 1989), 111–34.
8. Law, "Technology and Heterogeneous Engineering," 113.
9. I portray NASA in this essay as the "architect" who endeavors to keep the network of actors relevant to the robotic planetary exploration enterprise intact. This portrayal is rooted in the fact that the National Aeronautics and Space Act of 1958 charges NASA with executing space research programs; I do not mean to imply here that the robotic planetary exploration program is or ever has been a high priority at NASA. The human space flight program has always been the agency's priority, and even other areas of science have taken clear precedence over planetary science at times.
10. For a comprehensive programmatic history of NASA's planetary exploration program, see Amy Paige Snyder, "NASA and Planetary Exploration," in *Exploring the Unknown: Selected Documents in the History of the U.S. Civil Space Program, Volume V: Exploring the Cosmos*, gen. ed. John M. Logsdon with Amy Paige Snyder, Roger D. Launius, Stephen J. Garber, and Regan Anne Newport (Washington: U.S. Government Printing Office, 2001), 263–300.
11. Two books authored by leaders of NASA's early space science program provide excellent background to the effort's origins. See Homer E. Newell, *Beyond the Atmosphere: Early Years of Space Science* (Washington, DC: NASA SP-4211, 1980); and John E. Naugle, *First Among Equals: The Selection of NASA Space Science Experiments* (Washington, DC: NASA SP-4215, 1991).
12. National Aeronautics and Space Act of 1958, Public Law 85–568, 72 Stat., 429. Signed by the president on July 29, 1958.

13. Edward Clinton Ezell and Linda Neuman Ezell, *On Mars: Exploration of the Red Planet, 1958–1978* (Washington, DC: NASA SP-4212, 1984), 135.
14. John M. Logsdon, "The Survival Crisis of the US Solar System Exploration Program in the 1980s," chapter two in this volume.
15. Ibid.
16. NASA Advisory Council, Solar System Exploration Committee, *Planetary Exploration through the Year 2000: A Core Program (Executive Summary)* (Washington: US Government Printing Office, 1983), 6.
17. See also Stephanie A. Roy, "The Origin of the Smaller, Faster, Cheaper Approach in NASA's Solar System Exploration Program," *Space Policy* 14 (August 1998): 153–71.
18. NASA Advisory Council, Space and Earth Science Advisory Committee, *The Crisis in Space and Earth Science* (Washington, DC, 1986).
19. Robert A. Brown, Solar System Exploration Division, NASA, "Presentation to the Committee on Planetary and Lunar Exploration, Space Studies Board, National Research Council," March 7, 1990; Krimigis and Veverka, "Foreword: Genesis of Discovery," *Journal of the Astronautical Sciences* 43 (October–December 1995): 345–47.
20. Steven Isakowitz, interview with the author, December 22, 2011.
21. Howard E. McCurdy, *Faster, Better, Cheaper: Low-Cost Innovation in the U.S. Space Program* (Baltimore: Johns Hopkins University Press, 2001), 44.
22. See, e.g., Daniel S. Goldin, *Remarks to the Aeronautics and Space Engineering Board, National Research Council*, October 1, 1992.
23. Senate Report accompanying H.R. 2519, the Fiscal Year 1992 VA, HUD, Independent Agencies Appropriations Bill.
24. NASA, "Small Planetary Mission Plan: Report to Congress," April 1992, 1. This report addressed NASA's commitment to pursue low-cost projects not only for planetary science but also for Earth science, astrophysics, solar physics, and searches for planets around other stars. It also conveyed NASA's plans to consider low-cost missions for lunar exploration to support the Space Exploration Initiative as well as for outer planet exploration.
25. Jefferson S. Hofgard, interview with the author, December 1, 2011.
26. Steven Isakowitz, interview with the author, December 22, 2011.
27. See, e.g., L. Siegel, "At NASA, Red Planet or Red Faces?" *Washington Times* (August 25, 1993): A1.
28. NASA, Office of Space Science, Solar System Exploration Division, *Discovery Program Handbook*, November 1992, 5.
29. Earth and Mars are in a configuration relative to each other once every 26 months, which minimizes travel time between the two planets.
30. Daniel S. Goldin, *Remarks to Jet Propulsion Laboratory Workers*, May 28, 1992.
31. Diane Vaughan, *The Challenger Launch Decision: Risky Technology, Culture, and Deviance at NASA* (Chicago: University of Chicago Press, 1996), 209.
32. The thirteen new starts included eight Discovery missions: NEAR in 1994; Mars Pathfinder in 1994; Lunar Prospector in 1994; Stardust in 1995; Comet Nucleus Tour (CONTOUR) and Genesis in 1997; and Deep Impact and Mercury: Surface, Space Environment, Geochemistry, and Ranging (MESSENGER) in 1999. Mars Global Surveyor, designed to fly spares or rebuilds of five of Mars Observer's seven instruments, received a new start in

fiscal year 1995 as the first mission within the Mars Surveyor line, followed by orbiter-lander pairs to launch to Mars in 1998 and 2001.

33. The Jet Propulsion Laboratory also vied for the opportunity to develop a small asteroid-orbiting mission, but NASA ultimately chose APL to build NEAR given the latter institution's experience with building low-cost space systems for Strategic Defense Initiative programs and pressure from Maryland senator Barbara Mikulski. For more on this issue, see Howard E. McCurdy, *Low-Cost Innovation in Spaceflight: The Near Earth Asteroid Rendezvous (NEAR) Shoemaker Mission*, The NASA History Series, Monographs in Aerospace History Number 36, NASA SP-2005–4536 (Washington, DC, 2005), 5–15.
34. Japan's MUSES-A spacecraft had previously demonstrated aerobraking but did not rely on it for its operations.
35. The "skunk works" concept dates back to World War II, when the Lockheed Aircraft Corporation granted autonomy to a small team of workers apart from the rest of the company to complete advanced projects. McCurdy has noted that the success of Mars Pathfinder team members to execute the mission with so little oversight came in part from the fact that NASA and JPL were preoccupied with the human space flight program and the flagship Cassini mission to Saturn, respectively. See McCurdy, *Faster Better, Cheaper*, 99.
36. Brian K. Muirhead and William L. Simon, *High Velocity Leadership: The Mars Pathfinder Approach to Faster, Better, Cheaper* (New York: HarperCollins, 1999), 113.
37. Susan M. Niebur, "Principal Investigators and Project Managers: Insights from Discovery," *Space Policy* 26 (August 2010): 174–84.
38. S. Alan Stern, interview with the author, February 10, 2010.
39. The seven papers can be found in *Science* 278 (December 5, 1997): 1734–74.
40. NASA press release 97–204, "Mars Global Surveyor Detects Magnetic Field As Aerobraking Begins," September 17, 1997; six papers related to Mars Global Surveyor's findings can be found in *Science* 279 (March 13, 1998): 1671–98.
41. G. Scott Hubbard, William Feldman, Sylvia A. Cox, Marcie A. Smith, and Lisa Chu-Thielbar, "Lunar Prospector: First Results and Lessons Learned," IAF-98-Q.4.01, 49th International Astronautical Congress, September 28–October 2, 1998, Melbourne, Australia.
42. National Research Council, *The Role of Small Missions in Planetary and Lunar Exploration* (Washington, DC: National Academy Press, 1995), 12–14.
43. Robert S. Boyd, "Pathfinder a Picture of Success; Spacecraft Snaps Photos of Mars," *The Denver Post* (July 5, 1997): A1; Matt Crenson, "Mars: Giant Leap for U.S.; Barren Surface 'Paradise' to NASA Scientists," *The Herald-Sun* (July 5, 1997): A1; and Glennda Chui, "'Hallelujah! Back on Mars!' NASA: Scientists Thrilled with First Pictures," *San Jose Mercury News* (July 5, 1997): 1A.
44. Contributing to this excitement was the claim made by two NASA-funded scientists the year prior that they had found evidence of life in a Martian meteorite, Allan Hills 84001, found in Antarctica.

45. National Research Council, *An Integrated Strategy for the Planetary Sciences: 1995–2010* (Washington, DC: National Academy Press, 1994), 182.
46. See, e.g., Tony Reichhardt, "Does Low-Cost Mean Low-Value Missions?" *Nature* 389 (October 30, 1997): 899; and Leonard David, "Is Faster, Cheaper, Better?" *Aerospace America* 36 (September 1998): 42.
47. Jefferson S. Hofgard, interview with the author, December 1, 2011.
48. See, e.g., Kathy Sawyer, "Pathfinder Lands on Mars, Sends Back Surface Images; Cocooned Craft Bounces 50 Feet on Impact In First Touchdown on Planet in 21 Years," *The Washington Post* (July 5, 1997): A1; "Pathfinder Success Vindicates Faster-Better-Cheaper Approach," *Aerospace Daily* (July 8, 1997): 34; and William Jefferson Clinton, "Statement on the Landing of the Mars Pathfinder Spacecraft, July 4, 1997" in *Public Papers of the Presidents of the United States, William J. Clinton, 1997, Book 2, July 1 to December 31, 1997* (Washington, DC: Government Printing Office, 1997), 915.
49. WIRE Mishap Investigation Board Report, June 8, 1999, 11.
50. The string of "faster, better, cheaper" mission failures continued beyond this pivotal year. NASA and APL lost the CONTOUR Discovery mission in 2002 due to a technical problem traced to insufficient contractor oversight, and in 2004 the Genesis sample return capsule crash-landed on Earth when its drogue parachute did not deploy because the capsule's accelerometers had been wired incorrectly.
51. NASA had to date lost six out of sixteen spacecraft developed under "faster, better, cheaper." In addition to the five spacecraft lost in 1999, a small Earth science probe, Lewis, failed in 1997.
52. Mars Climate Orbiter Mishap Investigation Board, Phase I Report, November 10, 1999, 6–7.
53. Mars Program Independent Assessment Team Summary Report, March 14, 2000.
54. McCurdy, *Low-Cost Innovation in Spaceflight*, 38, 54.
55. The works by Howard McCurdy and Liam Sarsfield cited in this essay describe a narrow region in which cost, schedule, and complexity balance to enable a reliable space mission. Both analysts reference the work of David Bearden (The Aerospace Corporation) in this area. See, e.g., David A. Bearden, "When Is a Satellite Mission Too Fast and Too Cheap?" Presentation at the 2001 MAPLD International Conference, September 11, 2001.
56. G. Scott Hubbard, interview with the author, January 5, 2010.
57. Edward J. Weiler, interview with the author, December 29, 2009.
58. Mars Program Independent Assessment Team Summary Report, March 14, 2000. An internal study requested by Goldin and led by Mars Pathfinder project manager Tony Spear to assess the initiative also was optimistic about this approach's continuation, provided that NASA implemented certain changes. See Tony Spear, "NASA Faster, Better, Cheaper Task Final Report," March 18, 2000.
59. Robert Lee Hotz, "Mars Probe Lost Due to Simple Math Error," *Los Angeles Times* (October 1, 1999), http://articles.latimes.com/1999/oct/01/news/mn-17288 (accessed December 12, 2011).
60. David Perlman, "NASA Craft Lost in Orbit Approach/Blow to U.S. Space Program—NASA Stunned," *San Francisco Chronicle* (September 24, 1999), http://articles.sfgate.com/1999-09-24/news/17698592_1_mars-surveyor-red-planet-mars-climate-orbiter (accessed December 12, 2011).

61. "Reports: NASA's Approach to Exploration Not Working," *USA Today* (March 14, 2000): 3A.
62. James Oberg (*UPI*), "NASA Knew Mars Polar Lander Doomed," March 21, 2000.
63. Senate Committee on Commerce, Science, and Transportation, Subcommittee on Science, Technology, and Space, Hearing on NASA's Performance, March 22, 2000, transcript page 4.
64. Ibid., transcript page 2.
65. Ibid., 33.
66. Senate Committee on Appropriations, Subcommittee on Veterans Affairs, HUD, and Independent Agencies, Hearing on NASA FY 2001 Budget, April 13, 2000, transcript page 5.
67. James Sensenbrenner, "Sensenbrenner Statement on Mars Orbiter Error," September 30, 1999.
68. Aerospace Safety Advisory Panel, *Annual Report for 1999* (Washington, DC: NASA Headquarters, February 2000).
69. Steven Isakowitz, interview with the author, December 22, 2011.
70. Senate Committee on Appropriations, Subcommittee on Veterans Affairs, HUD, and Independent Agencies, Hearing on NASA FY 2001 Budget, April 13, 2000, transcript page 13.
71. National Research Council, *Assessment of Mission Size Trade-offs for NASA's Earth and Space Science Missions* (Washington, DC: National Academy Press, 2000), 3.
72. By the time of its 2011 launch, the Mars Science Laboratory had grown to $2.5 billion (FY 11$) in cost.
73. Scott Hubbard, *Exploring Mars: Chronicles from a Decade of Discovery* (Tucson: University of Arizona Press, 2011).
74. National Research Council, *New Frontiers in the Solar System: An Integrated Exploration Strategy* (Washington, DC: The National Academies Press, 2003).
75. Hubbard, *Exploring Mars*, 85–86.
76. Orlando Figueroa, interview with the author, January 8, 2010; and G. Scott Hubbard, interview with the author, January 5, 2010.

Part II

Developing New Approaches to Planetary Exploration

Chapter 4

Redefining Celestial Mechanics in the Space Age: Astrodynamics, Deep-Space Navigation, and the Pursuit of Accuracy

Andrew J. Butrica

Space exploration unquestionably has had a major and valuable impact on a range of scientific disciplines. Equipped with highly sensitive instruments, probes have discovered new bodies and phenomena and have contributed to our knowledge of the atmospheres, ionospheres, magnetospheres, and geologies of solar-system objects. Crucial to achieving these scientific advancements has been the careful placement of spacecraft-borne instruments in close proximity to their subjects. In addition, flights into the solar system made possible by the creation of the National Aeronautics and Space Administration (NASA) in 1958 engendered a concomitant shift in longstanding funding patterns, a shift that was especially visible in the field of astronomy.[1]

Astronomy differed from other space-related disciplines in that, in addition to growing with the increasing amount of data and discoveries that solar-system exploration provided as was happening in other fields, it also enabled the exploration that made possible those advancements through its relationship with space navigation. This essay focuses on the dynamic relationship between space navigation, the mathematics-based discipline that underlies all spacecraft guidance, and its evolving relationship with that branch of astronomy known as celestial mechanics, which concerns itself with the motions and gravitational effects of celestial objects and the drawing up of ephemerides.

The values assigned to physical constants are at the heart of both astronomy and navigation. Constants are a necessity in the sciences; agreement on their value assures the compatibility of what theory predicts and observation quantifies. In astronomy, the derivation of the values of such constants results from many years of observations and the fitting of observations to models called theories. These theories are analytical, mathematical approximations of a body's observed motion in space. In order to have a common basis for comparison over long time periods, astronomers require that star catalogs and ephemerides be based on a fixed and self-consistent set of astronomical constants. They therefore are reluctant to alter the values of those constants.

With the advent of space exploration, one could determine constants faster and to an increasingly greater degree of precision. In space navigation, it is of the highest importance that one use the most accurate and latest set of constants available. Thus, the astronomer's need for *consistent* values was set against the navigator's craving for the newest and most *precise* value. In a sense, one had a dual system of constants: (1) the astronomer's set of self-consistent constants not subject to speedy change and available for general scientific use, and (2) the navigator's set of continually updated constants prepared ad hoc for a specific mission. Samuel Herrick, an astronomy professor at the University of California at Los Angeles (UCLA), drew attention to this division by distinguishing between astronomy and what he called astrodynamics.[2]

Deep-space navigation made new demands on astronomy not unlike those made by deep-water navigation centuries ago. Seafaring navigators relied on instruments of increasing accuracy as well as the knowledge of the sky that astronomers possessed and packaged for navigators in the form of ephemerides and almanacs. The direction of knowledge was always from the theoretical to the practical: celestial mechanics informed the ephemerides, and astronomers in turn converted the ephemerides into almanacs and tables in a form handy for use by navigators. Observations made at sea never became the basis for improved ephemerides or advanced celestial mechanics, because they lacked the accuracy necessary to improve those ephemerides and almanacs.

With the dawn of space travel, a need similar to that created by ocean travel arose for the practical products of astronomy. At first navigators borrowed astronomers' ephemerides, but they quickly found that these lacked the accuracy requisite for guiding craft to the Moon and beyond. Unlike the case of oceanic navigation, navigators derived measurements of spacecraft motion that were far more precise than those made by astronomers. This tracking data—plus observations made with ground-based radar—became the basis for constructing improved ephemerides. Navigators quickly came to question the physical constants and assumptions regarding the motions of the Moon and other bodies that were at the heart of celestial mechanics. Thus, space travel turned the flow of knowledge from the practical back to the theoretical. Eventually, moreover, the astronomers who prepared the ephemerides and almanacs adopted the values and theories derived by navigators as their own.

Almanac Offices and Ephemerides

Celestial navigation, the term for navigating by means of the Sun and stars, is centuries old. It involved measuring the altitude of the Pole star or the Sun perhaps using a sextant, which gave only latitude information, that is, one's position relative to the equator. In order to guide one's way with accuracy, both latitude and longitude are necessary. The advent of accurate chronometers in the 1760s revolutionized navigation by enabling oceanic travelers to determine their longitude, as well.[3] The main rationale for founding the Royal Observatory at Greenwich was to acquire and to improve astronomical data for celestial navigation. Similar navigational needs gave rise to the establishment of the United States Naval Observatory and, in 1849, the Nautical Almanac Office.[4]

One of the key advances that prepared almanac offices for the Space Age was the technology that they used to formulate ephemerides and to prepare publications. Previously, everything was done "by hand," using slide rules, multiplication tables, and desk calculators.[5] During the 1920s and 1930s, Leslie J. Comrie, a leading exponent of the use of punched-card machines in astronomy, automated the British Nautical Almanac Office.[6] Wallace J. Eckert, aware of Comrie's work, pioneered the automation of astronomical computations in the United States. While teaching Celestial Mechanics at Columbia University, Eckert founded the Thomas J. Watson Astronomical Computing Bureau, operated jointly by Columbia University, the American Astronomical Society, and the International Business Machines (IBM) Corporation. This unique facility allowed researchers to solve the differential equations of planetary motion via numerical integration. Most of its early work consisted of two large astronomical projects undertaken in cooperation with Dirk Brouwer of Yale University, where Eckert had received his doctorate in astronomy.[7]

Starting in 1940, as the new director of the Nautical Almanac Office, Eckert simplified air navigation by supplying military pilots with precalculated data based on astronomical tables in the *American Air Almanac*. He used punched-card technologies not only to perform calculations, but also to produce and print the *Air Almanac* in a format that permitted its replication by photo-offset printing. While Eckert focused on the *Air Almanac*, Paul Herget helped to automate production of the *American Ephemeris and Nautical Almanac* for seafaring navigators, surveyors, and astronomers. An assistant professor of astronomy at the University of Cincinnati who began working at the Almanac Office in 1942, Herget, equally inspired by Comrie's work, had applied such techniques to his dissertation on the computation of orbits.[8]

World War II and the automation of astronomy at the Almanac Office forged the links of an important partnership between Eckert and Herget. To their number one must add Gerald Clemence, whom Eckert hired in 1940. Clemence worked with Eckert and Herget in introducing the new punched-card technology and in automating astronomical computations. In 1945, Clemence succeeded Eckert, who returned to the Watson Scientific Computing Laboratory.[9]

In his quest to improve planetary theories and astronomical constants, Clemence collaborated closely with Eckert and Dirk Brouwer at Yale

University. Beginning in 1947, their joint efforts received financial support from the Office of Naval Research, which awarded a long-term contract to Yale, the Naval Observatory, and the IBM Watson Laboratory to undertake work on a variety of solar-system problems. In addition to using automation to perform the computations, one of the novelties of the effort was to utilize numerical integration instead of traditional analytical techniques. Numerical integration used equations of motion and processed data in an iterative process. Because it was such a laborious process, numerical integration required access to a mainframe computer.[10]

Brouwer, Eckert, and Clemence thus took celestial mechanics to a new level that foreshadowed the new techniques and technologies of space navigation, namely, numerical integration and large-scale digital computers. These four astronomers accomplished more in celestial mechanics than anyone had achieved for many years previously, because they exploited the availability of punched-card and computer technology. Undoubtedly, their collaboration marked a major era in the history of celestial mechanics that was thriving at the dawn of the Space Age.[11]

From Comets to Spacecraft

The advances in celestial mechanics that Brouwer, Herget, Clemence, and Eckert forged prepared the astronomical community for the dawn of the Space Age. The launch of the Sputniks engendered an urgent need for the computation of orbits and the design of trajectories, which was the bailiwick and forte of the world's nautical almanac offices. Not surprisingly, then, Vanguard, an Earth-circling satellite to be launched as part of the International Geophysical Year, called on Clemence and his Almanac Office colleague, Raynor Duncombe, to serve as consultants along with Paul Herget who was back at the University of Cincinnati.[12]

The Almanac Office also lent assistance to the Apollo program. Beginning with Apollo 8, the first US human flight to escape the Earth's gravitational pull, the Almanac Office developed a series of star charts for navigation and visual orientation at NASA's request. Such voyages differed significantly from deep-space missions primarily because the human presence facilitated the application of traditional celestial navigation methods. Thus, in 1961, NASA called on Captain Philip Van Horn Weems for navigational assistance on Apollo. A graduate of the Naval Academy, Weems devised his eponymous system of simplified celestial navigation in the 1920s and wrote a book that became a standard text in navigation courses. For the Apollo astronauts, he developed a simplified mathematical navigational formula to assist them in determining their position in space within seconds by visual sighting.[13]

Although traditional celestial mechanics was not appropriate for deep-space navigation, calculating a spaceship orbit was not unlike the challenge of determining a comet's orbit, which similarly experiences gravitational perturbations. The idea that comets and spaceships shared certain characteristics was not new. Krafft Ehricke, a rocket-propulsion engineer, talked about spaceships as "comet craft,"[14] while Arthur C. Clarke likened navigating a spaceship to determining a comet's orbit.[15] The connection between the two led to the creation of the first university courses in space navigation.

The modern astronomical tradition of determining cometary orbits has its origins in Edmond Halley's application of Isaac Newton's laws to the movement of a comet that had appeared in 1531 and 1607 and his prediction of its return in 1758.[16] More recently, the study of comets (along with asteroids and meteors) received fresh attention between the two world wars, as astronomers in the United States studied the orbits of those objects and planets in an attempt to uncover clues as to their origin and the origin of the solar system itself.[17] Among those institutions performing investigations of cometary orbits were the Nautical Almanac Office and the University of California at Berkeley, where astronomy professors Armin O. Leuschner and Russell T. Crawford taught and conducted research. Leuschner's doctorate at the University of Berlin had been on the orbits of comets, and Crawford was a coauthor with Leuschner of an important 1930 work on determining the orbits of comets and asteroids.[18]

Samuel Herrick, who studied under Leuschner and Crawford, carried on the tradition. He wrote his dissertation on the determination of comet orbits and, as an astronomy professor at UCLA, he taught, among others, a course called "Determination of Orbits." In 1942 he established a new course, the first of its kind, in what he called "Rocket Navigation."[19] In order to disseminate his astronomy-based spacecraft orbit-determination methods, Herrick

Figure 4.1 Mariner 10 to Mercury was the first American spacecraft to perform a planetary gravity assist, a milestone in space navigation. This was made possible by increased timekeeping accuracy at tracking stations. Today gravity assist is a common and essential tool for space exploration. It requires that the position of the spacecraft be determined very accurately so a successful trajectory can be planned. (Credit: Courtesy of NASM, photo by Dane Penland.)

founded the Institute of Navigation in 1946 with Captain Weems. Herrick thus bound space navigation with traditional celestial navigation.[20]

He also compiled a book of tables useful for calculating both rocket and comet orbits with funding from the John Simon Guggenheim Memorial Foundation. The tables allowed the user to determine the position and velocity of a rocket or comet given the time. At the heart of the table was Herrick's theoretical work that sought to unify the calculation of rocket flight paths and comet orbits, In space navigation as with the compilation of almanacs, access to mainframe computers was a must. Herrick, unlike Herget, Eckert, Brouwer, and Clemence, did not have ready access to such computers. Therefore, in order to carry out the computations required for his tables, he had to rely on the staff of the National Bureau of Standards' Institute for Numerical Analysis and funding from the Office of Naval Research.[21]

Space Navigation Begins

Deep-space navigation began with the series of Pioneer probes launched as part of the Advanced Research Projects Agency's Lunar Program (and the heliocentric Pioneer 5) followed by the NASA Ranger, Surveyor, and Lunar Orbiter series to the Moon, Mariners 1 through 9 to Mars and Venus, and the interplanetary Pioneers 6 through 9, all of which launched between 1958 and 1971. During this period, NASA's JPL acquired and fostered the mathematical, astronomical, and computer resources necessary to meet the challenges of navigating spacecraft. These resources included a trajectory program, software for computing orbits, and the requisite ephemerides to support that software.

Key to the accuracy of navigation was the nature of the data utilized. The basic navigational data types were two-way Doppler readings and, beginning with the Lunar Orbiter series, range measurements.[22] Astronomers were more interested in the position of an object as viewed against a starry background than its distance or velocity, so they relied on angle measurements. Such positional readings had far less utility in navigation, because probes en route to the Moon and beyond scarcely moved against the background of fixed stars.[23] Navigators' reliance on radio data thus set them apart from traditional astronomers' reliance on positional (optical) measurements. Nonetheless, navigation relied heavily on astronomical methods and products, such as ephemerides, in the design of trajectories and the determination of orbits.

The development of the first JPL trajectory program, called the Space Trajectories Program or SPACE for short, was the virtually solo endeavor of Douglas B. Holdridge, a member of JPL's Computer Applications and Data Systems Section who had received an MS in mathematics from the California Institute of Technology (Caltech) in 1957.[24] Similarly, the individuals who first created the orbit determination program were neither expert nor experienced in astronomy. The three credited with crafting the program were R. Henry Hudson, a programmer in the Computer Applications and Data Systems Section who left JPL in 1964; Jack Lorell, a research specialist hired into Homer Joe Stewart's Research Analysis Section sometime during the 1940s; and Russell E. Carr, who worked in the laboratory's Space Sciences Division. Both Lorell and Carr during the mid-1950s had worked together at JPL on a propulsion project.[25]

The orbit determination program was far more complex than the trajectory software, which it ran as a subroutine. The fundamental mathematical process of this and later orbit determination programs was an iterative numerical integration of the Doppler data. Thus, from the start, the navigation software took advantage of the computing power available at the laboratory, purchased to support the NASA space program, and utilized numerical integration as the sine qua non for computing orbits.[26]

Both the orbit determination and trajectory programs utilized the same ephemeris information through various subroutines. As with the trajectory and orbit determination software, the creation of the ephemerides was not the work of an astronomer. The architect of the original ephemeris tapes was the computer programmer R. Henry Hudson, who also had helped to develop the orbit determination program. He assembled the data for the tapes in conjunction with personnel from Space Technology Laboratories who worked through a subcontract with JPL.[27]

The ephemeris tapes reflected heavy borrowing from the Almanac Office and common astronomical practices. For example, the tapes used the same frame of reference that astronomers utilized, and they gave all coordinates in astronomical units[28] with the exception of the distance to the Moon, which was expressed in Earth radii, a common astronomical practice that nonetheless made the ephemeris tapes inconsistent with the trajectory program. The adoption of astronomers' units of measurement reflected the origins of the ephemeris data in the publications of the Naval Observatory's Nautical Almanac Office, including the astronomical tables for the Sun, the Moon, Venus, and the outer planets. The Naval Observatory supplied much of this information in the form of punched cards.[29]

The imprecision of the constants underlying the Almanac Office ephemerides led navigation personnel at JPL—as well as at other NASA centers concerned with Earth-orbiting satellites and astronaut travel—to adopt their own self-consistent set of values for constants used in their ephemeris, trajectory, and orbit determination work. Without direction or approval from NASA Headquarters, they organized an ad hoc working group that agreed on the NASA Standard Set of Constants in 1961 for use by all concerned NASA centers and contractors. As a result, the JPL constants became equally those of Earth-orbiting and human lunar missions, and contractors on the Ranger, Mariner, and Surveyor programs implemented them, as well.[30] The JPL constants also proved to be of high value to astronomers.

JPL Development Ephemerides

One of the principal reasons for the eventual adoption of JPL values by astronomers was their high precision. That precision resulted from the types of data that JPL ephemeris developers had at their disposal. One such type was flight tracking information. During a mission, navigators fed values for constants into the orbit determination program in order to figure out a spacecraft's flight path. After the mission, they ran the program again, but solved for the constants and other values, because now the spacecraft's flight path was known to a high degree of accuracy. Their post-flight analyses yielded more precise measurements of, for example, the astronomical unit and the masses of the Earth and Moon and their gravitational constants (i.e., their mass times the universal value of gravity).

The other data type that provided the JPL ephemeris with more precise measures of physical constants came from radar astronomy. Benjamin S. Yaplee and others at the Naval Research Laboratory's Radio Astronomy Branch along with Dirk Brouwer computed new values for both the radius of the Moon and the distance between Earth and the Moon from radar observations of the Moon collected in 1959 and 1960.[31] A group of researchers at MIT's Lincoln Laboratory led by Irwin I. Shapiro refined the astronomical unit and other ephemeris values by combining radar and optical measurements.[32] Meanwhile, JPL computed a more precise value for the astronomical unit from its own radar studies of Venus.[33] In this way, radar quickly became a fundamental tool of deep-space navigation to the Moon, Venus, and Mars, because it furnished more accurate values for the basic physical constants that lay behind the ephemerides on which navigators had to rely.

The JPL ephemeris group also underwent a transformation that contributed in no small way to the growing sophistication and precision of its main product, the ephemeris tapes that NASA used to design trajectories and to determine spacecraft orbits. The personnel constituting the ephemeris group started to change with the addition of more and more astronomers trained in celestial mechanics. Indeed, as a result, the members of the ephemeris group came to resemble a miniature version of the larger celestial mechanics community that Brouwer, Clemence, Herget, Eckert—and Herrick—had been fostering.

Initially, the development of the ephemerides remained in the hands of individuals trained in mathematics and familiar with computer programming who worked in JPL's Computer Applications Section. The earliest leader of the group was Neil L. Block, better known by his astrologer alias Gary Duncan. P. Kenneth Seidelmann, at one time director of the Almanac Office, characterized him as "a very competent sort of numerical person." His fellow astrologers saw him as "a computer scientist and professional astronomer" who "lectured in celestial mechanics, numerical analysis and statistics." Block left JPL in 1967 to return to school.[34]

In addition to Douglas B. Holdridge mentioned earlier, the group included Dr. Charles "Chuck" L. Lawson, who started at JPL in 1960 as an applied mathematician specializing in mathematical software. Lawson had MSc and PhD degrees in mathematics from UCLA. Between 1964 and 1967, Lawson took over as head of the ephemeris development team and began to hire a sizeable number of astronomers who had studied under Brouwer, Herget, and Herrick. The bulk of them came from Yale.[35]

Between 1964 and 1966, for example, a number of Yale students spent their summers working with the JPL ephemeris group, including Carol Williams, Jay H. Lieske, Brian G. Marsden, and David Dunham, while Lieske became regular staff after completing his dissertation. In 1972, Lieske, who now led the group, hired fellow Yale astronomy graduate E. Myles Standish. Douglas A. O'Handley, a Yale graduate working at the Naval Observatory, joined the group at that time, as well.[36] The ephemeris group also included J. Derral Mulholland, a 1965 PhD in astronomy from Herget's University of Cincinnati program, and a good number who had studied astrodynamics under Samuel Herrick at UCLA. The first Herrick students were John D. Anderson and Theodore D. Moyer, followed by Satorios S. Dallas and Vladimir John Ondrasik, Jr.[37]

This miniature version of the celestial mechanics community improved the JPL ephemeris one iteration after another. The product of the group was known as a Development Ephemeris ("DE") followed by a number that indicated the particular iteration. The first ephemeris tapes distributed outside JPL in 1964, however, were known simply as E9510, E9511, and E9512, but received the designation DE3 upon the distribution of the next iteration, known as DE19, in 1967. A not insignificant number of copies went out to government and industry users. A December 1964 Interoffice Memorandum, for example, listed 18 companies using the ephemeris tapes.[38] Later versions were available from the NASA Computer Software Management and Information Center (COSMIC) at the University of Georgia.[39]

The creators of the Development Ephemeris relied on the work of the Almanac Office, but not without having to make important changes. For example, they had to deal with astronomers' units of measurement that, from a navigation perspective, were sometimes unwieldy. While astronomers measured the distance from the Earth to the Moon in terms of Earth radii, navigators expressed distances in kilometers, a unit of length based on an international laboratory standard, the meter. The Paris 1896 Congress of Directors of National Ephemerides adopted a value for the astronomical unit based on the solar parallax, which measured the distance indirectly. Here, too, navigators worked in kilometers, so they had to devise conversion factors.[40]

Many of the improvements in the JPL Development Ephemeris did not derive from the accuracy of tracking or radar data. One of the more far-reaching contributions came from a portion of Jay Lieske's doctoral dissertation[41] published as a JPL technical report and dubbed DE28. The ephemeris covered a greater span of time than any previous JPL ephemeris, from 1800 to 2000. Lieske emphasized the importance of basing numerical integrations on long time-spans of observations. The incorporation of the treasure trove of past astronomical observations into the JPL ephemeris database marked a major milestone in the program's evolution.[42] Additionally, unlike the numerical integrations of the widely distributed DE3 and DE19 that solved for a single body, Lieske's software was an adaptation of an N-body (multiple-body) computer program developed for asteroids and comets by Joachim Schubart and Peter Stumpff of the Heidelberg Astronomisches Rechen-Institut.[43]

Just as Lieske's DE28 provided a new conceptual approach to generating JPL ephemeris products, a new computer program, the Solar System Data Processing System (SSDPS), furnished the means for realizing them. The SSDPS, whose formulation Chuck Lawson oversaw, was actually a series of programs that determined the motions of the planets through simultaneous numerical integrations. The first such simultaneous integration of all the planets took place in the spring of 1968. In contrast, the existing DE19 consisted of successive single-body integrations for each of the planets. The SSDPS also compared optical, radar, and spacecraft observations and performed a number of other key navigational functions.[44]

The combination of Lieske's approach, the new software, and observations from an even greater time-span enriched subsequent iterations of the Development Ephemeris, DE69 (1969), DE96 (1976), and DE102 (1983). Reflecting the new emphasis on numerical integration of increasingly more observations, DE69 and DE96 used thousands of optical observations from the Naval Observatory's Six-Inch and Nine-Inch Transit Circles collected between 1910 and 1971, a period of six decades, while DE102 spanned the gamut of astronomical observations considered to have usable accuracy, namely, from 1411 BC to 3002 AD.[45]

In addition, DE69 incorporated radar data from JPL and Lincoln Laboratory, while DE96 added measurements from Cornell University's giant Arecibo dish in Puerto Rico. DE69 marked another milestone as the first JPL ephemeris to use spacecraft range data, specifically, the planetary range points collected in 1967 from the Mariner 5 mission to Venus. For DE96, the Mariner 9 Navigation Team contributed Earth-Mars ranges obtained between November 1971 and October 1972, while the navigation teams for Pioneers 10 and 11 provided Earth-Jupiter ranges. DE102 added lunar laser ranging to the database. The first laser ranges to the Moon became a reality in 1969, thanks to the so-called Laser Ranging Retro-Reflectors set up at Tranquility Base in the Sea of Tranquility by Apollo 11 astronauts.[46] In addition, the SSDPS software allowed navigators to integrate the dynamic movements of the solar system, including for the first time the asteroids Ceres, Pallas, and Vesta, over the entire twentieth century using n-body equations of motion.[47]

Measure for Measure

The success of space navigators in improving their ephemerides and in obtaining increasingly more precise values for the physical constants on which those ephemerides relied had an immediate impact on astronomers interested in celestial mechanics, particularly those working in the world's almanac offices, who saw a need to reevaluate such key astronomical constants as the astronomical unit. Already, in 1950, the time seemed to be ripe to discuss adopting new constants, but almanac office directors in Europe and America meeting in Paris recommended no immediate changes in those constants to the International Astronomical Union (IAU).[48]

The IAU met again in May 1963 at the Paris Observatory to discuss, among other issues, the appropriate value for the astronomical unit. According to Gerald Clemence, the recent radar-determined value of the astronomical unit and the precise value of the Sun-Venus mass ratio computed from Mariner 2 tracking data "provided the immediate incentive" for holding the meeting. In fact, were it not for the radar value, Mariner 2 would have passed Venus without acquiring any useful data.[49] (See Figure 4.2)

The attendees of the 1963 IAU meeting included the usual directors and other representatives of national almanac offices and astronomers interested in celestial mechanics. Among them were Dirk Brouwer and Gerald Clemence, now scientific director of the Naval Observatory; both had helped JPL prepare the ephemeris used in their Venus radar experiment. For the first time, the attendees included members of the radar astronomy and space communities, such as Eberhardt Rechtin from JPL; Duane O. "Dewey" Muhleman, a former JPL employee enrolled in the Harvard Astronomy Department; and

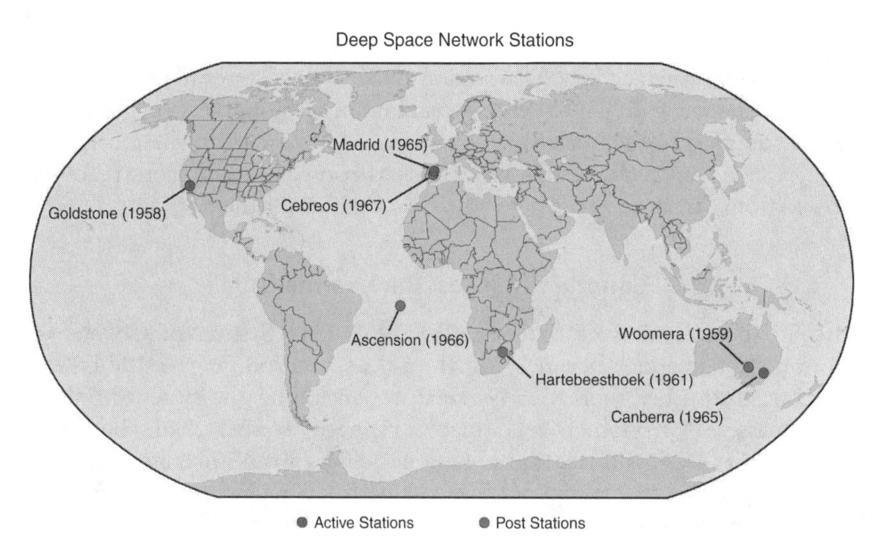

Figure 4.2 The Deep Space Network map. The Deep Space Network was constructed to communicate and command spacecraft across the solar system. This map shows the three active stations in blue, inactive stations in red, and the first year of operation for each station. Stations for the Deep Space Network have been constructed at sites that allow for constant communication throughout the solar system. The Goldstone station in the Mojave Desert of California was the first to begin operations. Stations in Australia, South Africa, Spain, and Ascension Island were later added. Today the stations at Goldstone, Madrid, and Canberra make up the network. The position of the three stations on the Earth's surface has been surveyed very accurately. The distance between stations, or the baseline, must be known when using two stations to determine the location of a spacecraft. (Credit: Courtesy of NASM.)

Irwin Shapiro, the "guru" who calculated the ephemeris for MIT Lincoln Laboratory's radar research.[50]

Astronomers expressed concern about the validity of the various radar values computed for the astronomical unit because of the discrepancy between the initial radar measurements made in 1958 and 1959 and those made during the 1961 conjunction. The IAU recommended that a working group study the astronomical unit and other constants. The members of the working group comprised representatives from US, French, German, English, and Soviet observatories and almanac offices. Nobody from the space navigation or radar astronomy communities participated in the working group. After soliciting comments from some 80 astronomers, the working group met again in January 1964, at the Royal Greenwich Observatory, Herstmonceux Castle, and drew up a list of constants, including the astronomical unit, for consideration by the IAU general assembly, which met in Hamburg in August 1964.[51]

Despite some reservations, the working group recommended adopting the radar value for the astronomical unit, which the IAU general assembly subsequently adopted.[52] It was now a matter of incorporating the new radar-based value into the various national almanacs and ephemerides. The accuracy of the radar data clearly surpassed the accuracy of optical measurements made by astronomers and used by almanac offices in formulating ephemerides. The technologies on which space navigation relied simply were inherently more accurate, and navigation into deep space required such high accuracy.With the IAU's adoption of the radar value for the astronomical unit, the movement of knowledge between celestial mechanics (theory) and navigation (practice) flowed from practice to theory for the first, but not for the last, time.

Planetary Masses: Back to the IAU

Despite the adoption of the so-called 1968 IAU System of Constants, meaning ephemerides began using the values in 1968, at the 1964 IAU meeting, unfinished work clearly remained, including the system of planetary masses. Astronomers recognized the need to alter them, but they believed that fresh information about the masses would become available in the coming years mainly, but not totally, from tracking spacecraft. The next meeting to address the question of the planetary masses—among other issues—was the Colloquium on the IAU System of Astronomical Constants held in Heidelberg in August 1970. Both Kenneth Seidelmann and Raynor Duncombe from the Naval Observatory contributed to the discussion on planetary masses, as did William Melbourne from JPL.[53]

JPL already deemed the IAU constants to be "out of date" and "of insufficient accuracy for astrodynamic purposes" in its System of Astronomical Constants released in 1968. Of all the IAU constants, the planetary masses were "the most seriously out of date." JPL claimed to have realized a "dramatic improvement in astronomical constants," refining the accuracies of the mass values of Venus, Mars, the Earth, and the Moon by two or three orders of magnitude thanks to direct measurements of their gravitational fields by spacecraft.[54]

At the Heidelberg meeting, Clemence introduced a general discussion on the desirability of adopting revised values for the planetary masses. Because better values would be available in four or five years, he argued for deferring the question until at least 1973. In the end, IAU members voted against adopting new values for planetary masses and other constants in 1973, because new data would become available. They also voted in favor of updating the system of planetary masses at a time closely linked to the production of the next fundamental star catalog (FK5) due in 1978. FK5, which stood for *Fundamental Katalog* ("Fundamental Star Catalog") number 5, was a German star catalog considered to be the definitive celestial coordinate system of its day.[55] The idea was to make a number of basic changes all at the same time.

The 1970 meeting also recommended setting up working groups, including one for the purpose of specifying, "in time for consideration in 1973, the basis for the planetary ephemerides to be published in almanacs for 1980 onwards." The 1970 IAU General Assembly meeting in Brighton, England, formally established the working group, and a Working Meeting on Constants and Ephemerides took place in October 1974 at the Naval Observatory in Washington, DC, to draft the group's proposed report. The chairmen of all working groups met in September 1975 and June 1976 in Herstmonceux, England, and Washington, DC, respectively, and the IAU Sixteenth General Assembly adopted their report and recommendations at its August 1976 meeting in Grenoble, France.[56]

The result of these deliberations and resolutions was a major fundamental readjustment of the quantitative and theoretical basis for the ephemerides. The IAU voted to introduce the IAU (1976) System of Astronomical Constants in 1984 in conjunction with the FK5 star catalog, so that national almanacs would incorporate both at one time. In addition, the 1984 almanac ephemerides would feature new theories, replacing current ones that were from three to eight decades old.[57] In order to base the almanac ephemerides on the latest and most accurate constants as well as on the best theories, the IAU chose to make JPL Development Ephemeris 200 (DE200) the basis for those constants as well as for the theories and tables printed in the Naval Observatory's *American Ephemeris and Nautical Almanac*, renamed *The Astronomical Almanac* with the 1981 edition. DE200 was a high-precision ephemeris created by numerical integration based on multiple data types spanning more than 44 centuries, from 1410 BC to 3002 AD Ken Seidelmann, an astronomer with the Naval Observatory, had recommended the adoption of DE200 to his working group because, he concluded, it was the most accurate available, and nobody else was going to have that level of accuracy.[58]

What Had Happened?

How, then, did the navigators at JPL overthrow the astronomers of the Almanac Office and become the authority for the national almanac and ephemerides? What were the institutional decisions, actions, and changes that led to this revolution? In his thorough and insightful history of the Naval Observatory, Steven J. Dick finds fault at the observatory itself, especially in the person of Gerald Clemence, scientific director of the Naval Observatory from 1958 to 1963. (See Figure 4.3)

Clemence deliberately limited the observatory's role in the space program, because he felt that space belonged to NASA, not the observatory, plus the observatory just did not have the budget to compete against NASA. "It was not our charter or our mission at the Naval Observatory," Duncombe recalled, "and he [Clemence] didn't see any purpose in trying to do this. Here was NASA, newly created to handle all this. Why should a little place like the Naval Observatory try to spread out its funds to compete with something

Figure 4.3 The Goldstone station, located in the Mojave Desert of California. Goldstone contains several individual sites, each with its own set of large dish antennas. Each site is commonly named for the mission or destination for which it was first used. The first to begin operation was the Pioneer site, used to receive data from the Pioneer probes to the Moon. The Uranus site was first used to receive data from the Voyager 2 flyby of Uranus. (Credit: NASA/JPL.)

like that? It made a lot of sense not to do that." According to Dick, this "was a decision that affected the Observatory for decades."[59]

A second reason for the shift in expertise to JPL, according to Dick, was the Naval Observatory's belated adoption of new, more accurate data types—such as radar and laser range measurements, very-long-baseline interferometry (VLBI) readings, and spacecraft range and Doppler information—to the formulation of the ephemerides. "This lack of data was crucial," Dick argues, "whereas modern transit-circle observations determined the position of Mars to within about 100 km, radar ranging to the planets determined distances to a precision of a few hundred meters." Furthermore, he calls attention to the fact that the observatory lacked JPL's access to spacecraft data, a fact that Seidelmann, a former observatory astronomer, confirmed. The lack of such data was critical, but on at least one occasion tracking data was made available to scientists. Mariner 4 (launched in November 1964) set a precedent, when navigators published the flight tracking data in a form suitable for use by researchers.[60]

Dick also highlights the production of competing ephemerides. Stimulated by a variety of space application needs, each sought to achieve greater accuracy than that of the Almanac Office's products. In addition to the JPL undertaking, a group at MIT's Lincoln Laboratory had its own ephemeris program, called the Planetary Ephemeris Program or PEP. It combined optical data from the Naval Observatory back to 1850 with current radar observations of Venus, Mars, and Mercury made by the laboratory's own facilities and the Arecibo dish, which relied on PEP for pointing its radar. PEP eventually also analyzed lunar laser range and VLBI data. Irwin Shapiro, who ran the project, exploited the laboratory's computing power and designed the program so that it numerically integrated the equations of motions of the planets.[61]

Although planetary radar astronomers at Arecibo continued to use PEP, pulsar observers and other astronomers shifted to the JPL ephemeris program. In the end, a lack of funding left the PEP just able to keep up with the Arecibo ephemeris work.[62] In contrast, the JPL ephemeris program enjoyed a more abundant source of manpower and funding, because it supported NASA spacecraft missions.

NASA's Goddard Space Flight Center and the Naval Surface Weapons Center at Dahlgren, Virginia, also had ephemeris projects. Unlike JPL and the Almanac Office, however, they were not necessarily vying to create ephemerides for general scientific purposes. Instead, each was addressing a niche need peculiar to that institution's mission. MIT's Lincoln Laboratory, for instance, focused mainly on serving radar astronomers; its chief client was the Arecibo radar. Goddard, on the other hand, concentrated on Earth-orbiting satellites, ceding deep space to JPL, a reflection of the types of missions carried out at the two centers. The Naval Surface Weapons Center served the missile-related needs of the navy. According to Dick, Dahlgren eventually became the navy's agency for orbit computations, rather than the Naval Observatory, because of its more general use of numerical integration.[63]

On top of these causes originating within the Naval Observatory that Dick enumerates, one can point equally to the death of Dirk Brouwer in 1966 as precipitating the end of an era in dynamic astronomy marked by the collaboration of Brouwer with Gerald Clemence, Wallace Eckert, and Paul Herget—and their computers—in the field of celestial mechanics. As Dick observes, "The subsequent dissolution of the Brouwer-Clemence-Eckert-Herget collaboration constituted the end of an era in celestial mechanics."[64] If that era ended, its legacy certainly lived on at least at the JPL ephemeris group.

The shift in astronomical authority from the Almanac Office to the Jet Propulsion Laboratory resulted not just from decisions and actions at the Naval Observatory, but also from the buildup of expertise in general celestial mechanics within the JPL ephemeris group. Significantly, as leaders of the ephemeris group, Chuck Lawson and his successor Jay Lieske hired individuals trained as astronomers from Dirk Brouwer's Yale Astronomy Department, and recruited several students from Samuel Herrick at UCLA and Paul Herget at the University of Cincinnati. It was their work and that of their colleagues—combined with the inherent accuracy of radio tracking and radar data—that made the JPL Development Ephemeris succeed in becoming the basis for the US and world almanacs. Among the secondary contributors were the more extensive use of numerical integration, the availability of high-speed digital computers at JPL—made possible in turn by the relative largesse of NASA funding levels—and the fact that JPL created an export version of the Development Ephemeris for specialized use by government and industry researchers.

Notes

1. Joseph N. Tatarewicz, *Space Technology and Planetary Astronomy* (Bloomington: Indiana University Press, 1990).
2. William G. Melbourne, J. Derral Mulholland, William L. Sjogren, and Francis M. Sturms, Jr., *Constants and Related Information for Astrodynamic Calculations, 1968*, Technical Report 32–1306 (Pasadena: JPL, July 15, 1968), p. 33; Samuel Herrick, "Astronomical and 'Astrodynamical' Values of Constants and Ephemeral Data," in Jean Kovalevsky, ed., *The System of Astronomical Constants* (Paris: Gauthier-Villars, 1965), pp. 105–108; John D. Anderson, *Determination of the Masses of the Moon and Venus and the Astronomical Unit from Radio Tracking Data of the Mariner II Spacecraft*, TR 32–816 (Pasadena: JPL, July 1, 1967), pp. 6–7.
3. Steven J. Dick, *Sky and Ocean Joined: The U.S. Naval Observatory 1830–2000* (New York: Cambridge University Press, 2003), pp. 15–16. On the early history of navigation, see E. G. R. Taylor, *The Haven-Finding Art: The History of Navigation from Odysseus to Captain Cook* (New York: Abelard-Schuman, 1957); and Charles H. Cotter, *A History of Nautical Astronomy* (London: Hollis & Carter, 1968). On the instruments, see J. A. Bennett, *The Divided Circle: A History of Instruments for Astronomy, Navigation, and Surveying* (Oxford: Phaidon, Christie's, 1987). For the history of longitude and the chronometer, see Dava Sobel, *Longitude: The True Story of a Lone Genius Who Solved the Greatest Scientific Problem of His Time* (New York: Walker, 1995).

4. For the history of the Greenwich Royal Observatory, see *Greenwich Observatory...the Story of Britain's Oldest Scientific Institution, the Royal Observatory at Greenwich and Herstmonceux, 1675–1975*, 3 vols. (London: Taylor & Francis, 1975). On the Naval Observatory, see Dick, *Sky and Ocean Joined*, pp. 28–70, 119, 121–22. The evolution of national almanacs is taken up by P. Kenneth Seidelmann, Paul M. Janiczek, and Ralph F. Haupt, "The Almanacs—Yesterday, Today, and Tomorrow," *Navigation: Journal of the Institute of Navigation* 24 (1976–1977): 303–12.
5. Paul Herget, interview by David DeVorkin, Cincinnati Observatory, April 19–20, 1977, conducted for the American Institute of Physics, Center for History of Physics, transcript, James Melville Gilliss Library, US Naval Observatory, Washington, DC (hereafter, Gilliss Library), pp. 16, 64; Paul Herget, "Numerical Integration with Punched Cards," *Astronomical Journal* 52 (1946): 115–17; LeRoy E. Doggett and Steven J. Dick, *Oral History Interview with Raynor L. Duncombe and Julena S. Duncombe: on June 18, 1983 and on Jan. 11, 1988* (Washington, DC: US Naval Observatory, 1988), pp. 5–6; Dick, *Sky and Ocean Joined*, pp. 517–20.
6. Herget, interview, p. 16; Donald E. Osterbrock and P. Kenneth Seidelmann, "Paul Herget, January 30, 1908-August 27, 1981," National Academy of Sciences, *Biographical Memoirs* V.57 (1987): 63; Comrie, "The Application of the Hollerith Tabulating Machine to Brown's Tables of the Moon," *Monthly Notices of the Royal Astronomical Society* 92, 7 (1932): 694–707. On Comrie, see Mary J. Croarken, "L. J. Comrie and the Origins of the Scientific Computing Service," *IEEE Annals of the History of Computing* 21, 4 (1999): 70–71; Croarken, "L. J. Comrie: A Forgotten Figure in the History of Numerical Calculation," *Mathematics Today* 36, 4 (August 2001): 114–18; H. S. W. Massey, "Leslie John Comrie, 1893–1950," *Obituary Notices of the Royal Society* 8 (November 1951): 97–105.
7. Jean Ford Brennan, *The IBM Watson Laboratory at Columbia University: A History* (New York: IBM, 1970), pp. 3–4, 7–8; Dorritt Hoffleit, *Astronomy at Yale, 1701–1968* (New Haven: Yale University, 1992), p. 197.
8. Wallace J. Eckert, "Air Almanacs," *Sky and Telescope* 4 (1944): 4–8; Brennan, *The IBM Watson Laboratory at Columbia University*, p. 10; Osterbrock and Seidelmann, "Paul Herget," pp. 63, 65–66; Kaj Strand, interview by David DeVorkin and Steven Dick, December 8, 1983, and January 3, 1984, Space Astronomy Oral History Project, National Air and Space Museum, Smithsonian Museum, transcript, Gilliss Library, pp. 53–54; Herget, interview, p. 16; Herget to Prof. Jan Shilt, January 20, 1940, "National Research Fellowship," drawer 2, Paul Herget Papers, Gilliss Library (hereafter, Herget Papers); Dick, *Sky and Ocean Joined*, pp. 521–22. Wallace Eckert and Ralph Haupt, "The Printing of Mathematical Tables," *Mathematical Tables and Aids to Computation* 2 (January 1947): 187–202, traces the evolution of the production of the *American Air Almanac* during the war years. Herget's dissertation was: "The Determination of Orbits," PhD thesis, University of Cincinnati, 1935. In the same year, he published "The Determination of Orbits," *Astronomical Journal* 44 (1935): 153–61, based on the dissertation and reprinted as "A Method for Determining Preliminary Orbits Adopted to Machine Computation" *Publications of the Cincinnati Observatory* no. 21 (Cincinnati: Cincinnati Observatory, 1936).
9. Dick, *Sky and Ocean Joined*, pp. 523–25.
10. Ibid., pp. 526–27.

11. Raynor and Julena Duncombe, interview, p. 17; Dick, *Sky and Ocean Joined*, p. 530.
12. Raynor and Julena Duncombe, interview, pp. 18, 32–34. On Vanguard, see Constance McLaughlin Green and Milton Lomask, *Vanguard: A History* (Washington: Smithsonian Institution Press, 1971).
13. Dick, *Sky and Ocean Joined*, p. 532; Lawrence H. Aller, John L. Barnes, and George O. Abell, "Samuel Herrick, Engineering, Astronomy: Los Angeles, 1911–1974," University of California, *In Memoriam* (March 1976): 58–59; various items in folder 2534, NASA Historical Reference Collection, NASA Headquarters, Washington, DC (hereafter NHRC).
14. See, e.g., Ehricke, "Instrumented Comets—Astronautics of Solar and Planetary Probes," ARS Paper #493–57, Proceedings of the 8th International Astronautical Congress, Barcelona, October 1957, which also appeared as Convair Report AZP-019, July 24, 1957, cited in Ben Evans with David Michael Harland, *NASA's Voyager Missions: Exploring the Outer Solar System and Beyond* (New York: Springer, 2004), p. 40.
15. Arthur C. Clarke, *Interplanetary Flight: An Introduction to Astronautics* (New York: Harper & Brothers, 1951), p. 124.
16. Peter Broughton, "The First Predicted Return of Comet Halley," *Journal for the History of Astronomy* 16 (1985): 123–33. See also Curtis Wilson, "Clairaut's Calculation of the 18th-Century Return of Halley's Comet," *Journal for the History of Astronomy* 24 (1993): 1–15.
17. Ronald E. Doel, *Solar System Astronomy in America: Communities, Patronage, and Interdisciplinary Science, 1920–1960* (New York: Cambridge University Press, 1996), pp. 16, 123–25.
18. Paul Herget, "Armin Otto Leuschner," National Academy of Sciences, *Biographical Memoirs* 49 (1978): 132; Russell T. Crawford, Armin O. Leuschner, and Gerald Merton, *Determination of Orbits of Comets and Asteroids* (New York: McGraw-Hill 1930); Maud Worcester Makemson, "Russell Tracy Crawford, 1876–1958," *Publications of the Astronomical Society of the Pacific* 71, 423 (December 1959): 503.
19. Samuel Herrick, *The Laplacian and Gaussian Orbit Methods* (Berkeley: University of California Press, 1940); Aller, Barnes, and Abell, "Samuel Herrick," pp. 58–59.
20. Aller, Barnes, and Abell, "Samuel Herrick," pp. 58–59.
21. John Simon Guggenheim Foundation, "Fellows Whose Last Names Begin With H," http://www.gf.org/hfellow.html (accessed October 29, 2007); Samuel Herrick, *Tables for Rockets and Comet Orbits* (Washington: National Bureau of Standards, 1953), pp. esp. v and xxii.
22. William R. Corliss, *A History of the Deep Space Network* (Washington: NASA, 1976), pp. 9–10; Nicholas A. Renzetti, "DSIF in the Ranger Project," in *The Ranger Program*, TR 32–141 (Pasadena: JPL, September 1961), p. 37; JPL, *Space Programs Summary 37–10, Volume I, for the Period May 1, 1961, to July 1, 1961* (Pasadena: JPL, August 1, 1961), p. 64; "Mark I Ranging Subsystem," 13 in *Space Programs Summary 37–20, Vol. III, The Deep Space Network, for the Period January 1 to February 28, 1963* (Pasadena: JPL, March 31, 1963).
23. Nonetheless, angle data obtained from cameras and other optical equipment have remained an essential component of near-Earth tracking systems. In fact, launches of NASA's space shuttle rely on both military and civilian tracking elements, with the air force following Cape Canaveral launches

during the initial preorbital phases of flight. All shuttle launches are tracked by the air force's Eastern Range tracking system, also known as the Air Force 45th Space Wing. In addition, during the Columbia accident investigation, the Air Force Space Command reviewed observations made by its Space Surveillance Network including 3,180 separate radar and optical observations of the Orbiter from sites at Eglin, Beale, and Kirtland Air Force Bases, Cape Cod Air Force Station, the Air Force Space Command's Maui Space Surveillance System in Hawaii, and the Navy Space Surveillance System. Columbia Accident Investigation Board, *Report*, Vol. 1 (Washington: NASA, August 2003), p. 62; Return to Flight Task Group, *Final Report* (Washington: NASA, July 2005), pp. 49–54.

24. John E. Ekelund, "History of the ODP at JPL," date unknown, page 2, manuscript, copy provided to author, states that Holdridge created the trajectory program "almost single-handedly." Thomas W. Hamilton, William L. Sjogren, William E. Kirhofer, Joseph P. Fearey, and Dan L. Cain, *The Ranger 4 Flight Path and its Determination from Tracking Data*, TR 32–345 (Pasadena: JPL, September 15, 1962), p. 82, supports that statement declaring that the program was "developed almost completely" by Holdridge.
25. Ekelund, "History of the ODP at JPL," pp. 2–3; Melba W. Nead, "Reminiscences of California Institute of Technology Guggenheim Aeronautical Laboratory, GALCIT No. 1 later JPL," memorandum from Nead to Kyky Chapman, November 5, 1991, 4, HC 16–7, JPL History Collection, JPL Archives, Pasadena, California. Lorell and Carr were studying the combustion of ethanol and hydrazine. See Jack Lorell and Henry Wise, "Steady-State Burning of a Liquid Droplet: I. Monopropellant Flame" *Journal of Chemical Physics* 23 (1955): 1928–32; and Jack Lorell, Henry Wise, and Russell E. Carr "Steady-State Burning of a Liquid Droplet: II. Bipropellant Flame," *Journal of Chemical Physics* 25 (1956): 325–31.
26. Nicholas A. Renzetti, Joseph P. Fearey, Justin R. Hall, and B. J. Ostermier, *Radio Tracking Techniques and Performance of the United States Deep Space Instrumentation Facility*, TR 32–87 (Pasadena: JPL, March 24, 1961), pp. 10–12; Carr and Hudson, *Tracking and Orbit Determination Program of the Jet Propulsion Laboratory*, Technical Report 32–7 (Pasadena: JPL, February 22, 1960), pp. 26, 43; Jack Lorell, Russell E. Carr, and R. Henry Hudson, *The Jet Propulsion Laboratory Lunar-Probe Tracking and Orbit-Determination Program*, Technical Release No. 34–16 (Pasadena: JPL, March 10, 1960), pp. 4, 5–6.
27. Renzetti, Fearey, Hall, and Ostermier, *Radio Tracking Techniques*, pp. 12–13; Lorell, Carr, and Hudson, *The Jet Propulsion Laboratory Lunar-Probe Tracking and Orbit-Determination Program*, pp. 4, 5; R. Henry Hudson, *Subtabulated Lunar and Planetary Ephemerides*, TR 34–239 (Pasadena: JPL, November 2, 1960), p. 1; Douglas B. Holdridge, *Space Trajectories Program for the IBM 7090 Computer*, TR 32–223 (Pasadena: JPL, September 1, 1962), p. 2.
28. The astronomical unit is the mean distance between the Earth and the Sun that astronomers use as a basic measuring stick of distance.
29. Hudson, *Subtabulated Lunar and Planetary Ephemerides*, p. 1–2; Holdridge, *Space Trajectories Program for the IBM 7090 Computer*, p. 2; Herget, *Solar Coordinates, 1800–2000*, Astronomical Papers Prepared for the Use of the American Ephemeris and Nautical Almanac, Vol. 14 (Washington: Government

Printing Office, 1953); Lorell, Carr, and Hudson, *The Jet Propulsion Laboratory Lunar-Probe Tracking and Orbit-Determination Program*, p. 5; Paul R. Peabody, James F. Scott, and Everett G. Orozco, *Users' Description of JPL Ephemeris Tapes*, TR 32–580 (Pasadena: JPL, March 2, 1964), pp. 9, 10; Victor C. Clarke, Jr., *Constants and Related Data for Use in Trajectory Calculations as Adopted by the Ad Hoc NASA Standard Constants Committee*, TR 32–604 (Pasadena: JPL, March 6, 1964), p. 10.

30. Clarke, *Constants and Related Data*, pp. 1–2.
31. Benjamin S. Yaplee, Stephen H. Knowles, Allan Shapiro, K. J. Craig, and Dirk Brouwer, "The Mean Distance to the Moon as Determined by Radar," in Kovalevsky, ed., *The System of Astronomical Constants*, p. 82; *The Mean Distance to the Moon as Determined by Radar*, NRL Report 6134 (Washington: Naval Research Laboratory, 1964); Clarke, "Earth Radius/Kilometer Conversion Factor for the Lunar Ephemeris," *AIAA Journal* 2, 2 (February 1964): 363; William L. Sjogren, David W. Curkendall, Thomas W. Hamilton, William Kirhofer, Anthony Liu, Donald W. Trask, Robert A. Winneberger, and Wilber R. Wollenhaupt, *The Ranger VI Flight Path and its Determination from Tracking Data*, TR 32–605 (Pasadena: JPL, December 15, 1964), p. 58.
32. Michael Ash, Irwin Shapiro, and William B. Smith, "Astronomical Constants and Planetary Ephemerides Deduced from Radar and Optical Observations," *The Astronomical Journal* 72 (1967): 338–50.
33. Andrew J. Butrica, *To See the Unseen: A History of Planetary Radar Astronomy*, NASA SP-4218 (Washington: NASA, 1996), 36–41; "In Conjunction with Venus," *IEEE Spectrum* 34, 12 (December 1997): 31–38.
34. Steven J. Dick, *Oral History Interview with P. Kenneth Seidelmann: July 20, 2000* (Washington, DC: US Naval Observatory, 2000); Ekelund, "History of the ODP at JPL," p. 4. On Neil Block/Gary Duncan astrologer, see Michael Erlewine, "Gary Duncan (Neil Llewellyn Bloch) 1931–1988," 2001 at http://www.solsticepoint.com/astrologersmemorial/duncan.html (accessed February 8, 2009); and Michael Erlewine, "Remembering Gary Duncan (1931–1988)," January 7, 2009, ACT Astrology, http://actastrology.com/viewtopic.php?f=30&t=25 (accessed February 8, 2009).
35. Ekelund, "History of the ODP at JPL," p. 4; Steven J. Dick, *Oral History Interview with Douglas A. O'Handley, E. Myles Standish, and Henry F. Fliegel: December 4, 1999* (Washington: US Naval Observatory, 1999); and The History of Numerical Analysis and Scientific Computing, "Charles L. Lawson" http://history.siam.org/oralhistories/lawson.htm (accessed February 4, 2009).
36. Charles L. Lawson, "JPL Ephemeris Development, 1960–1967," February 23, 1981, p. 3, box 13, folder 159, Daniel J. Alderson Collection, JPL Archives; E. Myles Standish, interview by author, JPL, April 16, 2007, pp. 3–4, tape and transcript, NHRC; Hoffleit, *Astronomy at Yale*, 199; Ekelund, "History of the ODP at JPL," p. 4.
37. Lawson, "JPL Ephemeris Development, 1960–1967," p. 2; "J. Derral Mulholland" at http://setas-www.larc.nasa.gov/LDEF/MET_DEB/IDE/TEAMVITA/JDMVITA.HTM (accessed February 6, 2009). Herrick students' dissertations in boxes 4 and 5, Samuel Herrick Papers, Archives of American Aerospace Exploration, Ms78–002, Special Collections

Department, University Libraries, Virginia Polytechnic Institute and State University, Blacksburg, VA. On Moyer's arrival at JPL, see "About the Author: Theodore D. Moyer," http://descanso.jpl.nasa.gov/Monograph/bios/moyer.cfm?force_external=0 (accessed February 6, 2009).

38. J. Derral Mulholland and Neil O. Block, *JPL Lunar Ephemeris Number 4*, Technical Memorandum 33–346 (Pasadena: JPL, April 15, 1967); Charles J. Devine, *JPL Development Ephemeris Number 19*, Technical Report 32–1181 (Pasadena: JPL, November 15, 1967); Lawson, "JPL Ephemeris Development, 1960–1967," p. 2; Peabody, Scott, and Orozco, *Users' Description of JPL Ephemeris Tapes*, pp. 1–2.
39. Douglas A. O'Handley, Douglas B. Holdridge, William G. Melbourne, and J. Derral Mulholland, *JPL Development Ephemeris Number 69*, Technical Report 32–1465 (Pasadena: JPL, December 15, 1969), p. 1; E. Myles Standish, Michael S. W. Keesey, and X. X. Newhall, *JPL Development Ephemeris Number 96*, Technical Report 32–1603 (Pasadena: JPL, February 29, 1976), p. 7.
40. Hudson, *Subtabulated Lunar and Planetary Ephemerides*, 1; Dirk Brouwer, "An Assessment of the Present Accuracy of the Value of the Astronomical Unit," *Navigation* 9, 3 (Autumn 1962): 206; Anderson, *Determination of the Masses*, 7; Clarke, "Earth Radius/Kilometer Conversion Factor," p. 363.
41. Jay Henry Lieske, "A Dynamical Determination of the Solar Parallax from the Motion of (433) Eros," 1968, cited in Hoffleit, *Astronomy at Yale*, p. 199.
42. Lawson, "JPL Ephemeris Development, 1960–1967," p. 3; Standish, interview, p. 8; O'Handley, Holdridge, Melbourne, and Mulholland, *JPL Development Ephemeris Number 69*, pp. iv, 3–5.
43. Jay Henry Lieske, *Newtonian Planetary Ephemerides 1800–2000: Development Ephemeris Number 28*, Technical Report 32–1206 (Pasadena: JPL, November 15, 1967), pp. 1–2; Joachim Schubart and Peter Stumpff, *On an N-Body Program of High Accuracy for the Computation of Ephemerides of Minor Planets and Comets*, ARI-Heidelberg Veröffentlichungen, No. 18 (Karlsruhe: Verlag G. Braun, 1966).
44. Lawson, "JPL Ephemeris Development, 1960–1967," p. 3; Standish, interview, p. 8; O'Handley, Holdridge, Melbourne, and Mulholland, *JPL Development Ephemeris Number 69*, pp. iv, 3–5.
45. X. X. Newhall, E. Myles Standish, and James G. Williams, "DE 102: A Numerically Integrated Ephemeris of the Moon and Planets Spanning Forty-For Centuries," *Astronomy and Astrophysics* 125 (1983): 150–51.
46. Carroll O. Alley, "Story of the Development of the Apollo 11 Laser Ranging Retro-Reflector Experiment," *Adventures in Experimental Physics* (1972): 132–49.
47. O'Handley, Holdridge, Melbourne, and Mulholland, *JPL Development Ephemeris Number 69*, pp. iv, 4, 5; Lawson, "JPL Ephemeris Development, 1960–1967," p. 2; E. Myles Standish, S. W. Keesey, and X. X. Newhall, *JPL Development Ephemeris Number 96*, Technical Report 32–1603 (Pasadena: JPL, February 29, 1976), pp. 2, 5.
48. Gerald M. Clemence, "The System of Astronomical Constants," *Annual Review of Astronomy and Astrophysics* 3 (1965): 96–97. The need for new

constants moved Clemence to write "On the System of Astronomical Constants," *The Astronomical Journal* 53 (May 1948): 169–79.

49. Clemence, "The System of Astronomical Constants," pp. 97–98; Nicholas A. Renzetti, *Tracking and Data Acquisition Support for the Mariner Venus 1962 Mission*, Technical Memorandum 33–212 (Pasadena: JPL, July 1, 1965), 9, 17, 75–76; Corliss, *A History of the Deep Space Network*, p. 29.
50. Kovalevsky, *The System of Astronomical Constants*, 1; Butrica, *To See the Unseen*, p. 47.
51. Irwin Shapiro, "Radar Determination of the Astronomical Unit," in Kovalevsky, ed., *The System of Astronomical Constants*, pp. 177–215; Duane O. Muhleman, "Relationship between the System of Astronomical Constants and the Radar Determinations of the Astronomical Unit," in Kovalevsky, ed., *The System of Astronomical Constants*, pp. 153–75; Kovalevsky, *The System of Astronomical Constants*, pp. 314 and 323; "Joint Discussion on the Report of the Working Group on the IAU System of Astronomical Constants," in J. C. Pecker, ed., *Proceedings of the Twelfth General Assembly* (New York: Academic Press, 1966), p. 600.
52. "Joint Discussion," p. 606; "Report to the Executive Committee of the Working Group on the System of Astronomical Constants," in Pecker, ed., *Proceedings of the Twelfth General Assembly*, p. 594.
53. Kovalevsky, "Introductory Remarks," *Highlights of Astronomy* 3 (1974): 209; B. Emerson and G. A. Wilkins, eds., "The IAU System of Astronomical Constants," *Celestial Mechanics* 4 (1971): 128, 136.
54. Melbourne, Mulholland, Sjogren, and Sturms, *Constants and Related Information* pp. 33, 35.
55. Emerson and Wilkins, "The IAU System of Astronomical Constants," pp. 138–39, 144, 147; P. Kenneth Seidelmann, "The Ephemerides: Past, Present, and Future," in Raynor L. Duncombe, ed., *Dynamics of the Solar System* (Boston: D. Reidel Publishing Company, 1979), p. 99; Dick, *Sky and Ocean Joined*, p. 430.
56. Emerson and Wilkins, "The IAU System of Astronomical Constants," pp. 128–49, 147–48; Seidelmann, "The Ephemerides," pp. 99, 101; Dick, *Sky and Ocean Joined*, p. 538.
57. Seidelmann, "The Ephemerides," pp. 99, 106, 108, 111; Dick, *Sky and Ocean Joined*, p. 539.
58. Standish, "The JPL Planetary Ephemerides," *Celestial Mechanics* 26 (1982): 181–86; Newhall, Standish, and Williams, "DE 102: A Numerically Integrated Ephemeris of the Moon and Planets," pp. 150–67; E. Myles Standish, "The Observational Basis for JPL's DE 200, the Planetary Ephemerides of the Astronomical Almanac," *Astronomy and Astrophysics* 233, 1 (July 1990): 252–71; Notes, telephone interview, Seidelmann, with author, February 16, 2009.
59. Dick, *Sky and Ocean Joined*, pp. 533–34; Raynor and Julena Duncombe, interview, pp. 37–38; Seidelmann, interview, pp. 15–16.
60. Dick, *Sky and Ocean Joined*, p. 533; Seidelmann, telephone interview; Appendix A, "Listing of Tracking Data," in George W. Null, Harold J. Gordon, and Dennis A. Tito, *The Mariner IV Flight Path and Its Determination From Tracking Data*, TR 32–1108 (Pasadena: JPL, August 1, 1967), pp. 57–170.

61. Butrica, *To See the Unseen*, p. 124; Shapiro, interview with author, Harvard-Smithsonian Center for Astrophysics, Cambridge, MA, September 30, 1993, tape and transcript NHRC, p. 18.
62. Newhall, Standish, and Williams, "DE 102: A Numerically Integrated Ephemeris of the Moon and Planets," p. 150; Butrica, *To See the Unseen*, p. 125; Shapiro, interview, pp. 20–22.
63. Dick, *Sky and Ocean Joined*, p. 533.
64. Ibid., p. 530.

Chapter 5

Big Science in Space: Viking, Cassini, and the Hubble Space Telescope

W. Henry Lambright

Since World War II, the US federal government, as well as some governments in other nations, have supported "big science."[1] Big science is shorthand for a broad range of research and development (R&D) projects that have certain characteristics. They are extremely costly; are large-scale in the number of scientists, engineers, and technicians employed; entail multi-institutional government, industry, and university relationships; have complex management systems; are both visible and often controversial; and last a decade or more from concept to completion. A wide range of projects fit under big science, most of which are big because of engineering requirements for large and complex machines. The purposes of big science are varied, but scientific discovery is only one and in some instances not the dominant rationale for governmental expenditures.

Within the United States a handful of agencies support big science. NASA is one of these. In fact, the history of NASA is in many ways a saga of large-scale science and technology projects. Apollo is the most notable by far; and the glamorous manned projects frequently are classified under big science. However, the heart and soul of those big science projects whose purpose is primarily discovery are found in the Science Mission Directorate, a major program office that supports robotic spaceflight. This office, whose name has changed over the years, is the home today of planetary spaceflight and space telescopes, as well as projects that turn the eyes of satellites on Earth. The purpose of this chapter is to discern the dynamics of big science supported by this office, using a comparative case history method. The cases studied are: Viking, Hubble, and Cassini. As in the case of the Science Directorate, the names of projects can change over time. For purposes of clarity, we will refer to the projects by their established names.

Viking had roots going back to the beginning of NASA but got its official start in 1968. It ended in 1982.[2] It was a $1-billion project whose main goal was to find life on Mars. The conception of the Hubble Space Telescope is usually accorded to astronomer Lyman Spitzer in 1946. It did not get officially started until 1977, was launched in 1990, repaired in late 1993–early

1994, and serviced again periodically afterward. It was almost terminated in 2004, and then rescued in 2006. It may extend five years beyond the repair date of 2009, assuming a successful servicing. The cost of developing and launching Hubble is usually put at $2 billion.[3] Cassini is a big science project in an international mode. Hubble had international involvements, but Cassini was special in this respect. Conceived in 1982, it was launched in 1997 to Saturn. Once in orbit around Saturn in 2004, it released a European-built Huygens probe to Titan, Saturn's largest moon. Like Hubble, Cassini is still returning information into the second decade of the twenty-first century. The cost of Cassini and the Titan probe was $3.3 billion, with the United States contributing $2.6 billion.[4]

These three projects are all costly and long-term. They illuminate the dynamics of big science in space well. In examining these projects we ask how they were born, implemented, evaluated, perhaps reoriented, and what have been their results. We are particularly interested in who moved them forward from stage to stage in the policy process: from agenda setting to adoption, to execution. Who were the champions? Opponents? What barriers did the advocates face? How were they overcome?

Most observers would regard these projects as relatively successful in outcomes. But that does not mean they had smooth rides through their lifetimes or do not provide cautionary lessons about policy and management. In all cases, they in fact faced a host of issues that could have stopped them, or at least derailed them for a time. In examining these projects, we will especially focus on the role of NASA. NASA is the prime institutional mover behind big science in space. But who or what moves NASA?

Viking

Agenda Setting and Adoption

The first step for big science in space is to get on NASA's agenda. It can be on the front-burner or a back-burner depending on who is pushing, who is opposing, and the skill with which they do either. The dominant idea of Viking—to send a robotic probe to Mars, land on the Red Planet, and search for life—goes back to the early 1960s and a previous project called Voyager. This project, which used a Saturn rocket and was driven by both space science and manned space interests, was killed by Congress in 1967. Administrator James Webb directed the Science Directorate to propose a less expensive version he could sell to the president and Congress. Without a major new start, the planetary science program was in jeopardy. The Science Directorate did so in a matter of weeks and developed the Viking proposal.[5] The most important difference was that it used an intermediate-sized rocket rather than a Saturn, and was otherwise substantially scaled back in price. Both President Johnson and Congress endorsed the project in 1968. Viking thus can be said to have had a remarkably short gestation period (figure 5.1). It could also be argued that it was a reorientation of an earlier project.

Figure 5.1 This "glamour" photograph of the Viking lander from 1976 displays its major features, including its scoop and chemistry instruments. (Credit: NASA.)

Viking was promoted as costing approximately $400 million in contrast to Voyager, which had been projected at well over $2 billion. The intent was to launch Viking in 1973, preceded by Mariner 9, in 1971. Mariner's purpose, in part, was as a scouting mission to help find a good landing site for Viking.

Implementation

Webb got Viking adopted but it was up to his successor, Tom Paine, who became administrator in October 1968, to implement it. Paine did so by making Viking NASA's top science priority. This gave Viking a means of protection even as budget cuts took their toll elsewhere. A visionary, Paine embraced Viking as both important for science (the search for life) and as a precursor for manned Mars flight. He also was spurred on by Cold War competition. The Soviet Union, which lost the Moon race in 1969, still had robotic flights to Mars on its agenda.[6]

Paine sought to sell a comprehensive post-Apollo program to President Nixon in 1969 and 1970. This program would have had Viking as a first step in a series of robotic flights culminating in human exploration of the Red Planet by the end of the twentieth century. He also sought to get a space station and space shuttle adopted. Paine desperately wanted to lead NASA in a bold program that would build on what had gone before.

Nixon was not interested in a bold program. Paine collided with what space-policy observer John Logsdon has termed the "Nixon Doctrine" on space issued March 7, 1970. Nixon declared that:

> Many critical problems here on this planet make high priority demands on our attention and our resources. By no means should we allow our space program to stagnate, but—with the entire future and the entire universe before us—we should not try to do everything at once. Our approach to space must be bold—but it must be also be balanced.

He went on to explain that "space expenditures must take their place within a rigorous system of national priorities."[7] In other words, NASA was no longer "special." This was a doctrine that lowered the priority of space. Under the Nixon doctrine, the Office of Management and Budget (OMB) had power vis-à-vis NASA it never had at the height of Apollo. Access to the president was no longer privileged for Paine as it had been with Webb under Kennedy and Johnson. With the NASA budget under enormous pressure, Paine had to decide what was most important to save. He protected Viking, but had to delay the launch date from 1973 to 1975. Frustrated, Paine resigned in September 1970. George Low, his deputy, served as acting administrator until April 1971, when a permanent NASA administrator, James Fletcher, took the helm. Low and then Fletcher affirmed the importance of Viking not just to space science, but to NASA.[8] What this meant was that NASA continued to make Viking a top priority even as its overall agency budget fell—and Viking's costs for execution rose. In 1972, Fletcher got Nixon to accept the space shuttle as NASA's prime agency goal for the 1970s, a critical decision that finally gave a measure of stability to the agency. What was essential for the agency as a whole was also important for Viking. Without a clear and large-scale manned goal after Apollo, NASA was in jeopardy. Those at the Viking project level could now worry more about getting their job done than survival.

Roger Launius has argued in an essay on Fletcher that one of the reasons he was so supportive of Viking was because of his Mormon roots, a tradition that encouraged his belief in life on planets beyond Earth.[9] There is no question Fletcher cared about Viking. He was personally involved in its administration.

Organization

The organization for Viking had the Science Mission Directorate in charge, with much power delegated to the lead center, Langley, in Virginia. The Viking project manager, James Martin, at Langley was totally committed to the project and it was a major management challenge. Viking was a huge scale-up organizationally and budgetarily from the predecessor Mariner series. It involved other centers, especially the Jet Propulsion Laboratory (JPL), and also Ames and Lewis. It had a number of contractors and subcontractors, with the prime contractor being Martin Marietta. The chief scientist was Gerald Soffen of Langley. He presided over 13 teams of scientists representing various scientific disciplines and crosscutting needs, such as choosing landing sites for the two Viking orbiter/landers. The principal scientists numbered 70 and included both academic and government researchers. Among the scientific groups, the exobiologists stood out as first among equals. It was their task to design unprecedented equipment and experiments to search for life. No one really knew how to do that. At its height, Viking employed over ten thousand people in a government, industry, and university team.[10] Viking wound up costing $1 billion. Considering the value of 1970s money today, Viking may have been the biggest big science project in NASA's planetary exploration history. Its prime rival would probably be Cassini.

As noted, Fletcher was personally engaged and an active participant from the management perspective. Like the Science Mission Directorate and its head, John Naugle, he was worried about the pace and cost of Viking. Even though slowed down from 1973 to a 1975 launch, Viking ran into innumerable technical problems. The window for efficient Mars launches opened every 26 months. If NASA missed a 1975 opportunity, it would have to wait another two years. The Science Directorate and Martin needed Fletcher to help ride herd on contractors that were behind schedule or not performing. Fletcher engaged CEOs and got their attention. He also listened to Joshua Lederberg, a Nobel-Prize winner and Viking exobiologist, who lobbied him on the landing site issue. The major issue here was how to balance safety in landing with the potential of a site for life. Lederberg wanted NASA to tilt in favor of life, even if that meant taking greater risks. Fletcher indicated to Naugle that he felt as did Leaderberg that the search for life was the main reason for Viking.[11]

Fletcher wanted to emphasize life for personal and institutional reasons. He knew that if NASA found evidence of life on Mars it would reawaken interest in the space program, at a time in US history when space had receded as a national interest. Carl Sagan, a member of Viking's site selection team, fanned the flames of media and public interest in stressing optimism in the life search. The president, Gerald Ford, was among those eagerly following the project as it reached its scheduled launch date in 1975. NASA's plan was for the first Viking to land on July 4, 1976, America's two hundredth birthday, and the second Viking somewhat later. The fact that the Soviet Union had tried to get to Mars with landers and send back pictures in 1971 and 1973 and failed added to the drama and suspense of Viking.

Completion

Big science projects in space go through various phases: development, testing, spaceflight, and operations. Development and testing for Viking ended in 1975 with the launch and trip to Mars. In 1976, space flight transitioned to operations as the two Vikings went into orbit and began providing information on the Red Planet and where to land. NASA and its team had preliminary sites selected based on Mariner, but as the Viking scientists looked at the landing site chosen for the first Viking, they realized that the place they had picked was too dangerous. The decision to delay landing beyond July 4 was made by Martin and endorsed by managers up the line to Fletcher. Finally, on July 17, Viking safely landed at an alternate site. The Soviets had had a landing a few years earlier, but their probe survived only a few seconds and transmitted no images back to Earth. When Viking communicated and sent back the first extraordinary images, NASA knew it had succeeded in a secure landing. President Ford congratulated NASA. *The New York Times* heralded the event as a "miracle" achievement.[12]

But the real test was whether Viking would find signs of life. The initial results of the exobiology experiments proved potentially exciting, then puzzling, then inconclusive. The media followed every nuance of NASA's actions at this point, putting scientists in a fish bowl, as day by day they tried to determine what their information was saying.

Viking 2 entered orbit August 7. After another last-minute shift in site based on new information provided by the orbiter, Viking 2 landed September 3. The first Viking was relatively near the Mars equator, the second closer to one of the poles. Like Viking 1, Viking 2, could not find any indicators that were clearly favorable to life. The scientists agreed among themselves that they would not make strong positive statements about life unless they had incontrovertible evidence. That, they did not have. What they found could be explained by chemistry, rather than biology. One of the key scientists in the biology experiments believed he did find life, but his Viking colleagues said he was wrong. At a NASA press conference, the NASA official in charge of the biology team, Chuck Klein, announced that the scientific consensus was that Viking had not found life at these two places on Mars.[13] Fletcher wrote President Ford, explaining that Viking had advanced knowledge about Mars tremendously in meteorology, seismicity, geology, and other fields, but the "search for life remains inconclusive."[14] Viking continued to send information from its orbiters or landers for years after 1976, the last piece of equipment dying in 1982.

Viking was a great scientific and engineering success, but its advocates had raised expectations about finding life, and the media, public, and most scientists saw Viking as falling short of that goal. Fletcher spoke of "what might have been" in the wake of the disappointment. He believed success would have created a momentum for NASA and particularly Mars exploration, even manned expeditions. Advocates like Sagan did not give up, pointing out that failure in two places chosen in large part for safety reasons did not mean "no life" on Mars. But many others in the core coalition that had promoted Viking, and lived with it day-in and day-out, at least since 1968, drifted off, exhausted and frustrated. The advocacy coalition for Mars fragmented, and was unable to agree on what to do next. Project Manager Martin, who left NASA for a high-paying executive position in industry, admitted that Viking would probably be the highlight of his career. Maybe a man could only have one Viking in his life, he said.[15]

There were others scientists, inside and outside of NASA, who wanted to visit other planets and spend money on different priorities. They became the force that demanded to have their time of glory. While some argued against Mars, the more telling rhetoric was about "balance." After many years of giving Mars top priority, NASA and the Ford administration decided that it was time to look elsewhere. As the United States moved away from Mars, so also did the USSR, which emphasized Venus. US priorities went to the outer planets and the development of a large new telescope in space. The United States did not return to Mars until the 1990s.

Hubble Space Telescope

Agenda Setting

The idea that became the Hubble Space Telescope (figure 5.2) was conceived by 31-year-old Lyman Spitzer in a report he did for the Rand Corporation in 1946.[16] It got on NASA's agenda in the 1960s in a serious way as NASA considered its post-Apollo options. Nancy Roman, director of astronomy

Figure 5.2 Flyaround of the Hubble Space Telescope (HST) after deployment on this second servicing mission (HST SM-02) on February 19, 1997. Note the telescope's open aperture door. (Credit: NASA.)

programs in NASA's Science Mission Directorate, was extremely interested in a "Large Space Telescope," as it was then called, that could be placed above the distorting effects of Earth's atmosphere. She heard from Spitzer, now a senior astronomer at Princeton, other astronomers, and various advisory committees advocating the telescope. Two NASA centers, Goddard Space Flight Center and Marshall Space Flight Center, vied to be lead center on a project to build such a telescope. By the early 1970s, Marshall had maneuvered itself into the lead role, with Goddard also heavily involved. Roman provided planning funds to nurture telescope ideas, and kept the concept alive at a time when funds were tight and getting tighter.

Adoption

The adoption process for the Hubble Space Telescope was exceedingly elongated. Part of the reason was that it took a while for the advocates to make it a high enough priority to reach an agency decision. Fletcher, as noted earlier, joined NASA in April 1971 and was beset with the need to sell a shuttle decision to the president and find money for Viking and other major projects he inherited. In December 1972, he and Naugle discussed the telescope project. Fletcher asked how much it would cost. Naugle said somewhere between $500 million and $900 million. Fletcher told him to hold the line at $300 million.[17]

Over the next year, Naugle's office and the telescope advocates in the astronomy community worked to find a way to bring costs down while proposing a worthwhile telescope. In late 1973, the advocates had developed a concept that seemed acceptable at $400–500 million. Fletcher went along, as did the White House.

A modest $6.2 million was put into NASA's budget to get the project started with definitional studies. The congressional subcommittee reviewing NASA's budget in March 1974 asked about total costs and Naugle responded with the $400–500 million estimate. The subcommittee noted that in 1972 a National Academy of Sciences review of astronomy needs headed by Jesse Greenstein had placed a space telescope third in priority. The subcommittee wound up rejecting NASA's proposal on the grounds it was not a top priority.

Hubble's backers in the astronomy community led by John Bahcall of Princeton, ably assisted by Spitzer, reacted strongly. They got the Greenstein panel to "clarify" its position in such a way to support the launching of a large telescope in space. In late 1975, the Hubble advocates felt they had sufficient scientific backing to try again with Congress. This time, however, it was NASA that the protagonists found to be a barrier. Fletcher told Noel Hinners, now the associate administrator for space science, to hold off on funds planned for the telescope. Hinners, who favored the telescope, eliminated all funds for Hubble, knowing Bahcall and his allies would fight the decision.[18]

That they did, and they managed to persuade Congress in 1976 to direct NASA to release requests for proposals to industry on how to build the telescope. This decision to authorize a limited action was in anticipation of later full congressional approval of the Hubble Telescope when NASA leadership was amenable to moving forward. In late 1976, agency leadership put Hubble into a budget submission approved by the Ford administration. In 1977, Congress considered Hubble as a new start. The problem for NASA (and Congress) was that the agency was at the same time requesting adoption of a Jupiter probe, what would be called Galileo. The debate in Congress was not so much about the telescope, as about big science at NASA in general. Key Congressional forces told NASA it could have its telescope, but Galileo would have to be sacrificed. Galileo, however, had its own constituency of support, and Galileo proponents got their mission approved along with Hubble.[19]

The long and tortuous process of adoption for Hubble pointed up the degree to which the political environment for big science was constricted for NASA in the 1970s. The shuttle decision had stabilized NASA's overall budget and brought some modest growth. But to give shuttle what it needed in funds, NASA squeezed other program offices, including the Science Mission Directorate. NASA could afford only so many big science projects at one go. As there were windows of opportunity for spacecraft launches, so there were political windows to start major projects.

Implementation/Evaluation

Implementation of the Hubble Space Telescope got underway in 1977. NASA chose contractors for Hubble—Perkin-Elmer for the mirror, and Lockheed for the structure. Marshall was lead center, with Goddard the chief partner. The Science Mission Directorate was in charge. The project manager at Marshall was Fred Speer. In 1980, with an eye to utilization, NASA created the Space Telescope Science Institute at Johns Hopkins. It was run via a contract with the Association of Universities for Research in Astronomy (AURA).

At the time Congress authorized NASA to build the Hubble Space Telescope, it agreed to a $475-million cost estimate. As development proceeded, however, it became clear that the costs were going well beyond that figure. Hence, a time for evaluation came in 1981 when James Beggs took office as NASA's administrator. Confronted by reality, Beggs decided to maintain the telescope's development. In 1981 and in 1982 he provided additional funds, mainly through reprogramming actions. There had to be modifications in Hubble's design, and the project manager, Speer, was replaced in a management shake-up.[20]

Another major change in Hubble policy came in 1983, guided by Roman's successor as director of the astronomy program, Ed Weiler. Weiler argued that servicing should take place in space, using the shuttle and astronauts. There was too much risk to delicate mechanisms in Hubble to repair it through periodic landings and relaunchings, a procedure that was then the servicing plan. Also, he persuaded NASA to build a duplicate of the primary mirror—just in case it was needed.

These reviews also resulted in decisions to reschedule Hubble from a 1983 launch to 1988. When the Challenger shuttle accident intervened in 1986, the launch was postponed again, to 1990.

From Setback to Success

Development ended, and operations were set to begin in 1990. Preceded by a wave of publicity about how Hubble would revolutionize astronomy, Hubble was carried aloft by a shuttle. Unfortunately, the much-anticipated pictures came back blurred. There was a "spherical aberration," NASA sadly announced. The media had a field day and late-night TV comedians ridiculed the space agency. An inquiry found that a manufacturing error early in the development process by Perkin-Elmer was to blame. NASA's credibility for technical management was hurt greatly, and Senator Barbara Mikulski, a senior Democrat on the agency's appropriations committee, called Hubble a "technoturkey."[21]

Dan Goldin, who became NASA administrator in April 1992, made Hubble repair a top priority. He realized he could not sell the construction of a space station, NASA's flagship for the 1990s, if NASA could not repair a telescope. He gave it his personal attention and made sure it got the oversight and resources it needed. He appointed a manager specifically to coordinate the repair effort on his behalf and told him the agency's fate rode on the success of the repair. He said the same to the astronauts whose job it was to actually do the work in space. In late 1993–early 1994 astronauts accomplished the repair through an unprecedented set of activities in space. Following repair, the pictures from Hubble came back and were spectacular. "The trouble with Hubble is over," Mikulski beamed at a press conference. The repair, which was seen on television, helped turn a setback into a great public relations success for NASA.[22]

Service missions were carried out in 1997, 1999, and 2002. Hubble increasingly captivated the public with awesome views of the cosmos.

Termination versus Completion

In February 2003, the Columbia Shuttle disintegrated. Once again, a crew was lost and NASA was in crisis. A Columbia Accident Investigation Board (CAIB) was established and found numerous management errors in NASA risk-taking with an aging shuttle. The NASA administrator, Sean O'Keefe, declared that NASA would not only implement CAIB's safety recommendations "but we are also seeking ways to go beyond their recommendations."[23]

As NASA put together its upcoming budget in November and December 2003, O'Keefe deleted funds for a servicing mission. CAIB had called for NASA's developing a capacity to repair spacecraft (e.g., Hubble) away from the safe haven the space station provided. The kind of in-space repairs needed in a repeat of the Columbia accident, which involved damage to insulating tile, was extremely hard to do and dangerous in NASA's view.

O'Keefe decided the risk to astronauts from a scheduled servicing mission was too great and he had to live up to his words about going the extra mile in averting risk.

News of the decision leaked in early 2004 in a news story about President Bush's decision to return to the Moon and go on to Mars. The Hubble astronomy community erupted in anger and disbelief. AURA and the director of the Space Telescope Science Institute mobilized the astronomy community and lobbied lawmakers to overturn O'Keefe's decision. The critics painted the Hubble decision as a sacrifice to Moon-Mars. Senator Mikulski, herself a strong Hubble supporter, asked O'Keefe for a "second opinion." O'Keefe turned to Harold Gehman, chair of CAIB, who he hoped would back him. Instead, Gehman said a decision to terminate Hubble required a substantial study of risk. O'Keefe was forced to go to the National Academy of Sciences to conduct the study. He also asked NAS to determine the possibility of a robotic repair of Hubble.

It was clear that the termination decision hit a public nerve. Hubble was not only a scientific icon, but a public one as well. O'Keefe stood virtually alone. The NAS came back with a report saying the risks of astronaut repair were acceptable and robotic repair not feasible in the time needed given Hubble's deteriorating condition. In early 2005, O'Keefe, refusing to change his position on Hubble, left NASA for the presidency of Louisiana State University.[24] Michael Griffin became NASA's administrator in April of that year. He said he would revisit the Hubble decision once the shuttle had returned to flight and flown safely for at least two occasions. He also undertook an intense risk assessment study of the repair mission. On October 31, 2006, he announced a reversal of the O'Keefe decision. Hubble underwent a repair mission in May 2009. This was the last repair, and would hopefully keep the telescope operating for perhaps several more years. The completion of this project would thus come at approximately the same time that Hubble's successor, the James Webb Space Telescope, would take over.

Cassini

Agenda Setting

Cassini, like Viking, is an example of planetary big science. But its dynamics are very different. It took place at a later time in history; went to an outer planet, Saturn and its Moon, Titan; and was an international project. Viking was spurred on by *competition* with the Soviet Union. Cassini was catalyzed by *cooperation* with Europe.

Cassini was a logical outgrowth of earlier missions to the outer planets, particularly Voyagers 1 and 2, launched in 1977, and Galileo, mentioned earlier, which was launched in 1989. The Voyager missions were flybys of Jupiter and Saturn (1979 and 1981, respectively) whereas Galileo was an orbiter, which arrived at Jupiter in 2003. Voyager 2 passed by Saturn's largest Moon, Titan, and found views blocked by dense orange clouds. The fact that it was one of the few planets or moons in the solar system to have an atmosphere particularly intrigued planetary scientists. The concept of a mission that would orbit Saturn and also send a probe beneath Titan's clouds evolved along with the planetary program of NASA.

However, the impetus for action on the Saturn-Titan front came initially from Europe. The European Space Agency (ESA), an organization to which several European nations contributed funds to enable larger-scale missions than individual European nations could undertake, was considering its agenda in 1982 and issued a call for proposals. Two European scientists, Daniel Gautier of L'Observatoire de Paris-Meudon, France, and Wing Ip, working at Germany's Max Planck Institute, wanted to respond with the Saturn-Titan concept and discussed a possible NASA-ESA project with an American colleague, Tobias Owen of SUNY Stony Brook. The Europeans did in fact submit a proposal embodying the idea, while Owen began to advocate the concept of a joint project to his peers in the United States and at NASA.[25]

The timing was propitious, as the National Academy of Sciences-Space Science Board in the same year had formed a Joint Working Group with

Figure 5.3 Jet Propulsion Laboratory (JPL) technicians clean and prepare the upper equipment module for mating with the nuclear propulsion module subsystem of the Cassini orbiter in the Payload Hazardous Servicing Facility at KSC in 1997. (Credit: NASA/JPL.)

the European Science Foundation-Space Science Committee. Gautier and Ip were members of the European body, and Owen had contacts on the NAS-Space Science Board.

The Joint Working Group recommended Saturn-Titan as an option for international collaboration and in 1983 a key NASA advisory group, the Solar System Exploration Committee, also did the same (figure 5.3). If there was a push from outside scientists for the mission, there also was impetus from inside-government scientists. NASA and ESA agreed in 1984 on a joint assessment of the proposed Saturn-Titan mission. The assessment went ahead in 1984–1985. Two key leaders of the assessment were Wesley Huntress and Ron Draper of NASA-JPL and Jean-Pierre Lebreton and George Scoon from ESA. The assessment group recommended NASA fund the launch vehicle and Saturn orbiter, and ESA the Titan probe.[26]

Adoption

What the advocates of the Saturn-Titan project needed was a "new start" decision. By this time the project had a name: Cassini-Huygens, for two seventeenth-century astronomers who made discoveries relevant to Saturn and Titan. ESA sought to get the collaborative adoption process moving. In 1988, its science advisory committee urged ESA to sponsor Huygens, on the condition the United States moved forward with Cassini. ESA awaited action by the United States.

One of the reasons the United States was slower in decision-making than Europe was that the Cassini mission had come to be connected in NASA planning with another project, the Comet Rendezvous Asteroid Fly-by (CRAF). The two projects were both aimed at the outer solar system and shared certain launch facilities and equipment. There were strong advocacy groups behind both projects and Len Fisk, NASA's associate administrator for space science, wanted to accommodate both groups. In 1989, the new George H. W. Bush administration was giving a priority to space and accepted the CRAF-Cassini project in its proposed budget. Congress went along with the White House and NASA, but put a cap of $1.6 billion on CRAF-Cassini.[27]

In 1990, ESA approved an international agreement under which NASA and it would commence the Cassini-Huygens project. The targeted date for launch was 1996. ESA had the power to commit funds for the duration of development. NASA had to get funding year to year, but it had funding to get started. NASA's appropriation was for CRAF-Cassini, but it was understood that Cassini was linked with Huygens.

Implementation

As with other big science projects discussed here, NASA put the Science Mission Directorate in charge of CRAF-Cassini. JPL was responsible for technical management. The project manager for Cassini was Dick "Spe"

Spehalski. The agency selected a number of scientific investigators to plan for Cassini instruments and research to be undertaken. The Cassini prime contractor was Lockheed Martin for the launch vehicle. NASA also had an agreement with the Department of Energy to supply radioisotope thermoelectric generators (RTGs). RTGs use plutonium as a power source for spacecraft too far from the sun for solar energy and too distant also for chemical fuels. Nuclear fuels provide required range and duration. In 1991, NASA entered into another agreement with Italy (separate from that with ESA) for Italy to provide several instruments for Cassini.

Early on, it was obvious that NASA would have trouble living within the $1.6 billion cap. In the budget announced at the outset of 1992, the Bush administration killed CRAF, with the intent that money could thereby shift entirely to Cassini. OMB, however, warned NASA that the economic pressures on the Bush administration were such that not even Cassini was secure.[28]

The insecurity of Cassini (and thus Huygens) increased in April 1992 when Bush appointed Dan Goldin NASA administrator. Goldin proclaimed himself a change agent. Big science was past, he said, and the future belonged to "faster, better, cheaper" (i.e., smaller) missions. When Fisk resisted his proposed changes, he no longer was Goldin's associate administrator for space science. The new head was Huntress.[29]

Evaluation by the Administrator

Retained by President Bill Clinton in 1993, Goldin took his reforms to a new level of intensity. He evaluated all science missions in accord with his criteria. He believed huge missions took too long to build, cost too much and were technically obsolete by the time they launched. Moreover, they were too risky politically, for if they failed, NASA took tremendous criticism. He pointed to a $1-billion Mars Observer that failed in 1993 as an example. He wanted to kill Cassini, which he called "Battlestar Galactica."[30] His new associate administrator, Huntress, had been an advocate of Cassini when at JPL and remained such. His position was that Space Science required faster, better, cheaper projects *and* larger projects ("flagships") like Cassini. The scale depended on science requirements. Unlike Fisk, however, Huntress was careful not to challenge Goldin directly on policy. He worked behind the scenes to shore up support for Cassini.[31]

The issue came to a head in 1994. Convinced that Goldin was seriously threatening Cassini (and thus Huygens), the Europeans acted. In addition to other moves, the director general of ESA contacted Vice President Al Gore, who had space policy as part of his portfolio, calling "unilateral withdrawal" from the joint project "unacceptable," and a threat to any future US-European collaborative efforts in science and technology.[32] Goldin got the message—Cassini was too far along to cancel, too integrated with Huygens. Goldin did indeed back off, but told JPL and others connected with the project that they had better not fail!

Cassini was rescheduled to launch in 1997 rather than the year earlier planned, 1996, but otherwise continued without interruption.

Protests by the Public

As the date of launch approached, NASA ran into confrontational protests from antinuclear activists who worried that the plutonium-based RTGs could spread radiation over the population near the launch site in Florida if Cassini crashed. NASA had used RTGs on many previous flights with minimal opposition. Galileo had run into some resistance. Cassini used more RTG power sources than any spacecraft in history. It consequently ran into more opposition than any launch in history.

NASA carefully prepared an environmental impact statement that called the risk negligible and launched a public information campaign in southern Florida. To no avail. The opposition had its own campaign and it gathered steam as the launch date of October 13, 1997, approached. NASA increased security at Kennedy Space Center to thwart possible disruption or sabotage. Activists lobbied the Florida governor and other politicians and received considerable local and national media attention. Fifteen hundred protestors showed up at the gates of KSC to make their views known.[33]

The White House gave NASA the official go-ahead as required for nuclear-powered flight. On October 13, Cassini roared into space, carrying its Huygens probe.

Completion

Seven years later, in June 2004, Cassini saw Saturn. The riskiest part of the flight was near. The rings around Saturn were beautiful, but deadly. The spacecraft had to maneuver between the rings to get to Saturn. There were gaps between the rings, but the danger of a being hit by a chunk of ice or rock was sufficient to cause all associated with the project and especially those watching at mission control/JPL to hold their breath as the moment of insertion came.

On June 30, Cassini sailed through the gap in the rings and settled into orbit, its planned four-year observational mission commencing. Then, on January 14, 2005, Huygens left the mother ship toward Titan. It used a parachute so its descent would be slow and it could take photographs of the mysterious moon. It found a mix of methane and nitrogen in the atmosphere, and an active and diverse weather system based on liquid natural gas. It rained liquid methane on the frigid moon. Huygens performed splendidly before hitting the surface and dying.[34] Meanwhile, Cassini continued to study Saturn, providing much more information about the ringed planet than ever before known. The US-European team was ecstatic. They regarded the mission a great success, technically and organizationally. It was also a success politically in the sense of merging the space programs of the United States, ESA, and Italy. The mission involved over 250 scientists, 5,000 engineers and other professionals, and 19 nations.[35] It did so over many years and involved a journey of 2.2 billion miles. It continues to return information at the into the twenty-first century, and may be extended.

Conclusion

The three big science projects discussed illuminate factors that influence the course of major R&D efforts at NASA and other governmental institutions. Seven interrelated factors are of particular importance: goals, constituency, expertise, esprit, competition, cooperation, and leadership.

Goals

Each of these projects gained by having clear goals. The aim of Viking was to send an orbiter to Mars, land on the Red Planet, contribute new science about Mars, and particularly to find if life existed. This last goal was by far the most important in driving the mission. It was also a Viking goal to beat the Russians in landing successfully and finding life on Mars. The aim of Hubble was to place a telescope in Earth orbit and see further into space and back in time than man has ever done before. The Cassini goal was to orbit Saturn and land on its mysterious moon, Titan, and thus contribute to knowledge about this most spectacular of outer planet systems. Titan would also, it was hoped, produce knowledge about the conditions of life before it arose on Earth.

Viking certainly revolutionized understanding of Mars, but fell short of finding life or determining conclusively that it existed or did not exist on Mars. Hubble started out a failure in reaching its goal because of blurred images, but the technical problems were corrected and Hubble stands as one of the most successful and popular projects in NASA's history. Cassini has been successful in its goals and, like Hubble, has in some ways exceeded its aims.

Big science needs clear goals and they have to be bold to justify the investment. Nations do not spend billions on science unless clarity and excitement are conveyed. Big science has to attract many scientific and engineering contributors, and these individuals have to work toward something important they understand and share in common. The goal has to be challenging technically—that is part of the attractiveness—but attainable, or at least believed to be attainable at the time a project gets underway.

Constituency

Big science projects need constituencies who will support them in obtaining the financial support they require. This means, in the United States: NASA, scientists, the president, White House offices such as OMB, Congress, industry, the media, US public, and sometimes other countries. As the cases show, it takes a while to build a constituency that is broad enough to sustain a project over years that costs a billion dollars and more. It takes advocacy on the part of a core constituency, one that typically embraces an "inside-outside coalition" of NASA officials and academic scientists. The bigger the science project, the bigger the constituency needed. Getting public interest in big science requires goals the public can understand or images the public can see and appreciate. In Cassini's case, the constituency included international partners. Hubble did also, but to a much lesser degree than Cassini. More and more, it can be expected that big science will be international in constituency-base.

Expertise

Because big science projects push the state of the technical art and have goals that are first of a kind in many ways, they must have outstanding scientists, engineers, and technical managers associated with them. Such individuals must be willing to devote years of their careers to these projects. It is clear that Viking, Hubble, and Cassini have been blessed by having some of the "best and brightest" technical personnel associated with them. However, it appears that Hubble was guilty of poor technical management in the development of the telescope. It is rare for projects at the frontier of science and technology not to have some technical glitches. However, these problems should be discovered early. The Hubble case points up the fact that technical management is itself an expertise critical in big science.

Esprit

Esprit refers to morale, an intangible that keeps teams of personnel engaged in a project over a long period even when there are setbacks. Projects inevitably have not only technical problems, but political/budgetary ones as well. Individuals involved have to not only believe in the goals of the project, but be willing to work with others in a team. The "team" concept is difficult for many scientists to embrace. But big science requires cohesion among disparate individuals and in the best cases, the whole is greater than the sum of the parts. Viking, Hubble, and Cassini all revealed this concept of esprit. It drives people to work in concert with others long hours, often far from their families, toward a common goal.

Competition

Competition is always a driver among scientists, and there was competition among ideas and NASA centers within each project studied. However, the only project studied where international competition was a driver in its conception and implementation was Viking. Viking was born during the Cold War. The Russians had an active Mars program. They were not successful, but competition was a spur adding to the other factors influencing the project.

Cooperation

For big science, cooperation across institutions is essential. This always means domestic institutions (e.g., government, industry, universities) and increasingly international institutions (e.g., NASA, ESA, and the Italian Space Agency). It is unlikely that Cassini would have been possible without the cooperation between the United States and Europe. The fact that Europe had a distinct and prestigious part of the overall project (the Titan probe) made ESA more of an equal partner than it would have been if its work were absorbed into a larger entity for which NASA got all the glory. That being the case, Europe fought especially hard to keep Cassini going when it was threatened with cancellation in the early and mid-1990s by the NASA administrator.

Leadership

There has to be leadership throughout big science projects for them to succeed. Such leadership needs to be found at every level: project managers, center directors, program associate administrators, and agency administrators. The agency administrators are at the top, and at the boundary between the technical/administrative side of the project and its political environment. They have potentially the most crucial role in a project's success, failure, or mixed results. Administrators choose other leaders in the agency and are most responsible for getting the resources to make the project happen. Administrators can help or hurt a project by their decisions.

In the case of Viking, the most influential NASA administrators were Webb, Paine, and Fletcher. Webb got Viking started. He pushed for Viking to be born as Voyager died. He energized his managers and got the president and Congress to keep planetary exploration going in the post-Apollo era. Paine supported Viking as the top planetary priority and kept it viable when NASA's budget fell by stretching its launch date. Fletcher also made its implementation a top priority, not just for science, but for the agency as a whole.

In contrast, Fletcher did not seem particularly enamored of Hubble, which arose on NASA's agenda as Viking ended. He gave his managers (and Hubble advocates) the green light to start but ordered them to keep the budget to a level that was extremely low and that proved wholly unrealistic. The shortfall in funding became too obvious to ignore for Beggs. He provided the funds Hubble needed and these saved the project at a critical evaluation stage. In doing so he exacted a price in change of leadership at the project manager level. Goldin understood the importance of Hubble repair to NASA's future and provided it resources and top-leadership attention that made it not only a priority, but an urgent priority. The repair had to succeed for the agency's overall credibility and especially in getting support for the International Space Station.

O'Keefe, on the other hand, saw astronaut safety as the priority he wanted to promote after the Columbia tragedy. He tried to cancel a scheduled Hubble service mission. His successor, Griffin, reinstated the mission.

O'Keefe's failure to achieve his intent on Hubble points up the fact that projects can get constituencies with influence sufficient to defeat the NASA administrator. Ordinarily for top-priority projects, the NASA administrator seeks to build constituencies that will sustain a project after he has gone. But occasionally he wants to end a project. Big science projects stretch in time over several administrators, Congresses, and presidents. Cassini got started under Fletcher in his second tour at NASA's helm and was officially adopted under his successor, Richard Truly. Goldin, pushing "faster, better, cheaper," wanted to kill Cassini. But it had a constituency within NASA and especially the international community. The "inside-outside" advocacy coalition made an end-run around Goldin to save the project. O'Keefe presided over Cassini's fruition.

If there is a central lesson about leadership in big science it is that it has to be broad and shared. There has to be an advocacy coalition with influence at various stages of the policy process, from agenda setting to completion. Who is in the coalition can change, but there needs to be one or the process ceases to move. Another lesson about leadership derives from the Viking case. To get external support, advocates of big science sometime go too far in their claims. By elevating the quest for life so much, Viking's leaders made the failure to find life a sign of failure generally to the media and public. That was not the case, as Viking added enormously to the store of knowledge about Mars. The evidence about life was "inconclusive" in just two places chosen under emergency circumstances. But failure overall was what was widely perceived and Mars exploration was abandoned for many years. Perhaps the most difficult aspect of leadership in big science, whether by a NASA administrator or an insider-outsider coalition, is how to sell a project to the public and political world. Without advocacy, highly visible, expensive big science projects do not take place. But overselling can have a negative effect on scientific and Washington decision-making and determine whether a particular project is a beginning or end of a program.

Notes

1. Peter Galison and Bruce Hevly, eds., *Big Science: The Growth of Large-Scale Research* (Stanford, CA: Stanford University Press, 1992).
2. Edward Ezell and Linda Ezell, *On Mars: Exploration of the Red Planet* (Washington, DC: NASA SP-4212, 1989).
3. Eric Chaisson, *The Hubble Wars* (New York: Harper Collins, 1994); Robert W. Smith, "The Biggest Kind of Big Science: Astronomers and the Space Telescope," in Galison and Hevly, ed., *Big Science*, pp. 184–211; Robert Zimmerman, *The Universe in a Mirror* (Princeton, NJ: Princeton University Press, 2008).
4. Bram Groen and Charles Hampden-Turner, *The Titans of Saturn* (London, UK: Marshall Cavendish, 2005), p. 4.
5. Ezell and Ezell, *On Mars*, p. 134.
6. Ibid, pp. 151–52.
7. Cited in John Logsdon, with Linda J. Lear, Jannelle Warren Findley, Ray A. Williamson, and Duane A. Day, eds., *Exploring the Unknown: Selected*

Documents in the History of the U.S. Civil Space Program, vol. I (Washington, DC: NASA SP-4407, 1995), p. 385..

8. Ezell and Ezell, *On Mars,* p. 191.
9. Roger Launius, "A Western Mormon in Washington, D.C.: James C. Fletcher, NASA and the Final Frontier," *The Pacific Historical Review* (May 1995): 217–41.
10. Beverly Orndorff, "Viking Manager Moves Team," *Richmond Times-Dispatch,* June 20, 1973, pp. 9–10.
11. Ezell and Ezell, *On Mars,* p. 312.
12. "The Viking Miracle," Editorial, *New York Times,* July 25, 1976.
13. Andrew Chaikin, *A Passion for Mars* (New York: Abrams, 2008), p. 168.
14. James Fletcher, letter to the president, September 2, 1976, NASA Historical Reference Collection, NASA Headquarters, Washington, DC.
15. Bill Delany, "Manager of Viking Project Pays Final Visit to Langley," *Hampton Daily Press,* December 18, 1976.
16. Zimmerman, *Universe in a Mirror,* p. 13.
17. Ibid., pp. 51–52.
18. Ibid., p. 69.
19. Ibid., p. 76.
20. Ibid., pp. 112–14.
21. Ibid., pp. 143–44.
22. Joseph Tatarewicz, "The Hubble Space Telescope Servicing Mission," in Pamela Mack, ed., *From Engineering Science to Big Science* (Washington, DC: NASA, 1998), p. 394.
23. Zimmerman, *Universe in a Mirror,* p. 188.
24. W. Henry Lambright, *Executive Response to Changing Fortune: Sean O'Keefe as NASA Administrator* (Washington, DC: IBM, 2005), pp. 34–36.
25. Groen and Hampden-Turner, *Titans of Saturn,* pp. 15–18.
26. Ibid., pp. 19–20.
27. Ibid., p. 27.
28. Ibid., p. 29.
29. W. Henry Lambright, *Dan Goldin and the Remaking of NASA* (Washington, DC: IBM, 2001).
30. Groen and Hampdon-Turner, *Titans of Saturn,* p. 31.
31. Wesley Huntress, Oral History, NASA Historical Reference Collection.
32. Groen and Hampden-Turner, *Titans of Saturn,* p. 32.
33. Ibid., p. 116.
34. Guy Gugliotta, "Scientists Detail Diverse Weather System on Titan," *Washington Post,* January 22, 2005, A2.
35. Groen and Hampden-Turner, *Titans of Saturn,* p. 6.

Chapter 6

Visual Imagery in Solar System Exploration

Peter J. Westwick

On January 3, 2004, a small robotic car, snug inside an aeroshell, plunged through the sky above Mars trailing fire. Parachutes and rockets slowed it down, and an airbag cushioned its final 15-meter drop to the Martian surface. Tense engineers waited nervously in the control room at the Jet Propulsion Laboratory in Pasadena for the radio signal of a safe landing. When the signal arrived the assembled engineers erupted in cheers, but the evening's highlight came a couple of hours later, when the Spirit rover unfolded its cameras and returned a panoramic image of its landing site. Breathless scientists declared that further images promised major discoveries, and, indeed, pictures from Spirit and its twin rover Opportunity soon provided persuasive evidence that Mars once had extensive standing bodies of salty water.

This essay explores the role of visual images in the communication of scientific knowledge about the planets, focusing on the American space program. It highlights two interconnected strands in the story: the development of imaging technology, and the place of imaging in solar system exploration. Space scientists did not wholeheartedly embrace images from the outset, and the pursuit of visual images competed against other interests, scientific and otherwise, on planetary missions. Over time, increasingly advanced imaging technologies enabled the return of striking images from every major planet, revolutionizing human understanding of the solar system.[1] These pictures revealed a diverse, dynamic system with weather and tectonics even at the coldest reaches of the solar system, and strong evidence of water and hence the possibility of life beyond Earth.

Planetary exploration provided particularly fertile ground for visual images. Astronomy itself had a long tradition of visual representation, which included planetary astronomy.[2] Planetary geology similarly drew on a long visual tradition in terrestrial geology.[3] Whereas such tradition may have appeared as a "craft" in earlier periods, in planetary science it appears as a highly technical pursuit, and one bringing in new disciplines such as electrical engineering and computer science. Imaging science involved exceedingly complex mathematics and technologies, but it did retain one element of earlier craft traditions: namely, an aesthetic sensibility that appreciated images for their visual appeal.[4]

Historians of science have increasingly appreciated the importance of visual representation in science. Whereas earlier historians viewed visual images and aesthetic considerations as marginal, if not contrary, to quantitative and literary forms of scientific discourse, historians now highlight the central role of graphs, photographs, and digital images and the complex process of interpreting such images. Pictures do not necessarily convey understanding by themselves; rather, researchers manipulate and analyze them to reveal particular sorts of information. From Galileo sketching the moon with techniques of *chiaroscuro* to highlight features newly visible through his telescope; to nineteenth-century spectroscopists using tricks of photographic developing and retouching to control contrast, sharpen objects outside the focal plane, and correct for frequency response in images of the solar spectrum; to twentieth-century high-energy physicists parsing tracks in bubble chambers to reveal invisible subatomic particles—the production and interpretation of pictures has entailed decisions about which features to emphasize and which to ignore, and has thus determined what people would see.[5]

Imaging Technology

Astronomers had long used telescopes to furnish images of distant planets unavailable to the naked eye, going back to Galileo's first sketches of Saturn's rings. Spacecraft provided a more close-up platform, but required technologies to capture images from the spacecraft, return the image to Earth, and then develop or process it. These technologies evolved over time in a mutually reinforcing ratcheting-up of capabilities, which in turn helped ensure the wide embrace of visual images in the planetary community and the general public.

All planetary spacecraft before the Galileo mission used vidicons to capture images. In a vidicon, light falls on a photoconductive surface, which builds up an electric charge proportional to the light received at each point; an electron beam then scans the charge-density pattern to record the image. The Galileo mission, planned in the late 1970s and finally launched in 1989, switched to charge-coupled devices (CCDs), which convert light falling on individual solid-state pixels into electrical signals; CCDs were invented at Bell Labs in late 1969 but took several years to develop into a reliable technology. Spacecraft designers recognized their potential for planetary imaging, since CCDs did not have the scanning electron beam and were hence smaller and used less power than vidicons.[6]

Spacecraft cameras also involved several other technologies, and increasing emphasis on images—and on ever-higher image resolution—ratcheted up these technology requirements. Engineers built microstep actuators for the turntable-like platforms that held the camera, for precise pointing of the camera. Images also contained many data bits, which strained spacecraft communication systems, especially as succeeding missions produced ever more images and at higher resolution. The Mariner spacecraft had maximum rates of 16 kilobits per second; the higher resolution Voyager images were each over 5,000 kilobits, so Voyager specifications included data rates of up to 115 kilobits per second. That in turn entailed larger antennas on the spacecraft and in the Deep Space Network, faster spacecraft computers and more data storage capacity, and new software routines for telecommunications coding.[7]

The increasing requirements for imaging technology derived from increasing capabilities of electronic computers, which in particular enabled the development of digital image processing, the second key step in obtaining visual images from planetary spacecraft. It was often not enough to just take a picture and send it back to Earth; rather, like a photographer developing film in a darkroom, engineers had to apply digital processes to reveal images contained in the streams of digital bits.

Processing data from space-borne cameras took two stages, one before launch, and one after. Vidicons needed careful calibration on the ground: first, to correct for photometric distortion caused by uneven response in different portions of the photoconductor; second, to correct geometric distortion caused by warping of the electron beams owing to external electric fields; and last, to remove residual images on the photoconductor. CCDs similarly required calculations—such as carefully comparing closed-shutter exposures to the open-shutter image—to correct for thermal noise and the effects of cosmic rays.[8]

The second stage involved processing of images returned from the spacecraft. Images came to earth in streams of telemetry, for the early missions as analog waves, later as digital bits. Processing required first extracting imagery data from the telemetry, relating it to cartographic coordinates, and, for analog signals, digitizing it. The digital data were then arrayed in a two-dimensional grid and translated to a gray-scale image, based on the intensity ascribed to each pixel. Much of this work took place at the Jet Propulsion Laboratory, the main NASA center for planetary spacecraft, which developed the Ranger and Mariner spacecraft in the 1960s and many subsequent planetary missions. The first Ranger pictures proved too fuzzy to count or size craters, let alone pick out finer features. Robert Nathan of JPL, who had worked with pattern recognition in crystallography, began cleaning them up. At first he focused on correcting distortions and removing signal noise, such as a particular frequency superimposed on an image by vibration of the camera. The process entailed applying a Fourier transform from spatial to frequency variables, which revealed a bright spot for noise at particular frequencies; wiping out the spot and transforming back from frequency to spatial variables produced a clean image. Similar digital filters could enhance images in other ways—for instance, averaging the contrast in local areas around each image point, subtracting the averaged pattern, and then stretching the remainder had the cumulative effect of sharpening the overall picture. Nathan produced the first computer-enhanced images from Ranger 7 in 1964 and similarly enhanced the low-contrast, featureless images returned later that year by Mariner 4, and proceeded to evolve ever more sophisticated techniques over the decade for Surveyor and Mariner flights (figure 6.1).[9]

By the mid-1970s what was known as the Image Processing Laboratory at JPL had over 50 staff and a library of software and image processing algorithms; one of the most prominent algorithms was VICAR, or Video and Image Communication and Retrieval. Since the reconnaissance community was just switching from satellite film drops to real-time digital telemetry, JPL and the planetary program had perhaps the most advanced digital image processing capability in the country at the time. The software included routines for contrast enhancement, cartographic projection, constructing mosaics, motion compensation, foreshortening and topographic corrections, and full-color composition from black-and-white cameras and single-color filters.[10]

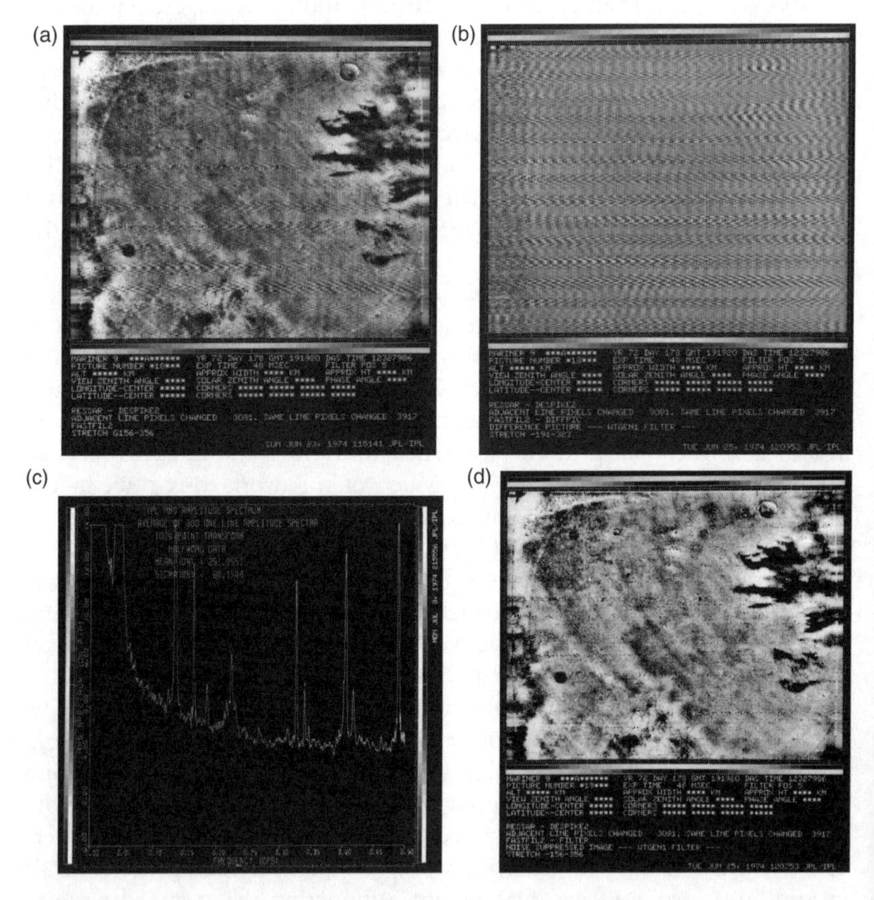

Figure 6.1 Four types of data from the Mariner 9 mission to Mars in 1971. (a) The original image of the surface; (b) the interference pattern superimposed from the spectrometer data; (c) the Fourier transform showing the noise frequency peaks; (d) the final image, with coherent noise suppressed. (Credit: NASA/JPL.)

Image processing capitalized on and helped drive the development of computers. JPL engineers viewed image processing as an information system, encompassing selection and targeting of features for photographs, on-board processing, telemetry and data compression, and ground-based analysis.[11] In particular image processing required fast processors. Fourier transforms were very calculation-intensive, and processing even simplified algorithms for the Fourier transform of a 256-by-256 image could still require 20 minutes on the computers of 1970.[12] An even tighter bottleneck was in the displays. Initially JPL engineers had no interactive digital displays; the processed images were printed on film, which meant that images went from analog format to digital for processing, then back to analog for output. The engineers developing new algorithms had to wait a day or two for film to return from the photo lab before they could see how their software worked. Using Polaroid snapshots taken at particular steps of the processing still forced them to use all the computer memory just to refresh the display. JPL then encouraged some of the small firms just coming out with digital image displays, which had their own memory for refreshing the image and so left the main computer memory free for processing. JPL engineers worked with these manufacturers to develop new digital displays and provided a key early market, since the $100,000 devices otherwise had few buyers.[13]

Digital images from planetary exploration thus helped drive the general emergence of digital image processing. JPL's work drew on earlier efforts in pattern recognition in crystallography and other fields as well as digital computing; and digital pictures themselves had been around since newspapers digitized photos for telegraph transmission in the 1920s.[14] The importance of digital planetary images was to realize the possibilities and by example spark subsequent applications far afield from space science. Several review articles on image processing acknowledged the inspiration provided by the planetary program; a 1975 review article called JPL's work "probably the most visible part of digital image processing to the general public," and a *Computer* magazine article two years later called the Ranger images "essentially the beginning of digital image processing technology."[15] Proliferating textbooks and special journal issues in the 1970s, including several standard texts and review articles by JPL engineers, helped to spread the techniques.[16]

JPL's image processing techniques spread beyond planetary science. In November 1976 JPL and its parent institution, Caltech, convened a conference on image processing attended by nearly 400 people, from academia, government agencies, and industry, who heard about the potential for fields such as geology, oceanography, astronomy, and biomedicine.[17] In the 1970s NASA made JPL's software available through a licensing program, and industry also began hiring away JPL staff. The availability of software, people, and publications helped to catalyze a commercial software industry for image processing starting in the late 1970s, especially with the emergence of higher-performance work stations and graphics accelerators starting in the 1980s. By the 1990s desktop packages such as Photoshop were deploying many of the same techniques—contrast enhancement, image stretching, color corrections, and so on—developed in the planetary exploration program.

Images influenced spacecraft design. Spacecraft designers weighed two competing approaches, spin stabilized versus three-axis stabilized. Spin-stabilized craft spun about their roll axis like a rifle bullet, which helped maintain their trajectory and also provided 360-degree sampling of the space environment for particles-and-fields experiments. Three-axis spacecraft did not spin but rather were stabilized in three dimensions in order to keep antennas pointed at Earth for communication and solar panels at the sun for power. Three-axis stabilization required complex guidance and control systems to keep proper attitude, but as an important side benefit also provided a stable platform for imaging experiments. The Explorer and Pioneer programs, starting in the late 1950s into the early 1970s, used the simpler spin-stabilized approach, but the Ranger and Mariner craft of the 1960s adopted three-axis stabilization.[18] The competing approaches reflected a disciplinary divide in the space-science community: atmospheric scientists and space physicists pushed for particles-and-fields experiments, while geologists preferred images of planetary surfaces.

The Galileo spacecraft highlighted the competition between the particles-and-fields scientists and those pushing visual images. The particles-and-fields camp preferred a spin-stabilized Pioneer design, while geologists and the imaging camp preferred the three-axis-stabilized Mariner platform.[19] James Van Allen, discoverer of the eponymous radiation belts and a senior figure in the field, led the particles-and-fields side, which resented the growing emphasis on three-axis stabilization for imaging; John Casani, a JPL spacecraft engineer, joked that atmospheric scientists had "been in the kitchen trying to get into the banquet hall for years and years." NASA at first sided with the Pioneer design, since it did not require Mariner's complicated sun and star sensors, gyroscopes, and thruster system, with their associated mass, cost, and risk. Imaging scientists protested, according to Casani, that "you cannot send a major spacecraft to Jupiter without cameras." Then, after NASA switched to the Mariner design, "the fields and particles guys are saying, 'over my dead body, this is our mission, we were the ones that put this together, you're going to come in and stabilize the damn thing and take half the value away.' "[20]

Van Allen suggested a compromise, a spacecraft that combined the two designs: half of it could spin and carry the particles-and-fields instruments and probe; the other half could stand still and carry the cameras. Hence JPL pursued what was called a spun/despun design for Galileo. That forced engineers to manage the momentum of two massive segments, each in the neighborhood of a thousand kilograms, with the despun part requiring very stable and precise pointing for the camera. The spin-bearing assembly connecting the two sat close to the center of the spacecraft, requiring engineers to route mechanical, electrical, and thermal subsystems through it. Galileo review boards suggested dumping the dual-spin design and going back to a three-axis spacecraft, but in the end they stuck with a spun/despun design. Space program managers were willing to accept higher engineering risk and cost in order to appease both the imaging advocates and particles-and-fields scientists.[21]

The Results

Even as the debate over Galileo played out in the late 1970s, the twin Voyager spacecraft were on their way to the outer planets. The spectacular images they returned starting in 1979 sealed the acceptance of visual images, but before then images had an uncertain role in solar system exploration. Although rockets and spacecraft gave scientists direct access to space and thus promised remarkable advances, the role of visual images for space science was far from clear at the outset.[22] Meanwhile, space also beckoned mankind with the fulfillment of primitive dreams of flight to the heavens, and here too images had an undefined role in exploration.[23]

This debate over the relation between science and exploration in the planetary program, and the role of images, played out over the first decades of the US space program. For example, the Ranger missions to the moon in the early 1960s were primarily intended to prepare the way for Apollo, not to produce scientific data; plans for several experiments on Ranger were dropped to concentrate on images to scout landing sites as well as spark public interest, to the chagrin of space scientists.[24] But accumulating scientific discoveries from visual images, enabled by the combination of high-resolution spacecraft cameras and ever more sophisticated digital image processing, would eventually settle the debate.

The first results came in the 1960s, when Mariner cameras revealed geological features on Mars dwarfing any on earth: volcanoes hundreds of kilometers across and almost twenty kilometers high, and a grand canyon thousands of kilometers long and five kilometers deep. Most surprising was evidence that Martian canyons had been carved by running water; the past presence of water rekindled speculation about life on Mars.[25] At Mercury, Mariner 10 images detected numerous craters, the remnants of what was called the "Great Bombardment," when a large number of planetesimals shot through the inner solar system early in its history. These results supported catastrophist theories of earth's geological and biological history.[26]

Starting in 1979, Voyager cameras returned a treasure trove of images from the outer planets—Jupiter, Saturn, Uranus, and Neptune—which fundamentally altered human understanding of the solar system. In addition to sharp images of Jupiter's and Saturn's weather systems, Voyager cameras revealed unanticipated complexity in Saturn's ring system, with images of dark spokes and intertwined or braided rings challenging the existing theories of ring dynamics. Voyager also detected several new moons at both Jupiter and Saturn and new idiosyncrasies in the known moons. Since the moons appeared in ground-based photos as small smudges, planetary scientists tended to extrapolate from Earth's moon and hence expected to find geologically dead satellites, pockmarked by impact craters. Murray and planetary geologists had struggled to persuade Voyager planners that the satellites deserved close scrutiny.[27]

The results justified the attention. Whereas scientists had anticipated craters on Io but found none, they expected none on Ganymede, whose ice surface was expected to absorb them, but found them in abundance; other regions on Ganymede exhibited a corduroy pattern indicating extensive ancient fault lines. Europa, another icy satellite, by contrast turned out smooth as a billiard ball, its icy crust obliterating craters but retaining a lacy pattern of dark cracks in the ice.[28]

Images of Jupiter's moon Io showed a surprising absence of impact craters, which implied that some active geological process had erased them. An image produced by JPL's navigation team soon showed a plume from an active volcano extending some three hundred kilometers above the edge of the planet. Io thus joined Earth as the only other body in the solar system to display active volcanoes. Volcanism explained the lack of craters, but in turn raised the question, where did the heat come from? Earth's hot interior drives plate tectonics; but Io occupies a much colder orbit, and any radioactive heat sources on the small moon would have long ago decayed. The heat engine was found instead in tidal forces from Jupiter's gravity and the other large moons, which combined to flex Io and thus heat it.[29] Voyager's later encounters with Uranus and Neptune produced similar visual images of unexpected energy at such great distances from the sun, whether it was meteorological activity as with Neptune's weather systems, or geological activity on moons, as with sharp canyons and cliff faces on Miranda (figure 6.2) or geysers venting material several kilometers into Triton's sky (figure 6.3).[30]

Voyager images rewrote the textbooks on the outer solar system and clinched the argument on the utility of images for planetary science.[31] Voyager cameras turned the outer planets and especially their moons from blurry smears on astronomer's plates to complex individual bodies undergoing

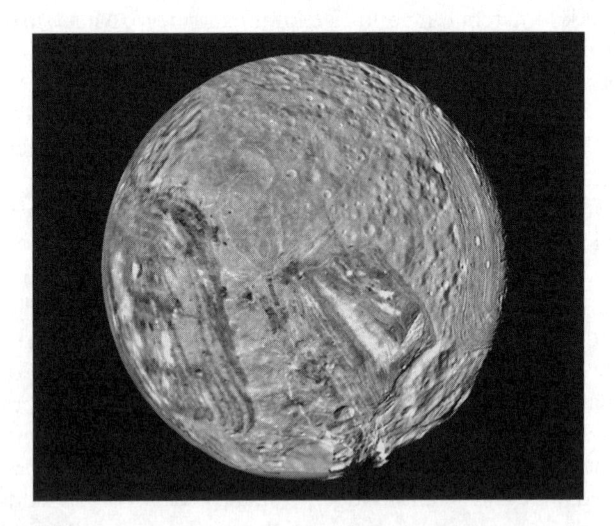

Figure 6.2 Uranus's moon Miranda is shown in a computer-assembled mosaic of images obtained on January 24, 1986, by the Voyager 2 spacecraft. Miranda is the innermost and smallest of the five major Uranian satellites, just 480 kilometers (about 300 miles) in diameter. Nine images were combined to obtain this full-disc, south-polar view, which shows the varying geologic provinces of Miranda. The bulk of the photo comprises seven high-resolution images from the Voyager closest-approach sequence. Data from more distant, lower-resolution images were used to fill in gaps along the limb. (Credit: NASA/JPL photo no. PIA01490.)

Figure 6.3 Global mosaic of Triton, taken in 1989 by Voyager 2 during its flyby of the Neptune system. (Credit: NASA/JPL/USGS photo no. PIA00317.)

dynamic processes; in particular, the weather systems on the large gaseous planets and the volcanism and tectonics on their moons surprised scientists who thought the frigid outer reaches of the solar system lacked the energy for such processes. The Voyager images emphasized the solar system's diversity and dynamism and thus helped to correct the geocentric perspective of planetary scientists.[32]

Later missions confirmed the role of visual images. As the first spacecraft to orbit an outer planet, Galileo provided pictures of Jupiter and its major moons from a variety of angles starting in 1995. The most important result concerned Io's colder cousin Europa. Pictures of curving cracks in Europa's ice surface, linked in long chains extending hundreds of kilometers, led to models of tides in a vast ocean of liquid water under the ice. The theory of extensive and perhaps accessible liquid oceans had profound implications: water is seen as a key ingredient for life, and Europa became another prime target in the search for extraterrestrial life.[33]

In early 2000 Mars Global Surveyor's camera, which could pick out objects 1.4 meters across from orbit, provided pictures of widespread gullies and canyons, apparently cut by recent melting ice, suggesting that liquid water might currently exist just below the surface.[34] The Spirit and Opportunity images from Mars in 2004 reinforced evidence for water on Mars. The same year, the Cassini spacecraft arrived at Saturn. Cassini's camera, with five times the resolution of Voyager's, provided fine-grained images of huge geysers of icy water shooting at supersonic speeds from the surface of the moon Enceladus, suggesting yet another source of liquid water in the solar system. The photographic evidence for liquid water on Mars, Europa, and Enceladus raised hopes of finding signs of life beyond Earth.

Visual Imagery and the Public

The planetary program pursued visual images not only for science, but also for public relations. NASA was highly attuned to public relations from the outset, evident in the celebrated *Life* astronaut profiles, and extended this orientation to the planetary program. In their first proposals for space projects after Sputnik, JPL mission designers ranked the priority of various goals: first was technical feasibility, but second was public relations, ahead of scientific and technical objectives.[35] Visual images provided a prime source of public interest in the planetary program.

Changes in the American and world media reinforced the importance of images—in particular, the emergence of television in addition to print media. In the 1960s public affairs for the planetary program was print dominated, but by the 1970s television had acquired a powerful role in American mass media, and news stories on television relied on snappy visuals.[36] Viking and then especially Voyager starting in the late 1970s capitalized on television interest. When Voyager reached Jupiter in 1979, public television aired a nightly "Jupiter Watch" hosted by JPL engineer Al Hibbs, beaming pictures of Jupiter nationwide together with commentary by Hibbs and guest scientists. Another television event the following year reinforced public interest in planetary images. "Cosmos," a 13-part series on astronomy for public television hosted by planetary astronomer Carl Sagan, was an instant hit, eventually reaching not quite billions and billions, but about a half billion viewers worldwide. The "Cosmos" series used Voyager images of Jupiter and other planetary images as visual stimulation.

The increasing emphasis on visual images shaped planetary science itself. For one, it encouraged "instant science," or science by press conference, where scientists announce their results not in academic journals after months or years of data analysis, but rather to the popular media within days or hours of getting the data. Scientists adapted slowly to the new mode, with some opposing quick release of photos from Mariner spacecraft and seeking instead to hold them back for scholarly publications. Such resistance seemed justified on Mariner 10, when the press seized on reports of a moon around Mercury; on closer examination the image turned out to be a distant star.[37] But by the time of Voyager scientists had come to accept and even relish the new mode of publicity, especially the media-friendly imaging team. Voyager scientists produced a new sort of peer review by convening meetings of the spacecraft science team to review incoming images and data before releasing them to the press. The process prevented major mistakes with instant science based on visual images, although a few small ones—such as the assertion that Saturn's rings contained thousands of separate ringlets (later analysis determined that there were waves in the rings, but not discrete gaps)—snuck through.[38]

Digital image processing supported the public-relations effort but raised questions about the epistemology of images. Image processing removed unsightly distractions, such as reticles or blank patches, and created color photos, in some cases expressly to wow the public.[39] Voyager images of Io, for example, revealed a colorful landscape of light and dark patches against a vivid orange background, provoking comparisons to a pepperoni pizza. Some images were perhaps too colorful and appealing. The extensive digital manipulation of images raised the possibility that the result was a subjective, idealized view of what humans *thought* the planets should look like, rather than the real thing. One JPL manager in the 1980s, describing the new field he called "Imageering," suggested this sensibility in declaring that "the idea is to generate an image which a perfect 'eye' would have seen."[40] The digital production of false-color images in astronomy has raised eyebrows among astronomers, who see such images as aimed more toward marketing than science—in particular, as a way to win taxpayer support for expensive space missions. As one astronomer and Hubble telescope supporter put it, "Big science requires big publicity," and jazzed-up photos were just part of the public-relations process.[41] Similarly, some scientists and journalists complained that false-color photos from Voyager were, indeed, false representations of the outer planets. JPL's public-relations staff replied that they always identified images as false-color, but that news media did not always convey this to the public.[42]

Visual images also provided fodder for the community of conspiracy theorists and UFO buffs, a constant presence for the American planetary program. These fringe groups focused especially on Mars, whose features had encouraged speculation about extraterrestrials for centuries, including Percival Lowell's famous canals. In 1976, cameras on the Viking mission captured an image of a hill resembling a sculpted face, which some theorists claimed was a remnant of an extraterrestrial civilization. Some conspiracy theorists claimed that NASA killed the Mars Observer spacecraft in 1993 after its images provided definitive proof of the "face on Mars." The high-resolution camera on Mars Global Surveyor mostly deflated such speculation by showing that the "face" was in fact an accident of light and shadow.

The emergence of television promoted not only still images, but also motion pictures. JPL staff combined still photos from Ranger and Mariner to produce motion pictures, and more elaborate simulated flights over Venus and Mars would later be produced using images from Magellan and Mars Global Surveyor in the 1990s. Meanwhile, the planetary program incubated pioneering work on computer animation. In preparation for Voyager, JPL hired a young computer programmer named James Blinn to generate three-dimensional animations simulating Voyager's encounters at the outer planets. Blinn developed new techniques of volume rendering, bump mapping, forward hierarchical kinematics, and other complex algorithms to model the motions of the spacecraft components and rotating planets against an accurate backdrop of planets and stars and lighting angles. The three-dimensional, textured color movies—in which the viewer rode along with Voyager as the spacecraft swooped past Jupiter and its satellites, and even more spectacularly two years later over Saturn's rings—proved a hit with TV news editors and viewers.[43] In 1980 Blinn took his techniques from the planetary program to George Lucas's special-effects studio, where he worked on the first "Star Wars" sequels; his Voyager images also sparked a collaboration with Ed Catmull and Alvy Ray Smith, who would go on to cofound Pixar.[44] Disney later raided JPL's image processing lab for digital animators in the 1990s; by that time JPL had a separate Digital Image Animation Laboratory funded mostly by IMAX, which produced several films for popular audiences, including "Cosmic Voyage," featuring dazzling visual images of outer space.[45]

By the 1980s, the cumulative results from Voyager had raised the popular status of images to a point where some members of the planetary community advocated dumping all other scientific goals. A 1987 proposal from the Planetary Society for a robotic Mars spacecraft exemplified the view that planetary missions should aim primarily at entertaining the public. The proposal stressed that "science should be given a low priority on this mission, if it is given any direct participation at all...Imaging, imaging, and more imaging is the name of the game...The aim here is to obtain images that are shameless crowd-pleasers and show Mars from a human perspective. If that's not good science, well then tough."[46]

Conclusion

Fifty years into the space age, scientists and the public have grown accustomed to viewing detailed images delivered from robotic spacecraft at distant planets. These images have transformed human understanding of the planets and their satellites, revealing diverse and dynamic environments throughout the solar system and reversing the human tendency to view Earth as exceptional. The evidence for extraterrestrial water raised the possibility of even more fundamental discoveries of potential life beyond Earth (figure 6.4).

To provide these images scientists and engineers developed several new technologies, including digital spacecraft cameras, high-data-rate telemetry, three-axis stabilized spacecraft, and digital image processing, and some of these technologies, such as image processing and computer animation, influenced fields far beyond planetary exploration. Acceptance of visual images in planetary programs, though, came after programmatic debates about science versus exploration and disciplinary squabbles between geology and space physics. Image processing meanwhile crossed disciplinary lines, bringing software developers together with hardware engineers and planetary scientists.

The aesthetic appeal of planetary images was not just for public relations. Or, rather, the reason the pictures captivated the public was because they were visually arresting. Visual artists such as Michael Light and Michael Benson demonstrated the aesthetic possibilities in shows and art books on the photographs from the moon and planetary programs.[47] Photos from planetary spacecraft were intended as works of science, but were also, indeed, works of art. Contrary to popular belief, science and art are not distant and disconnected. The perceived division between science and art is itself a historical construct, emerging first in the early modern period amid the

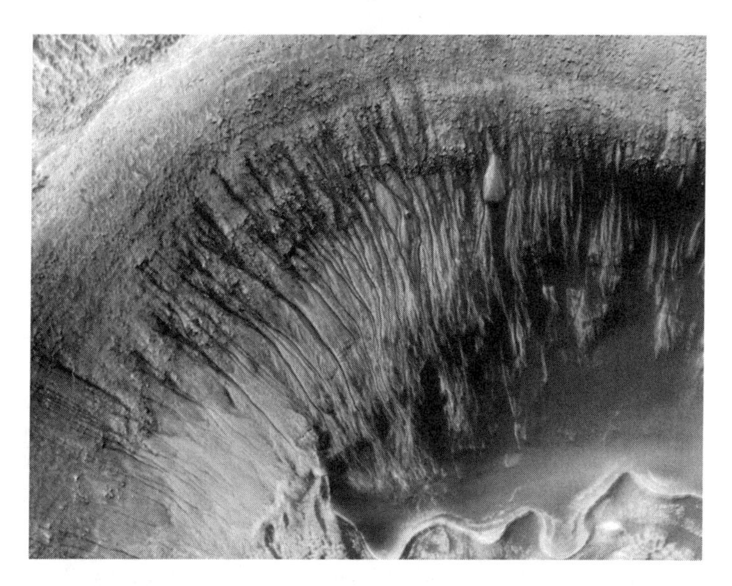

Figure 6.4 A small crater inside the large Newton Crater on Mars. Such images as this one from Mars Global Surveyor showing unusual channels on Mars excited the public about the possibility of the planet once being a watery place, and if so leading to speculation about life once evolving there. (Credit: NASA/JPL/MSSS.)

Scientific Revolution and the humanist project for art in the Renaissance, and then deepening in the modern period, as science self-consciously rejected aesthetic subjectivity while artists turned away from the industrialized, mechanized trappings of science. Historians have shown how this perceived division was constantly challenged in the twentieth century by connections from both sides.[48] Solar system exploration, literally the highest expression of scientific and technological mastery, represented another reconnection of science and art.

Notes

1. Spacecraft have not yet visited Pluto, which recently lost its official status as a planet.
2. Mary G. Winkler and Albert Van Helden, "Representing the Heavens: Galileo and Visual Astronomy," *Isis* 83 (1992): 195–217.
3. Martin J. S. Rudwick, "The Emergence of a Visual Language for Geological Science, 1760–1830," *History of Science* 14 (1976): 149–95; *Scenes from Deep Time: Early Pictorial Representations of the Prehistoric World* (Chicago: University of Chicago Press, 1995).
4. Michael Lynch and Samuel Y. Edgerton, Jr., "Aesthetics and Digital Image Processing: Representational Craft in Contemporary Astronomy," in Gordon Fyfe and John Law, eds., *Picturing Power: Visual Depiction and Social Relations* (London, UK: Chapman & Hall, 1988), pp. 184–220.
5. For an introduction to the growing literature on the history of scientific images, see Alex Soojung-Kim Pang, "Visual Representation and Post-Constructivist History of Science," *Historical Studies in the Physical and*

Biological Sciences 28:1 (1997): 139–71. On Galileo, see Erwin Panosky, “Galileo As a Critic of the Arts: Aesthetic Attitude and Scientific Thought,” *Isis* 47 (1956); Mary G. Winkler and Albert Van Helden, “Representing the Heavens: Galileo and Visual Astronomy,” *Isis* 83 (1992), 195–217; on spectroscopy, see Klaus Hentschel, *Mapping the Spectrum: Techniques of Visual Representation in Research and Teaching* (New York: Oxford University Press, 2002); on bubble chambers, see J. L. Heilbron, Robert W. Seidel, and Bruce R. Wheaton, *Lawrence and his Laboratory: Nuclear Science at Berkeley, 1931–1961* (Berkeley: University of California Press, 1981), pp. 91–93; and Peter Galison, *Image and Logic: A Material Culture of Microphysics* (Chicago: University of Chicago Press, 1997), pp. 370–431.

6. Gilbert F. Amelio, “Charge-Coupled Devices,” *Scientific American* 230.2 (February 1974): 23; Robert W. Smith and Joseph N. Tatarewicz, “Counting on Invention: Devices and Black Boxes in Very Big Science,” *Osiris* 9 (1994): 101–23.
7. Bruce Murray, *Journey into Space: Three Decades of Space Science* (New York: W.W. Norton and Co., 1989), pp. 103–109, 120.
8. William B. Green interview, February 12, 2002.
9. T. Rindfleisch, “Getting More Out of Ranger Pictures by Computer,” *Astronautics and Aeronautics* (January 1969): 70–74; Robert Nathan, interview by Cargill Hall, January 22, 1975 (JPL Archives).
10. *Proceedings*, Caltech/JPL Conference on Image Processing Technology, Data Sources and Software for Commercial and Scientific Applications, November 3–5, 1976, JPL SP-43–30.
11. A. R. Hibbs, “An Analysis of the Effect on JPL of NASA’s ‘Outlook for Space’ Study,” February 5, 1976 (JPL 142, 37/639); “Outlook for Space: A Synopsis,” report by NASA study group, January 1976 (JPL 142, 37/638).
12. Harry C. Andrews, *Computer Techniques in Image Processing* (New York, 1970), p. 33.
13. William Green interview.
14. Kenneth R. Castleman, “A History of Digital Image Processing at JPL,” in Castleman, *Digital Image Processing* (Englewood Cliffs, NJ, 1979), pp. 383–400; Azriel Rosenfeld, *Picture Processing by Computer* (New York, 1969); Maynard D. McFarlane, “Digital Pictures Fifty Years Ago,” *Proceedings of the IEEE* 60:7 (1972): 768–770.
15. B. R. Hunt, “Digital Image Processing,” *Proceeding of the IEEE* 63.4 (1975): 693–708; Esmond C. Lyons, Jr., “Digital Image Processing: An Overview,” *Computer* 10.8 (1977): 12–14; see also B. R. Hunt and D. H. Janney, “Digital Image Processing at Los Alamos Scientific Laboratory,” *Computer* 7.5 (1974): 57–61; M. P. Ekstrom, “Digital Image Processing at Lawrence Livermore Laboratory, Part I—Diagnostic Radiography Applications,” *Computer* 7.5 (1974): 72–80.
16. Special issues of *Proceedings of the IEEE* 60.7 (1972); *Computer* 7.5 (1974) and 10.8 (1977); Castleman, *Digital Image Processing*; and William B. Green, *Digital Image Processing: A Systems Approach* (New York, 1983).
17. R. B. Gilmore to Murray, November 29, 1976 (JPL 142, 23/384); *Proceedings*, Caltech/JPL Conference on Image Processing Technology, Data Sources and Software for Commercial and Scientific Applications, November 3–5, 1976, JPL SP-43–30.

18. Clayton R. Koppes, *JPL and the American Space Program: A History of the Jet Propulsion Laboratory* (New Haven, CT: Yale University Press, 1982), p. 119.
19. Meltzer, *Galileo*, chapter 2; John Casani interview, November 21, 2003.
20. Casani interview.
21. Norm Haynes interview, March 13, 2003; Casani interview.
22. On the space age and space science, see: David H. DeVorkin, *Science with a Vengeance: How the Military Created The U.S. Space Sciences after World War II* (New York: Springer Verlag, 1992); Joseph N. Tatarewicz, *Space Technology and Planetary Astronomy* (Bloomington: Indiana University Press, 1990); Stephen G. Brush, *Fruitful Encounters: The Origins of the Solar System and of the Moon from Chamberlin to Apollo* (Cambridge, UK: Cambridge University Press, 1996); Ronald E. Doel, *Solar System Astronomy in America: Communities, Patronage, and Interdisciplinary Science, 1920–1960* (Cambridge, UK: Cambridge University Press, 1996).
23. Howard E. McCurdy, *Space and the American Imagination* (Washington, DC: Smithsonian Institution Press, 1997).
24. Koppes, *JPL and the American Space Program*, pp. 121–22, 132–33, 150–51, 164.
25. Robert S. Kraemer, *Beyond the Moon: A Golden Age of Planetary Exploration, 1971–1978* (Washington, DC: Smithsonian Institution Press, 2000), pp. 44–61; Murray, *Journey into Space*, pp. 41–65; Koppes, *JPL and the American Space Program*, pp. 220–21.
26. Stephen G. Brush, *A History of Modern Planetary Physics*, vol. 3, *Fruitful Encounters: The Origin of the Solar System and of the Moon from Chamberlin to Apollo* (Cambridge, UK: Cambridge University Press, 1996), pp. 131–38; Ronald E. Doel, "The Earth Sciences and Geophysics," in John Krige and Dominique Pestre, eds., *Science in the Twentieth Century* (Amsterdam, 1997), pp. 391–416, on pp. 396, 406.
27. Stone interview by Dethloff and Schorn; Murray, *Journey into Space*, p. 136.
28. Henry C. Dethloff and Ronald A. Schorn, *Voyager's Grand Tour* (Washington, DC: Smithsonian Institution Press, 2003), pp. 160–61, 185–86; David Morrison and Jane Samz, *Voyage to Jupiter* (Washington, DC: NASA SP-439, 1980), p. 79; Mark Washburn, *Distant Encounters: The Exploration of Jupiter and Saturn* (Boston: Harcourt, 1983), pp. 109–10, 208–11; David Morrison, *Voyages to Saturn* (Washington, DC: NASA SP-451, 1982), pp. 146–47.
29. Stone interview by Dethloff and Schorn; Stone in David W. Swift, *Voyager Tales: Personal Views of the Grand Tour* (Reston, VA: AIAA, 1997), p. 50; Morrison and Samz, *Voyage to Jupiter*, p. 86; Washburn, *Distant Encounters*, pp. 115–17.
30. Dethloff and Schorn, *Voyager's Grand Tour*, pp. 209, 215–20; William E. Burrows, *Exploring Space: Voyages in the Solar System and Beyond* (New York: Random House, 1995), pp. 318–19, 399–417; John Noble Wilford, "Pictures of Triton, Neptune's Icy Moon, Show Signs of a Never-Seen Icy Volcanism," *New York Times*, August 26, 1989.
31. Dethloff and Schorn, *Voyager's Grand Tour*, pp. 147–50; Ronald E. Doel, *Solar System Astronomy in America: Communities, Patronage, and Interdisciplinary Science, 1920–1960* (Cambridge, UK: Cambridge University Press, 1996).

32. Stone interview by Dethloff and Schorn.
33. Gregory V. Hoppa, B. R. Tufts, R. Greenberg, and P. E. Geissler, “Formation of Cycloidal Features on Europa,” *Science* 285 (September 17, 1999): 1899–902; Torrence V. Johnson, “A Look at the Galilean Satellites after the Galileo Mission,” *Physics Today* (April 2004), pp. 77–83; David M. Harland, *Jupiter Odyssey: The Story of NASA’s Galileo Mission* (Chichester, UK: SpringerPraxis, 2000), pp. 230–32.
34. Arden L. Albee, F. D. Palluconi, and R. E. Arvidson, “Mars Global Surveyor Mission: Overview and Status,” *Science* 279 (1998): 1671–72; Laurence Bergreen, *Voyage to Mars: NASA’s Search for Life Beyond Earth* (New York: Riverhead Books, 2000), pp. 304–306.
35. Koppes, *JPL and the American Space Program*, p. 100.
36. Frank Colella interview, February 26, 2002; James L. Baughman, *The Republic of Mass Culture: Journalism, Filmmaking, and Broadcasting in America Since 1941* (Baltimore, MD: Johns Hopkins University Press, 1997), pp. 92–98, 108–15, 158–70; Sharon M. Friedman, “The Journalist’s World,” in Sharon M. Friedman, Sharon Dunwoody, and Carol L. Rogers, eds., *Scientists and Journalists: Reporting Science As News* (New York: Free Press, 1986), pp. 17–41, on 36–37.
37. On Mercury’s moon, Murray, *Journey into Space*, p. 119; and Washburn, *Distant Encounters*, p. 125.
38. Stone in Swift, *Voyager Tales*, pp. 39–41; Stone author interview, October 23, 2003.
39. For example, Murray, *Journey into Space*, pp. 115–16.
40. F. H. Felberg, “Imageering,” January 19, 1988 (JPL 230, 28/263).
41. Allison M. Heinrichs, “PR with Universal Appeal,” *LA Times*, September 5, 2003.
42. Joel Davis, *Flyby: The Interplanetary Odyssey of Voyager 2* (New York, 1987), pp. 121–22.
43. Colella interview; Kohlhase interview, July 12, 2002; James Blinn, telephone interview, July 20, 2007.
44. Natalie Angier, “It was Love at First Byte,” *Discover* (March 1981); David Salisbury, “Computer Art Takes Off Into Space,” *Christian Science Monitor*, July 20, 1979.
45. James A. Evans, “The Reimbursable Program,” February 25, 1995 (JPL 259, 50/553).
46. E. J. Gaidos, “Project Precedent,” July 31, 1987 (JPL 198, 36/528).
47. Michael Light, *Full Moon* (New York, 1999); Michael Benson, *Beyond: Visions of the Interplanetary Probes* (New York, 2003); see also Anthony Lane, “The Light Side of the Moon,” *New Yorker* (April 10, 2000).
48. Caroline A. Jones and Peter Galison, *Picturing Science, Producing Art* (New York, 1998).

Chapter 7

Returning Scientific Data to Earth: The Parallel but Unequal Careers of Genesis and Stardust and the Problem of Sample Return to Earth

Roger D. Launius

One of the most difficult tasks with which NASA has had to deal is how its space systems operate while transiting the atmosphere as they return to Earth. Coming home after a flight into space is a fundamental challenge, and research in aerodynamics, thermodynamics, thermal protection, simulation, guidance and control, stability, propulsion, and landing systems have proven critical to the success of these efforts from the beginning of the space age. Ablative heat shields and parachutes served well the task of delivering spacecraft back to Earth at their end of their missions during the Mercury, Gemini, and Apollo programs, and they have been used intermittently thereafter in other space projects. In the 1970s, however, NASA changed its methodology for reentry and recovery from space with the reusable Space Shuttle and its ceramic tile thermal protection system (TPS). While the ablative heat shields and parachute systems of early space missions were tailored to a wide array of scenarios and reengineered for each new flight regime, since the Viking lander program to Mars NASA ceased efforts to develop new reentry and recovery systems in favor of reusable systems supporting the Space Shuttle.

Not until the first part of the twenty-first century did NASA engineers return to the use of ablative heat shields and parachutes for coming back to Earth, and then with imperfect results. The Genesis spacecraft, launched in 2002, collected solar particles for two years and reentered the Earth's atmosphere on September 8, 2004, but the parachute failed to deploy causing the capsule to impact the ground at high velocity. Something as seemingly simple as a heat shield and parachute system similar to that used on earlier flights—albeit not for many years—had failed, and failed catastrophically. NASA had better success with the Stardust, launched in 1999 and recovered in 2006, also returning with an ablative heat shield and parachute system.

This was the first US space mission dedicated solely to returning extraterrestrial material from beyond the Moon, collecting samples both from Comet Wild 2 and the interstellar dust.

A Short History of Returning from Space

The atmosphere surrounding the Earth and supporting life here also makes spaceflight harder than it would be if it did not exist. It is said, only half-jokingly, that getting to orbit is halfway to anywhere because of the energy necessary to go beyond the gravity well of this planet.[1] Generally overlooked, however, is just how difficult it is to come home from orbit. All of the energy expended to get to orbit dissipates on the way back to Earth, usually in the form of extreme heating. In addition to the aerodynamic concerns with high-speed flight, there are serious thermodynamic issues with a 17,500 miles per hour plunge through the Earth's atmosphere.[2] Unlike the relatively long gestation periods of airplanes and launch vehicles, the technology needed to survive reentry matured quickly, largely in response to national security concerns. The warheads developed during the Cold War for ballistic missiles led directly to the capsules that first allowed humans and robotic systems to return from space. Of course, it is important to recognize that the requirements for human entry were much more severely restricted than for systems that did not have people aboard. The Mercury capsule, accordingly, could not simply be a warhead shell reconfigured for human occupancy. The heat shield configured for longer, lower-g entries drove the development of several new technologies.[3]

While most proposals for satellites between 1946 and 1957 avoided the difficult problem of reentry, it became obvious soon after World War II that returning to Earth represented a major step in flight. Beginning with the theoretical work of Germans Walter Hohmann and Eugen Sänger, as well as Americans such as Theodore Theodorsen and Robert Gilruth at the Langley Memorial Aeronautical Laboratory in the 1940s and early 1950s, it also included efforts to understand the heating of reentry during ballistic missile development in the 1950s. Three approaches dominated thinking at the time. The first was a heat sink concept that sought to move quickly from space through the upper atmosphere. Superheating proved a serious problem, however, and materials to protect the spacecraft a major concern. Initially the preferred thermal protection choice for ballistic missiles, as range and speed grew engineers realized that it was unacceptable for orbital reentry. The second approach, championed by Wernher von Braun and his rocket team in Huntsville, Alabama, called for circulating a fluid through the spacecraft's skin to soak up the heat of reentry. Von Braun's grandiose vision foresaw astronauts returning from wheeled space stations aboard huge spaceplanes, but when challenged to develop actual hardware he realized that there was no way for the heat to be absorbed without killing the occupants. For orbital flight, both of these concepts gave way to Julian Allen's and Alfred Eggers' blunt-body concept, which fundamentally shaped the course of spaceflight research and provided the basis for all successful reentry vehicles.[4]

All human spaceflight projects actually flown by the United States prior to the Space Shuttle employed a blunt-body reentry design with an ablative shield to dissipate the heat generated by atmospheric friction. This approach has also been used in reconnaissance, warhead, and scientific reentry successfully from the 1950s to the present. Additionally, the question of what materials to use to protect the spacecraft during blunt-body reentry emerged, and research on metallic, ceramic, and ablative heat shields prompted the decision to employ ablative technology.[5]

Once the orbital energy is converted and the heat of reentry dissipated, there is still the requirement to gently land the spacecraft in the ocean or on land. All of the American efforts until the Space Shuttle, and the Soviet and Chinese capsules, used from one to three parachutes for return to Earth. For the Americans, the capsules landed in the ocean and were recovered by ship. This exposed them to corrosive saltwater while waiting for expensive recovery efforts. Both the Soviet/Russian and Chinese spacecraft have always been recovered on land, which presented the crew with a harder landing than would be the case in the sea but obviated the need for naval deployments to recover the capsule and crew. For Project Gemini NASA toyed with the possibility of using a paraglider being developed at Langley Research Center for "dry" landings instead of a "splashdown" in water and recovery by the navy. The engineers never did get the paraglider to work properly, however, and eventually dropped it from the program in favor of a parachute system like the one used for Mercury.[6]

The United States also used parachutes to return film canisters from the nation's first reconnaissance satellite, CORONA, flown between 1960 and 1972. This program employed satellites with cameras and film launched into near-polar orbits to provide frequent coverage of the USSR. After the film was exposed, it was wound onto reels in a special reentry capsule that separated from the spacecraft at an altitude of about 100 miles (160 kilometers) and then at 60,000 feet (20,000 meters) jettisoned its heat shield and deployed a parachute. Air force planes flying over the Pacific then snagged the parachute and capsule, returning the film for processing and analysis.[7]

Even as Apollo was reaching fruition in the latter 1960s, NASA officials made the decision to abandon capsules with blunt-body ablative heat shields and recovery systems that relied on parachutes for its human spaceflight program. Instead, it chose to build the Space Shuttle, a winged reusable vehicle that still had a blunt-body configuration but used a new ceramic tile and Reinforced Carbon-Carbon for its TPS. Parachutes were also jettisoned in favor of a delta-wing aerodynamic concept that allowed runway landings. Despite many challenges, and the loss of one vehicle and crew due to a failure with the TPS in 2003, this approach has worked since first flown in 1981 through the end of the Space Shuttle program in 2011. The shuttle flew 135 orbital missions over the course of its career.[8]

The Soviet Union also built a space shuttle, the Buran, which flew only one mission without a crew aboard in 1988 before its retirement. Like its American counterpart, Buran landed on delta wings like an aircraft. With the demise of the Soviet Union in the latter 1980s, however, the Soviet Ministry of Defense realized that this system was expensive, and without a firm rationale, and was cancelled. Not until 1993 did NPO Energia head, Yuri Semenov, publicly announce the end of the project. As the twenty-first century has progressed, the preferred method for returning to Earth remains using an ablative heat shield for the dissipation of excess heat and speed and parachutes for soft landing. The Russians never made the decision, as did the Americans, to abandon ablative heat shields and parachutes for returning to the Earth from space. Generally, this has worked effectively but they did lose one mission, Vladimir Komarov's Soyuz 1 mission on April 24, 1967, when his reentry system failed.[9] The Russians likewise used the ablative heat shield and parachute approach for recovery of three lunar sample return missions, Luna 16, which successfully returned 101 grams of lunar soil in 1970; Luna 20, which returned 30 grams in 1974; and Luna 24, which returned 170.1 grams in 1976.

Bringing Back to Earth Deep-Space Probes: Genesis and Stardust

Because of the thirty-year diversion of NASA's efforts on reentry toward Space Shuttle efforts, a return to ablative technology with two small reentry vehicles for science sample collection and return to Earth in the 1990s proved a challenge. While the ablative heat shields of early space missions—especially Mercury, Gemini, and Apollo—were built specifically for the proposed use, tailored to a wide array of mission scenarios, and reengineered for each new flight regime, since the Viking Lander program to Mars in the mid-1970s NASA had gotten "rusty" in understanding and using this earlier reentry technology. As one analysis commented: "Over the past 30 years NASA adopted a 'risk averse' philosophy relative to TPS, i.e., use what was used before, even if it was not optimal, since it had been flight-qualified. An unintended consequence was that the ablative TPS community in the United States slowly disappeared." This capability had to be rebuilt with the Genesis and Stardust missions.[10]

When Genesis (figure 7.1) was first proposed, the manner in which it might return samples to Earth took several turns during the evolution of the mission. Some engineers proposed placing the spacecraft in Earth orbit and sending a Space Shuttle to retrieve it and bring it back in the payload bay. Others thought they could devise an approach whereby the spacecraft would rendezvous with the International Space Station and be captured by astronauts aboard. Still others believed they could park Genesis at L2, a location between the Earth and the Moon where it could remain with virtually no expenditure of energy indefinitely, from which it could be recovered

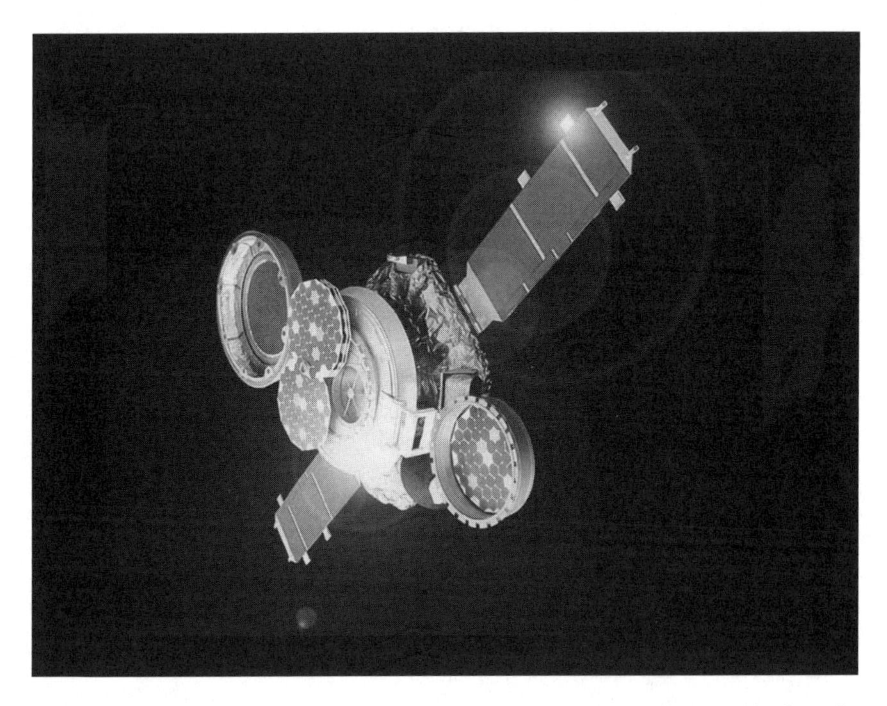

Figure 7.1 Genesis mission. The goal of the Genesis mission was to collect particles from the solar wind and return them to Earth for laboratory study. (Credit: NASA/JPL.)

at some point when NASA had the capability to do so. Not surprising, none of these options received much support, especially from the scientific community anxious to obtain the sample collected by the spacecraft during its mission.

Accordingly, when the mission lifted off in 2002 from the Kennedy Space Center, Florida, for its voyage to the L1 Lagrangian point between the Earth and Sun, mission planners had prepared for its return using an ablative heat shield similar to that on the Apollo spacecraft and equipped with parachutes for it to land safely. After two and a half years of collecting solar particles, it left the L1 halo orbit in April 2004, undertook a flyby of the Moon, and released its return capsule for the final descent back to Earth. The Genesis spacecraft, sans the return capsule, then returned to the L1 halo orbit. Everything appeared fine as the Genesis return capsule entered the Earth's atmosphere on September 8, 2004, and it safely made its way through the heavy ionization part of reentry. Unfortunately, the parachute failed to deploy allowing the capsule to impact the ground at a high velocity at the Utah Test and Training Range. Even so, scientists insisted that the majority of specimens returned were usable and the majority of the science goals were later met.[11]

Although the landing failed, the Genesis return to Earth had been planned as an extension of earlier efforts for recovery from superorbital missions. One statement laid out the manner in which the mission was to unfold:

> About four hours before Earth entry, the spacecraft reorients to the sample return capsule release attitude, spins up to 15 rpm and releases the capsule. Soon after release, the spacecraft reorients to joint its thrusters to Earth and performs a maneuver which will cause the spacecraft to enter Earth's atmosphere, but break- up over the Pacific Ocean. Following release from the spacecraft, the Genesis sample return capsule experiences a passive, spin-stabilized aero-ballistic entry, similar to that of the Stardust mission. When the capsule has decelerated to 1.4 times the speed of sound, the on-board avionics system fires a mortar to deploy the drogue parachute. The drogue is a disk-gap-band design, with heritage dating to the Viking program, and an extensive history of supersonic applications. It serves both to increase the deceleration of the capsule, and to stabilize it through the transonic phase. As the capsule descends into the airspace of the Utah Test & Training Range (UTTR), recovery helicopters are directed to fly toward the intercept point. The capsule's ballistic path is designed for delivery within an 84 x 30 km footprint with subsequent reduction to a 42 x 10 km helicopter zone... The first helicopter on-site will line up and match descent rate, then execute a Mid-Air Retrieval (MAR) capture. If a pass is aborted for any reason the pilot can line up and repeat.[12]

If only it had worked as intended; instead it crash-landed in the Utah desert.

While Genesis was built to return to Earth, the story of the Genesis recovery system was fundamentally shaped by an effort in the latter 1980s to develop an entry vehicle carried on the Galileo probe to Jupiter. That capsule was a strikingly different engineering challenge to be sure—it was intended to survive only a few minutes, Jupiter is a far cry from Earth in terms of pressures, atmospheres, and so on—but the knowledge gained in this R&D effort found application with Genesis as well. As Bernard Laub concluded:

> The Galileo probe to Jupiter was the most challenging entry mission ever undertaken by NASA. The probe employed a 45 deg blunt cone aeroshell and it entered the Jovian atmosphere at a velocity of ≈ 47.4 km/s. The forebody TPS employed fully dense carbon phenolic (= 1450 kg/m^3) that, at the time, was the best ablator available. The entry environment was very severe and estimates of the peak heating (combined convective and radiative) were on the order of 35 kW/cm^2 with a total integrated heat load of ≈ 200 kJ/cm^2.

Because of this experience, NASA engineers concluded that new materials to replace the carbon phenolic were appropriate for ablative TPS missions in the future.[13]

The heat shield for the Genesis return capsule consisted of a carbon-bonded carbon fiber (CBCF) insulating material as a base, an approach used with good success in several other NASA missions. The Genesis TPS also included PICA (phenolic impregnated carbonaceous ablator) and TUFROC (toughened uni-piece fibrous reinforced oxidation-resistant composite). As one study noted:

> CBCF utilized in the above mentioned TPS systems is an attractive substrate material because of its low density and high porosity, superior thermal performance, and compatibility with other components. In addition, it is low cost because of the commercial market it also serves. However, the current CBCF manufacturing process does not produce materials engineered to the specifications NASA desires to put in place. These emerging and highly innovative TPS designs require material manufactured to specification . . . The benefits derived include significantly improved flexibility for the TPS design engineer, as well as, more cost efficient CBCF derived TPS fabrication.[14]

As finally flown, the Genesis return capsule was an amalgam of existing technologies and newer concepts. Its backshell was made of SLA-561V "Mars" Ablator in an aluminum-honeycomb (Al-Hc) substructure. Its main structure possessed a carbon-carbon skin with continuous surface (multilayer fabric layup).[15] As reported by NASA:

> The heat shield is made of a graphite-epoxy composite covered with a thermal protection system. The outermost thermal protection layer is made of carbon-carbon. The capsule heat shield remains attached to the capsule throughout descent and serves as a protective cover for the sample canister at touchdown. The heat shield is designed to remove more than 99 percent of the initial kinetic energy of the sample return capsule.
>
> The backshell structure is also made of a graphite-epoxy composite covered with a thermal protection system: a cork-based material called SLA-561V that was developed by Lockheed Martin for use on the Viking missions to Mars, and have been used on several missions including Genesis, Pathfinder, Stardust and the Mars Exploration Rover missions. The backshell provides the attachment points for the parachute system, and protects the capsule from the effects of recirculation flow of heat around the capsule . . .
>
> The parachute system consists of a mortar-deployed 2.1-meter (6.8-foot) drogue chute to provide stability at supersonic speeds, and a main chute 10.5 by 3.1 meters (about 34.6 by 12.1 feet).
>
> Inside the canister, a gas cartridge will pressurize a mortar tube and expel the drogue chute. The drogue chute will be deployed at an altitude of approximately 33 kilometers (108,000 feet) to provide stability to the capsule until the main chute is released. A gravity-switch sensor and timer will initiate release of the drogue chute. Based on information from timer and backup pressure transducers, a small pyrotechnic device will cut the drogue chute from the capsule at about 6.7 kilometers altitude (22,000 feet). As the drogue chute moves away, it will extract the main chute. At the time of capture, the capsule will be traveling forward at approximately 12 meters per second (30 miles per hour) and descending at approximately 4 meters per second (9 miles per hour).[16]

The TPS worked well during reentry. This was an especially dicey task because of the concern about preserving the specimens on the Genesis return capsule while returning to Earth from interplanetary space and the resultant velocity at which it entered the Earth's atmosphere.

The fact that the parachute system failed was the dumbest of luck and the result was spectacular as the Genesis return capsule streaked into the ground. This was the case because the mortar that was to have fired at 100,000 feet releasing a drogue parachute to slow and stabilize the vehicle failed. Then at 20,000 feet the drogue was to have pulled out the main parafoil, and the spacecraft would then slow to 9 miles per hour, whereupon helicopters could capture the capsule from midair using a hook. As it was, the capsule failed to slow down, the helicopter failed to catch it, and the capsule plowed into the ground. The 5-feet-wide, 420-pound capsule burrowed into the desert floor at a speed of 193 miles per hour. In the end, NASA scientists were able to recover some of the solar wind particles from the Genesis sample return capsule and results from the analysis of these data have been appearing slowly thereafter.[17]

Why did the spacecraft's parachute fail? The accident investigation team released a statement on October 14, 2004, that implicated Lockheed Martin in faulty workmanship in constructing the system. It fixed blame on incorrectly oriented acceleration sensors—in essence a G-switch was installed backward—and a host of design reviews, checkers, and managers had failed to catch the error. It was the simplest of errors: the parachute system was designed to make an electrical contact inside the sensor at 3 *g* (29 m/s^2), maintaining it through the maximum expected 30 *g* (290 m/s^2), and breaking the contact again at 3 *g* to start the parachute release sequence. The sensors failed to make contact and the chute never deployed. Further evaluation revealed that a pretest for the system had been skipped by Lockheed Martin technicians.[18]

The Stardust sample return mission turned out much better, at least in the effectiveness of the reentry and recovery system. Stardust, launched in 1999 and recovered in 2006, was the first US space mission dedicated solely to returning extraterrestrial material from beyond the Moon (figure 7.2). It collected samples both from Comet Wild 2 and interstellar dust. The Stardust return system had six major components: a heat shield, back shell, sample canister, sample collector grids, parachute system, and avionics. The canister containing the samples was sealed in an exterior shell that protected them from the heat of reentry. The material Stardust returned may date from the formation of the solar system. Scientific studies of the samples may alter humanity's understanding of the universe. One major discovery is that ice-rich comets, the coldest and most distant bodies in the solar system, also contain fragments of materials that make up the terrestrial planets.

The Stardust TPS worked well during the mission. NASA realized that it required new technologies for reentry as its planetary exploration ramped up in the early/mid-1990s.[19] Accordingly, it pursued "development of two new lightweight ablators, Phenolic Impregnated Carbon Ablator (PICA) and Silicone Impregnated Reusable Ceramic Ablator (SIRCA). Owing to its performance in the ~ 1 kW/cm^2 heating environment and low heat shield

Figure 7.2 Stardust US space mission. Stardust was the first US space mission dedicated solely to returning extraterrestrial material from beyond the Moon. It collected samples from Comet Wild 2 and interstellar dust. Launched in 1999, it returned to Earth seven years later, parachuting to a landing in the Utah desert in 2006. (Credit: NASA/JPL.)

mass, PICA enabled the Stardust Sample Return Mission. SIRCA has been used on the backshell of Mars Pathfinder and the Mars Exploration Rover missions."[20] Furthermore, as one report elaborated:

> PICA was developed by NASA Ames in the early-mid 90s, is fabricated by Fiber Materials, Inc. (FMI) and was employed as the forebody TPS on the Stardust Return Capsule. It is currently the baseline forebody TPS for the Orion Crew Exploration Vehicle (CEV) and is being fabricated as the forebody TPS for the Mars Science Laboratory (MSL), scheduled for launch in mid-2009. PICA is a low density carbon-based ablator. Under both the CEV TPS Advanced Development Program and MSL, an extensive data base has been developed. The failure modes are well-understood (upper heat flux limits of ≈ 1500 W/cm2 and pressure of 1.0–1.5 atm) and validated design models have been developed. For Stardust, the PICA heatshield was fabricated as one piece. But that will not be possible for larger vehicles leading to a tiled design (such as the MSL design) that introduces significant design and fabrication complexities.[21]

PICA has been flight-qualified and used in the past and along with investments in new TPS materials offered acceptable, cost effective results.[22]

Because of the high velocity of reentry for Stardust, its forebody TPSs had to consist of ablative materials even as the afterbody's blunt cone aeroshell heating was much less hazardous to the spacecraft. Because of the stresses of forebody heating, NASA engineers used a complex analysis to arrive at the proper materials for its ablative heat shield:

> To meet the requirements for the Stardust mission, one of a family of lightweight ceramic ablator materials developed at NASA Ames Research Center was selected for the forebody heat shield of the Stardust sample return capsule. This material, phenolic impregnated carbon ablator (PICA), consists of a commercially available low density carbon fiber matrix substrate impregnated with phenolic resin...The Stardust program resulted in intensive material development, modeling, and testing efforts to provide a heat shield for the high convective heating conditions expected during Earth entry while under constraints of limited time and funding.[23]

As it turned out, the Stardust reentry system worked well, dissipating about 90 percent of the total energy via the bow shock heating of the atmospheric gases. And what was learned from this effort fed into the next major project NASA undertook for reentry and recovery from space.

So What?

This may well be the core question of all of history. It seems that one can offer three large conclusions from this story of the Genesis and Stardust missions. First, quite a lot of activity in the last decade of the twentieth century affected the course of reentry and recovery from space for Genesis and Stardust. After spending more than a generation using the reusable Space Shuttle and its unique TPS and runway landing capability, NASA needed to reinstitute the technologies necessary for ablative TPS and parachute recovery. This took some investment and some backtracking from the shuttle reentry and recovery approach.[24] Even so, the use of legacy technologies pioneered by NASA in some cases as far back as the Apollo program proved easier to reinstitute than might have seemed possible at first look. Thus far, it appears that only the Boeing TPS, developed for other systems in the recent past, pushed the envelope of knowledge about reentry. As reported in 2010, "Both capsules and space planes have their advantages, and neither has a spotless safety record. But it will be interesting to see which mode NASA eventually selects for the next generation of ISS missions."[25]

Second, the success of atmospheric reentry missions has long been constrained by the design of the TPS of the aerospace vehicles involved. During reentry into the Earth's atmosphere, space vehicles must operate in intense thermal stress regimes that require an effective TPS that ensures survival of the craft. Indeed, the TPS may be the most important system used on an spacecraft that must enter an atmosphere and land. Next most significant, of course, is the landing system, whether it be parachutes, rockets, glide systems, or Rogallo wings (figure 7.3).

Figure 7.3 Artist's rendering of the Stardust capsule returning to Earth. It shows the capsule coming home with samples of interstellar dust, including recently discovered dust streaming into our solar system from the direction of Sagittarius, under a parachute. (Credit: NASA/JPL.)

One recent study made the observation about the criticality of these systems in this particular way: "For vehicles traveling at hypersonic speeds in an atmospheric environment, TPS is a single-point-failure system. TPS is essential to shield the vehicle structure and payload from the high heating loads encountered during [reentry]...Minimizing the weight and cost of TPS, while insuring the integrity of the vehicle, is the continuing challenge for the TPS community."[26]

The modern TPS originated in the decade after World War II as part of the ballistic missile program. Required to ensure the reentry of nuclear warheads, TPS represented a fundamental technology for these missile systems. The ablative TPS protected the vehicle through a process that lifted the hot shock layer gas away from the vehicle, dissipated heat absorbed by the ablative material as it burned away, and left a char layer that proved remarkably effective as an insulator that blocked radiated heat from the shock layer.

Throughout the 1960s and much of the 1970s, engineers involved in ablative TPS development were quite active. The succession of programs, both human and robotic, kept them involved in making ever more capable TPSs. This changed in the mid-to-late 1970s, however, as research, development, and testing of ablative TPS materials significantly declined for both the nuclear missile programs and the robotic probes that NASA pursued. At the same time human programs that had used ablative systems were completed and NASA moved toward a nonablative, reusable TPS for the Space Shuttle program. Although this transition involved significant R&D, by the early 1980s the systems had been resolved, and only intermittent upgrades would follow. Accordingly, the ablative TPS community experienced a serious decline in capability. A return to ablative TPS for robotic entry probes in the last 20 years has required a reconstitution of the knowledge base lost in the 1970s. As a result, NASA invested in the development of two new lightweight ablators, PICA and SIRCA in the 1990s. It was PICA that enabled the Stardust Sample Return Mission. Additionally, SIRCA was used on the backshell of Mars Pathfinder and the Mars Exploration Rover missions. These efforts reached a serious crescendo when NASA began developing the Orion crew capsule to replace the Space Shuttle. In essence, we have come full circle in the context of TPSs, albeit with a spiral upward in capability and flexibility, and much of the resulting work critically based on the efforts of earlier systems dating back to the Apollo era.[27]

A key realization coming from the cyclical process of ablative TPS efforts was that despite the desire to use "off-the-shelf" materials, the knowledge base had so significantly atrophied that considerable effort had to be undertaken to reconstitute that knowledge base. Moreover, engineers working on TPS systems found that knowledge gained in other programs was profound. For example, several have concluded that without the investment in the 1990s in PICA TPS, the Mars Science Laboratory would have been even more behind schedule.[28] In essence, the space TPS community worked most effectively whenever it coordinated efforts to codevelop materials and technology that could be used in multiple projects and programs. As one study concluded: "The *very* important lesson learned here is that it is wise to have at least two viable candidate TPS materials in place for mission projects because although the selected TPS sufficed for a previous project, it may not be adequate for the next, even if the entry environments are only slightly more severe."[29]

The process of moving from ablative heat shields in the 1960s to the Space Shuttle TPS in the 1970s to the various types of TPS available at present has represented something of a back to the future approach to spaceflight. It raises important questions about the history of technology and its usually presumed progress. It looks more circular than progressive when considering the return and recovery from space during the history of the Space Age. It also points up the inexact nature of space technology and the lack of a clear line of development over time as there are loops back and forth, leaps in technology, and backtracks to earlier proven concepts.[30] It also suggests that there is very little finally decided in space technology, especially as older concepts are resurrected and brought to the fore years after their abandonment.

Finally, in the end this situation suggests that TPS technologies for many space missions are unique, challenging, and cross-cutting. Accomplishing them requires specialized efforts, but they offer multiple uses. In the end, the focus on thermal protection and landing systems cannot be deemphasized. As a recent study reported:

> NASA's ambitious exploration vision requires TPS *innovations*. Many future missions require TPS materials and/or concepts not currently available or, in some cases, new versions of old materials. New TPS materials, ground test facilities, and improved analysis models are required and will take some time to develop, Advances and improved TPS capabilities will benefit an array of missions (and *enable* some).[31]

Notes

1. Basic histories of rocketry include Roger D. Launius and Dennis R. Jenkins, eds., *To Reach the High Frontier: A History of U.S Launch Vehicles* (Lexington: University Press of Kentucky, 2002); David Baker, *The Rocket: The History and Development of Rocket and Missile Technology* (New York: Crown Books, 1978); Frank H. Winter, *Rockets into Space* (Cambridge, MA: Harvard University Press, 1990); Wernher von Braun, Frederick I. Ordway III, and Dave Dooling, *History of Rocketry and Space Travel* (New York: Thomas Y. Crowell Co., 1986 ed.); Eugene M. Emme, ed., *The History of Rocket Technology: Essays on Research, Development, and Utility* (Detroit, MI: Wayne State University Press, 1964); G. Harry Stine, *Halfway to Anywhere: Achieving America's Destiny in Space* (New York: M. Evans and Co., 1996).
2. Representative technical studies of the challenge of reentry and recovery include Patrick Gallais, *Atmospheric Reentry Vehicle Mechanics* (New York: Springer, 2007); Ashish Tewari, *Atmospheric and Space Flight Dynamics: Modeling and Simulation with MATLAB® and Simulink®* (Boston: Birkhäuser Boston, 2009); Frank J. Regan, *Dynamics of Atmospheric Reentry* (Reston, VA: American Institute for Aeronautics and Astronautics, 1993); Wilber L. Hankey, *Reentry Aerodynamics* (Reston, VA: American Institute for Aeronautics and Astronautics, 1988); John David Anderson, *Hypersonic and High Temperature Gas Dynamics* (Reston, VA: American Institute for Aeronautics and Astronautics, 2006).
3. *Reentry Studies*, 2 vols, Vitro Corp. report No. 2331–25, November 25, 1958; NACA Conference on High-Speed Aerodynamics, Ames Aeronautical Laboratory, Moffett Field, CA, March 18, 19, and 20, 1958, A Compilation of the Papers Presented, both NASA technical reports.
4. See Roger D. Launius and Dennis R. Jenkins, *Coming Home: Reentry and Recovery From Space*, forthcoming from NASA, 2012.
5. H. Julian Allen and Alfred J. Eggers Jr., "A Study of the Motion and Aerodynamic Heating of Ballistic Missiles Entering the Earth's Atmosphere at High Supersonic Speeds," NACA confidential report RM-A53D28, April 28, 1953 (this report was subsequently updated as TN-4047 in 1957 and Report 1381 in 1958); Edwin P. Hartmann, *Adventures in Research: A History of the Ames Research Center, 1940–1965* (Washington, DC: NASA

SP-4302, 1972), pp. 216–18. At the time, there was some dispute about exactly who had devised the blunt body concept. H. H. Nininger, the director of the American Meteorite Museum at Sedona, Arizona, claimed he first proposed the blunt nose for reentry vehicles in August 1952. Nininger, a recognized authority on meteorites, based his conclusion on studies of tektites and meteorites. He contended that the melting process experienced by meteorites during their descent through the atmosphere furnished a lubricant that protected them from aerodynamic friction. This letter evidently came to Ames some weeks after Allen and Eggers had completed their study. Despite the contention of Nininger, what Allen wanted to do was exactly the reverse: deliberately shape a reentry body bluntly in order to *increase* air resistance and dissipate a greater amount of the heat produced by the object into the atmosphere. See various letters in the file for H. H. Nininger 1935–1957, NASA History Office, Washington, DC.

6. Barton C. Hacker, "The Idea of Rendezvous: From Space Station to Orbital Operations, in Space-Travel Thought, 1895–1951," *Technology and Culture* 15 (July 1974): 373–88; Barton C. Hacker and James M. Grimwood, *On Shoulders of Titans: A History of Project Gemini* (Washington, DC: NASA SP–4203, 1977, rep. ed., 2002), pp. 1–26.
7. The best work on this subject is Dwayne A. Day, John M. Logsdon, and Brian Latell, eds., *Eye in the Sky: The Story of the Corona Spy Satellite* (Washington, DC: Smithsonian Institution Press, 1998); Robert McDonald, ed., *CORONA: Between the Sun & the Earth: The First NRO Reconnaissance Eye in Space* (Bethesda, MD: American Society for Photogrammetry and Remote Sensing, 1997); Curtis Peebles, *The Corona Project: America's First Spy Satellites* (Annapolis, MD: Naval Institute Press, 1997); Philip Taubman, *Secret Empire: Eisenhower, the CIA, and the Hidden Story of America's Space Espionage* (New York: Simon & Schuster, 2003).
8. Launius and Jenkins, *Coming Home: Reentry and Recovery From Space.*
9. Bart Hendrickx and Bert Vis, *Energiya-Buran: The Soviet Space Shuttle* (Chichester, UK: Springer-Praxis, 2007); Stephen J. Garber, "A Cold Snow Falls: The Soviet Buran Space Shuttle," *Quest: The History of Spaceflight Quarterly* 9, no. 5 (2002): 42–51.
10. Bernard Laub, "Development of New Ablative Thermal Protection Systems (TPS)," available online at http://asm.arc.nasa.gov/full_text.html?type=materials&id=1, accessed October 14, 2009.
11. M. Lo, B. Williams, W. Bollman, D. Han, Y. Hahn, J. Bell, E. Hirst, R. Corwin, P. Hong, K. Howell, B. Barden, and R. Wilson, "Genesis Mission Design," AIAA 98–4468, August 10–12, 1998; D. S. Burnett, B. L. Barraclough, R. Bennett, M. Neugebauer, L. P. Oldham, C. N. Sasaki, D. Sevilla, N. Smith, E. Stansbery, D. Sweetnam, and R. C. Wiens, "The Genesis Discovery Mission: Return of Solar Matter to Earth," *Space Science Reviews* 105 (January 2003): 509–34.
12. Lo, Williams, Bollman, Han, Hahn, Bell, Hirst, Corwin, Hong, Howell, Barden, and Wilson, "Genesis Mission Design," pp. 3–4.
13. Ibid.
14. Fiber Materials, Inc., "Advanced Thermal Protection Systems (ATPS), Aerospace Grade Carbon Bonded Carbon Fiber Material," PHASE-I CONTRACT NUMBER: NNA05AC11C, 2004.
15. Bill Willcockson, "Genesis Sample Return Capsule Overview," August 25, 2004, Lockheed Martin Space Systems, Denver, CO.

16. "Genesis: Search for Origins," June 24, 2008, available online at http://genesismission.jpl.nasa.gov/gm2/spacecraft/subsystems.html, accessed October 20, 2009.
17. Francis Reddy, "Salvaging Science from Genesis," *Astronomy*, September 18, 2004, available online at http://www.astronomy.com/asy/default.aspx?c=a&id=2444, accessed October 20, 2009; Ansgar Grimberg, Heinrich Baur, Peter Bochsler, Fritz Bühler, Donald S. Burnett, Charles C. Hays, Veronika S. Heber, Amy J. G. Jurewicz, and Rainer Wieler, "Solar Wind Neon from Genesis: Implications for the Lunar Noble Gas Record," *Science* 314 (November 17, 2006): 1133–35.
18. Maggie McKee, "Genesis Crash Linked to Upside-Down Design," *New Scientist*, October 15, 2004, available online at http://www.newscientist.com/news/news.jsp?id=ns99996541, accessed April 3, 2012, 3:04 P.M.; Associated Press, "Official: Genesis Pre-Launch Test Skipped," Space.com, January 7, 2006, available online at http://www.space.com/missionlaunches/ap_060107_genesis_update.html, accessed April 3, 2012, 3:07 P.M.
19. M. A. Covington, "Performance of a Light-Weight Ablative Thermal Protection Material for the Stardust Mission Sample Return Capsule," NASA Ames Research Center Technical Report, 2005, No. 20070014634.
20. Ethiraj Venkatapathy, Christine E. Szalai, Bernard Laub, Helen H. Hwang, Joseph L. Conley, James Arnold, and 90 coauthors, "Thermal Protection System Technologies for Enabling Future Sample Return Missions," White Paper to the NRC Decadal Primitive Bodies Sub-Panel, available online at http://www.psi.edu/decadal/topical/EthirajVenkatapathy.pdf, accessed October 22, 2009.
21. E. Venkatapathy, B. Laub, G. J. Hartman, J. O. Arnold, M. J. Wright, and G. A. Allen Jr., "Selection and Certification of TPS: Constraints and Considerations for Venus Missions," 6th International Planetary Probe Workshop (IPPW6), Atlanta, GA, June 23–27, 2008.
22. H. K. Tran, W. D. Henline, Ming-tu Hsu, D. J. Rasky, and S. R. Riccitiello, Patent 5,536,562 July 16, 1996, and Patent 5,672,389 September 30, 1997, "Low Density Resin Impregnated Article Having an Average Density of 0.15 to 0.40 g/cc."
23. M. A. Covington, J. M. Heinemann, H. E. Goldstein, Y.-K. Chen, I. Terrazas-Salinas, J. A. Balboni, J. Olejniczak, and E. R. Martinez, "Performance of a Low Density Ablative Heat Shield Material," *Journal of Spacecraft and Rockets* 45 (March–April 2008): 237–42.
24. Leonard David, "The Next Shuttle: Capsule or Spaceplane?" Space.com, May 21, 2003, available online at http://www.space.com/businesstechnology/technology/osp_debate_030521.html, accessed April 19, 2004; Leonard David, "NASA's Orbital Space Plane Project Delayed," Space.com, November 26, 2003, available online at http://www.space.com/businesstechnology/technology/space_plane_delay_031126.html, accessed April 19, 2004; White House Press Release, "Fact Sheet: A Renewed Spirit of Discovery," January 14, 2004, available online at http://www.whitehouse.gov/news/releases/2004/01/20040114-1.html, accessed April 4, 2004.
25. Clay Dillow, "Jumping into the New Space Race, Orbital Sciences Unveils Mini-Shuttle Spaceplane Design," *Popular Science*, December 16, 2010, available online at http://www.popsci.com/technology/article/2010-12/jumping-new-space-race-orbital-sciences-unveils-mini-shuttle-spaceplane-design, accessed February 27, 2011, 4:02 P.M.

26. Ethiraj Venkatapathy, Christine E. Szalai, Bernard Laub, Helen H. Hwang, Joseph L. Conley, James Arnold, and 90 coauthors, "White Paper to the NRC Decadal Primitive Bodies Sub-Panel: Thermal Protection System Technologies for Enabling Future Sample Return Missions," 2010, p. 2, NASA Historical Reference Collection, NASA History Division, Washington, DC.
27. Bernie Laub, "Ablative Thermal Protection: An Overview," presentation at 55th Pacific Coast Regional and Basic Science Division Fall Meeting, Oakland, CA, October 19–22, 2003.
28. Jean-Marc Bouilly, Francine Bonnefond, Ludovic Dariol, Pierre Jullien, and Frédéric Leleu, "Ablative Thermal Protection Systems for Entry in Mars Atmosphere: A Presentation of Materials Solutions and testing Capabilities," 4th International Planetary Probe Workshop, Pasadena, CA, June 27–30, 2006.
29. Venkatapathy, Szalai, Laub, Hwang, Conley, Arnold, and 90 coauthors, "White Paper to the NRC Decadal Primitive Bodies Sub-Panel," p. 2; emphasis in the original.
30. This is very effectively laid out in John Law, "Technology and Heterogeneous Engineering: The Case of Portuguese Expansion," and Donald MacKenzie, "Missile Accuracy: A Case Study in the Social Processes of Technological Change," both in Wiebe E. Bijker, Thomas P. Hughes, and Trevor J. Pinch, eds., *The Social Construction of Technological Systems: New Directions in the Sociology and History of Technology* (Cambridge, MA: The MIT Press, 1987), pp. 111–34 and pp. 195–222, respectively.
31. Bernard Laub and Ethiraj Venkatapathy, "Thermal Protection System Technology and Facility Needs for Demanding Future Planetary Missions," Presentation at International Workshop on Planetary Probe Atmospheric Entry and Descent Trajectory Analysis and Science, Lisbon, Portugal, October 6–9, 2003, p. 8; emphasis in the original.

Part III

Exploring the Terrestrial Planets

Chapter 8

Planetary Science and the "Discovery" of Global Warming

*Erik M. Conway**

On January 17, 2003, novelist Michael Crichton gave Caltech's Michelin Lecture. Its title was *Aliens Cause Global Warming*. "I am going to argue that extraterrestrials lie behind global warming. Or to speak more precisely, I will argue that a belief in extraterrestrials has paved the way, in a progression of steps, to a belief in global warming."[1] His paper begins with an attack on the Drake equation and SETI, then on Carl Sagan and nuclear winter, and then finally expands to an attack on all scientific modeling. Since one can choose any set of values to base a model calculation on (according to Crichton), any result is possible. Thus modeling is a religion, and global warming theology, not science.

Crichton's tale is a mere caricature of climate science. In reality, modelers draw upon physical quantities pulled from ice cores, sediment cores, tree rings, stalagmites, and other sources of climate information to constrain models. The United States spends more than $1 billion per year to measure current climate-related quantities, such as aerosols, to try to develop physical understanding of climate processes. All of this is discussed in great detail in the reports of the Intergovernmental Panel on Climate Change's Working Group I and in a long series of studies by the US National Academy of Sciences extending back to 1979.[2] Elsewhere, I have argued that this need to disbelieve scientific fact descends from late twentieth-century political ideology.[3]

But Crichton has one bit of his story almost right: extraterrestrials do have something to do with global warming. Or, to write more precisely, NASA's explorations of Venus and Mars during the 1960s in search of extraterrestrials helped focus scientific attention on the intersection of *chemistry* and *climate*. Whether or not Earth-like life can exist on another planet is fundamentally a question of climate; there are many environments in which our sort of life cannot exist. Early space age discoveries revealed planets so different than expected that they forced scientists to reconsider the role of biology in the evolution of our own Earth. It was no longer possible, by 1980, to argue with any scientific credibility that life had no impact on the physical Earth. Instead, scientists studying the Earth had to account for the role of biology in transforming the Earth into the cozy planet we enjoy.

This interplanetary context is an important, but generally neglected, thread of what Spencer Weart calls the *Discovery of Global Warming*.[4] During the same period as the Mariner missions, climate science itself was being remade. Originally a statistical endeavor linked to meteorology, a number of different research efforts began to provide a physical basis for climate-related research. Weart provides an excellent overview of the transition. James Fleming has examined the pre–World War II work of Guy Callendar on the carbon dioxide climate link, as well as examining some of the nineteenth-century and post–World War II discussions of climate change.[5] In more popular treatments, John Cox, Mark Bowen, and Richard Alley have all discussed the role of ice core analyses in the evolution of modern climate science. Mott Greene has recently surfaced the role of early computer model-based climate research. Only Weart raises the interplanetary connection, and in this essay, I want to reweave the threads of this now fairly well-told story to make that connection central.

Planetary Climates Prior to the Space Age

Three planets in the solar system count as "Earth-like" to planetary astronomers: Venus, Earth, and Mars. Venus, about 30 percent closer to the Sun than Earth is, happens to be about 80 percent Earth's mass. Mars is about a tenth the mass of Earth, and it is a little more than 30 percent further from the Sun. They were considered Earth-like in relation to the outer planets, which are many times Earth's size (Jupiter's mass is 317 times Earth's) and composed primarily of gases. These basic facts have been known for centuries; the Space Age has only allowed some reduction in error ranges.

In addition to these facts, planetary astronomers working prior to the Space Age had accumulated additional knowledge about them through observation, measurement, and informed speculation. The belief in life on Mars, for example, is well-documented in the historical literature. I need not recount the numerous stories of Lowell, canalae, and ancient Martian civilization.[6] Steven Dick has pointed out that by the late 1950s most of this speculation had been washed away by astrophysical research that suggested a much colder climate on Mars than would be expected simply from extrapolating by distance to the Sun. Yet an annual "wave of darkening" across the face of Mars was still widely interpreted as the seasonal flowering of plant life.[7] Mars was considered a living planet in 1957.

Venus's story was a bit different. Prior to the 1930s and the advent of radio, microwave, and infrared astronomy, Venus was often discussed as being a somewhat warmer Earth. Its proximity to the Sun suggested a current climate like that of Earth's Mesozoic period, when Earth was 7 or 8 degree Celsius warmer than today and the dinosaurs ruled. At least one series of children's books were written around that image of Venus.

The question of life, at least life like Earth's, is fundamentally a question of climate. Astrophysicists prior to the space age sought measurements of climatic factors in order to inform speculations about the presence or absence of Earth-like life. One could, by the late 1950s, look for the presence or absence of atmospheric oxygen and carbon dioxide with infrared and microwave spectrometers, which could (in theory) distinguish these molecules by their unique emissions. One could also make pretty good measurements of average temperature via spectroscopy as well, although one could not do much below the global scale. Spatial resolution did not permit examination of regional details. Measurement of these climate characteristics constrained theoretical models of planetary evolution while also restraining some of the more exotic speculations about Martian and Venusian life.

Thus astrophysicsts in the 1950s were very interested in planetary climates, although not yet what we now call global warming.[8]

Voyages of the Early Space Age

The first space vehicles of any kind were Earth orbiters. Sputnik 2 carried a dog named Laika aloft in November 1957, demonstrating that Earth creatures could live in space for at least a little while. The two US satellites Explorers 1 and 3 revealed the existence of the Van Allan radiation belts, a product of Earth's magnetic field. Their builder, the Jet Propulsion Laboratory (JPL) in Pasadena, California, used its success, and its connections to Caltech, to get out of its business of building rockets for the US Army and become the United States' primary planetary spacecraft center.[9]

The Space Science Board of the National Academy of Sciences had discussed the state of knowledge of the atmospheres of Venus and Mars during a set of meetings held at Caltech in late 1960 and early 1960. The panel had assembled a familiar group of astronomers, atmospheric scientists, and remote sensing specialists: Lewis Kaplan, Yale Mintz, and Carl Sagan among them.[10] With the Space Age just beginning and the possibility of gaining a closer perspective on the inner planets, the collected scientists hoped to provide a scientific strategy for answering some of outstanding questions about them. A key question was planetary climate, an interest because it spoke strongly about the possibility of finding life. Recent measurements by a Naval Research Laboratory scientist seemed to show the planet radiating strongly in the microwave region.[11] This suggested that planet's surface temperature averaged 600 degrees Kelvin, a clear refutation of the "Earth-like" Venus consensus that then dominated the profession.

One early adopter of this superhot Venus model was Sagan. In 1960, while still a student, he had made one of the first attempts at explaining how Venus could maintain such a surface temperature. He constructed a model atmosphere for Venus composed primarily of carbon dioxide, including small amounts of water vapor to explain the Venusian atmosphere's apparent absorbativity in the far infrared, based on the known properties of gases. Sagan calculated that Venus' needed to have a carbon dioxide rich atmosphere, with the mass of carbon dioxide equal to three–four times the mass of Earth's entire atmosphere; assuming that the ratio between nitrogen (the dominant gas in Earth's atmosphere) and carbon dioxide remained the same, Venus's atmosphere would be hundreds of times as dense as Earth's. This was a radically different concept of Venus, and left it, in Sagan's words, "a hot, dry, sandy, windy, cloudy, and probably lifeless planet."[12]

Sagan's model, however, had not been widely accepted within the scientific community and other models of the Venus atmosphere were presented. One was the "aeolosphere" model. In this model, the sun's energy was deposited high in the atmosphere, and intense winds carried heat to the surface while the strong winds kept the surface permanently shrouded in dust. The dust also served as a strong infrared absorber. Virtually no light at all would reach the planet's surface, rendering it extraordinarily hot and dark.[13] Spacecraft measurements might be able to distinguish between these measurements by examining the microwave emissions of the planet, and the JPL's Mariner 2 spacecraft, which was scheduled for launch in 1962, would carry a microwave spectrometer to test these hypotheses.

It was also possible that the surface was not as hot as the Naval Research Lab's observers had thought.[14] It was possible that the microwave radiation that Mayer had detected was not from the planet's surface but from a highly energetic ionosphere. This "hot atmosphere/cold surface" model might permit the surface to remain a more reasonable temperature.

The Mariner II mission (figure 8.1) to Venus dispelled the "hot atmosphere-cold surface model" for good. This was a "flyby" mission, with the vehicle passing about 22,000 miles from Venus during mid-December 1962. Mariner II carried both infrared and microwave radiometers for examining the planet. The microwave radiometer provided the telling surface measurement. The instrument had been designed to enable selection between the "hot surface/cold atmosphere" model and the "cold surface/hot atmosphere" model by scanning across the planet "disk." If the microwave radiation intensity increased toward the "limb" of Venus, the thin crescent of atmosphere between the planet surface and space, this would indicate that the atmosphere was the source of the radiation, not the surface. Similarly, if the limb "darkened," or showed less microwave intensity than the surface, then the source would be a hot surface. The Mariner data clearly indicated that the second case was true.[15] Venus had a very hot surface.

For the rest of the decade, scientists continued to argue over how Venus had become so dramatically different from Earth.[16] In 1969, for example, Andrew P. Ingersoll of the California Institute of Technology proposed that a water-vapor induced "runaway greenhouse effect" had occurred on Venus. On Earth, water vapor remains in equilibrium with surfaces of liquid water because, given the current composition of the atmosphere, incident solar flux is insufficient to cause continually increasing evaporation. Instead, the colder upper atmosphere forces water that evaporates from the surface to precipitate back out, maintaining a stable balance. On Venus, where the amount of sunlight reaching the top of the atmosphere is considerably higher than

Figure 8.1 Mariner 2 was the world's first successful interplanetary spacecraft. Launched August 27, 1962, on an Atlas-Agena rocket, Mariner 2 passed within about 34,000 kilometers (21,000 miles) of Venus, sending back valuable new information about interplanetary space and the Venusian atmosphere. This spacecraft showed that Venus was not an abode of life, but also opened questions that affected the study of Earth. (Credit: NASA/JPL photo no. PIA04594.)

on Earth, this might not have been true. In his calculations, water vapor outgassing from the young planet had never condensed as it largely had on Earth, producing a greenhouse effect that continually increased in intensity.[17] James Pollack, Carl Sagan's first graduate student, and head of Ames Research Center's atmospheric modeling group, made similar calculations the same year.

Scientists developed other hypotheses to explain Venus's scorching climate, too. One of these was a "deep convective" model. In this model, the majority of Venus's atmospheric energy was deposited in the high cloud layer on the sunward side of the planet and was carried to the planet's night side by large-scale zonal currents. There, of course, radiatively cooling air masses would descend, bringing energy to the lower atmosphere and surface. This hypothesis overcame the problem that many scientists believed that even the very thick carbon dioxide atmosphere of Venus could not have sufficient infrared opacity to generate Venus's high surface temperature through the classical greenhouse effect alone. Very little solar energy was thought to reach the surface through the cloud layers, and hence this small amount would have to be entirely retained by the atmosphere to maintain the high surface temperature. That did not seem possible, and this dynamical model was designed to overcome the limitations of Sagan's simple radiative model.[18]

Later spacecraft missions to Venus generated more observational data to help understand the Venusian climate. In 1967, the American probe Mariner 5 had made a "flyby" of Venus. This spacecraft was not equipped for extensive atmospheric investigation, although it did confirm the very high surface temperature via radio occultation. But it was the Soviet Union that carried out the most productive investigations of Venus, sending a series of atmosphere entry probes starting in 1967. The first of these was Venera 4, which successfully entered the Venusian atmosphere on October 18, the same day as the Mariner 5 flyby. Its measurements showed that the atmosphere was more than 90 percent carbon dioxide. The temperature and pressure reached 535K and 18 atmospheres before the probe failed.[19] A nearly continuous series of Venera spacecraft followed: two more atmosphere entry probes in 1969, the first successful Venus lander in 1970, a second lander in 1972, and still more in 1975, 1978, 1981, and 1984. The last pair carried a pair of French constant-level balloons as well, designed to float in the cloud system.[20]

An Alien Mars

The first successful robotic voyager to Mars, Mariner 4, destroyed belief in abundant Martian plant life. It flew by the planet on July 15, 1965, at a distance of about six thousand miles. The photographs sent back showed an unexpectedly crater-pocked, Moon-like Mars, effectively discrediting the vision of a planet inhabited by advanced sentient life and finally debunking the Martian canali. The cratered surface was one line of evidence that Mars had a very thin atmosphere. A more Earth-like atmosphere would have burned up most incoming meteorites prior to impact. And the craters' survival also showed that Mars no longer had a significant hydrologic cycle. On Earth, craters are erased by erosional processes led by liquid water; on Mars, they clearly lasted for billions of years.

Mariner 4's science team also used the vehicle's radio to obtain a better measurement of atmospheric pressure (figure 8.2). This radio occultation measurement indicated that Mars's atmospheric density was about 1 percent of Earth's, with a surface temperature of about 180K and a pressure between 4 and 7 mb.[21] This was, roughly, the same as an estimate one scientist had half-jokingly made several years earlier.

Two more Mariner "flyby" missions visited Mars in 1969. Mariners 6 and 7, equipped with cameras but no atmospheric instruments, flew past the planet in July and August of that year, returning additional photographs of a dead, and heavily cratered, surface. One produced the first close-up photographs of the southern polar cap, too. At the same time, several different astronomy groups confirmed the existence of water vapor in the Martian atmosphere using infrared spectroscopy. They were able to demonstrate that

Figure 8.2 First image from Mariner 4. A "real-time data translator" machine converted Mariner 4 digital image data into numbers printed on strips of paper. Too anxious to wait for the official processed image, NASA engineers attached these strips side by side to a display panel and hand colored the numbers like a paint-by-numbers picture. It showed a cratered surface that suggested that Mars was less habitable than previously thought. (Credit: NASA/JPL photo no. GPN-2003–00060.)

the water vapor content varied with latitude and season. This helped lead a JPL group under Crofton "Barney" Farmer to design a spacecraft-based instrument to provide more detailed measurements from Mars orbit; this became the Mars Atmospheric Water Detector and would fly on the 1976 Viking mission.[22]

The first American planetary orbiter reached Mars in late 1971. The spacecraft went into orbit November 13, and the images it returned showed a completely featureless Mars. A dust storm had covered the entire planet. The only visible features were Nix Olympica and three other peaks that protruded above the storm. This revealed to the camera team that they were the largest volcanoes in the solar system by far, dwarfing anything on Earth.[23]

After the planetary dust storm finally ended in late December, the imaging team began to produce a photographic map of the planet. In the process, they found that the images from the previous flyby missions had been misleading. While Mars certainly had Moon-like cratering, it also had vast volcanoes and still more interestingly, vast chasms and canyons.[24] Many of these looked like the natural drainage structures on Earth—extensive dendritic patterns, outwash plains, and deltas. The signs of past water were unmistakable, but the water itself was missing.[25]

The fascinating scientific questions then became what happened to it? And, how could Mars, with its thin wisp of an atmosphere, have once had a climate capable of supporting liquid water on the surface?

Mariner 9 made one more fascinating discovery about the Martian surface. The poles of Mars, like Earth's, are covered in ice. Unlike Earth's polar caps, Mars shows clear signs, even from space, of lamination. They consist of layers of ice separated by layers of what appear to be dust. The laminations were taken to be evidence of "quasi-periodic climatic changes" on Mars that had produced advance and retreat of the water-ice caps.[26] They helped lead many theoretically inclined researchers to speculate about what could cause such apparently rhythmic, and dramatic, changes. Orbital mechanics was an obvious place to start.

Speculative Climates

Mariner 9 ran out of maneuvering fuel and was shut down on October 27, 1972, having radically altered the scientific community's beliefs about Mars. The science teams turned to evaluating the reams of data it had sent them. One important line of research that emerged was the effort to conceive of a Martian climate that could have hosted an oceanic surface. In 1972, Sagan and George Mullen published a short paper in *Science* that encompassed the climate histories of both Earth and Mars. Their motivation was a thorny problem deriving from stellar astronomy and fusion physics. Specialists in both these fields believed the Sun had been substantially dimmer when it had first formed four and a half billion years ago. Known as the "Faint Early Sun" hypothesis, this meant that Earth should have been a frozen ball of ice for most of its history. But the geological evidence available in the 1970s was that it had never been one.[27] This required explanation.

Sagan and Mullen began their analysis of the Earth's past climate with a discussion of the greenhouse effect that maintains Earth's current average surface temperature about 30K higher than it would be in the absence of an atmosphere. They then turned to the Faint Sun hypothesis, reviewing the various estimates of the Sun's evolution, which ranged from luminosity changes of 30–60 percent. They chose 30 percent to be conservative, and then ran the Sun "backward" to evaluate the effect on surface temperatures. Average temperatures in their simple radiative model dropped below the freezing point of seawater 2.3 billion years ago. The geologic evidence available in the 1970s was relatively clear, however, that the Earth was largely ice-free as far back as 3.2 billion years, and life clearly existed in the form of algal "mats" called stromatolites at 2.8 billion years. There was thus a substantial conflict between what should have happened and what had happened.

They believed that only a much stronger greenhouse effect than the current Earth's atmosphere provides was necessary to keep the Earth from freezing under the faint early Sun. They postulated that Earth's early atmosphere was a mixture of carbon dioxide, water vapor, and ammonia, a strong greenhouse gas, with additional minor greenhouse contributions from methane and hydrogen sulfide. Their model required that small amounts of ammonia remained in the atmosphere up to the Precambrian-Cambrian boundary (570 million years before present), but this they found plausible even with the evolution of small amounts of atmospheric oxygen somewhere between one and two billion years ago. The origin of photosynthetic life in the early Cambrian, and subsequent production of the modern oxygenated atmosphere, would then have resulted in the removal of the ammonia. They concluded that "the evolution of green plants could have significantly cooled off Earth."[28]

Then in early 1973, Sagan and two of his students, Brian Toon and P. J. Gierasch, published a paper in *Science* that reported on a study they had done of the effects of orbital obliquity, solar luminosity, and polar cap albedo on Martian climate.[29] Using a general circulation model of the Martian atmosphere that had been adapted from Yale Mintz's Earthly GCM, for the 1973 paper they examined poleward heat transport in the present atmosphere as a means of estimating the overall stability of the present climate. Then, assuming varying amounts of carbon dioxide in the polar caps, then thought to be primarily carbon dioxide with a lesser amount of water ice, they had evaluated the effect of larger amounts of atmospheric carbon dioxide on Martian surface temperatures. An all-carbon dioxide atmosphere of 40 millibar surface pressure seemed enough to permit water at the equator; if much larger amounts of carbon dioxide had been available in the remote past, as the giant Martian volcanoes suggested was possible, a water-enhanced greenhouse effect of 30K could have permitted average equatorial temperatures to exceed water's freezing point.

They then turned to an examination of what could have caused the transfer of sufficient carbon dioxide from the polar caps to the atmosphere and back, starting with orbital variations. Their circulation model indicated that a 15 percent increase in energy absorbed at one pole was necessary to create the necessary conditions. They turned to orbital mechanics for their answer. A Czech scientist, Milutin Milankovich, had argued in a series of papers prior to World War II that Earth's ice ages had been driven by very slow, small changes in the Earth's orbit. His work had not received wide acceptance prior to the 1970s, however. This is partly because the four large glaciations known to have happened during the last 20 million years were not enough—Milankovich's analysis suggested that there should have been many more.[30] But Sagan and his colleagues, who were astronomers, not Earth scientists, chose to apply Milankovich's reasoning to Mars.

The two major components of orbital climate forcing are distance from the Sun and obliquity (the "wobble" of the spin axis). Planetary orbits are not perfectly circular; instead, they are elliptical and vary slowly. Sagan's analysis of the Martian orbit found it too stable and circular for distance from the Sun to have changed enough to produce the necessary changes in insolation. The observed obliquity, however, was quite sufficient to raise polar temperatures and produce the necessary outgassing of whatever carbon dioxide was present.[31] This would occur on a cycle of about one hundred thousand years. Hence Sagan's team concluded that "the atmospheric pressure on Mars has been both much larger and much smaller than present values during a considerable portion of Martian history."[32] Current models of solar luminosity changes suggested that the Sun was at its long-term minimum intensity; if correct, its gradual increase over the next several million years could also produce a warmer Mars. Finally, decreasing the albedos of the poles (making them darker and more absorptive) also appeared capable of producing such changes. In fact, their model was most sensitive to albedo change. Only a 4 percent reduction in polar albedo was necessary to produce the same effect as a 15 percent increase of insolation. There were, then, at least three ways to construct a warmer, wetter, and possibly living Mars that were physically realistic.

The three were arguing that only relatively small changes were necessary to bring about a radical shift in Martian climate. This was a direct attack on a long-held belief that the planets, once their catastrophic period of formation was over, were basically stable. Mars not only had changed since its formation, it would change again. Humans had the misfortune to have evolved when Mars and Earth were in "cold" modes (at least Sagan thought it a misfortune, as he would not get to witness a habitable Mars), but a warmer Mars would occur, eventually.

The result of the early Mariner missions was the opening up of new lines of argument about the evolution of the inner planets and of their climates. While I've focused on a handful of scientific actors, they were not the only interested parties. Many other scientists participated in these efforts to understand why Venus and Mars had turned out so differently, and those arguments continue—unresolved, although we have more evidence developed in ensuing years.

The Sagan and Mullen 1972 paper is still frequently cited (11 times in 2007 as of October), reflecting ongoing scientific interest in the intersection of *chemistry* and *climate* in Martian science. NASA's Mars program since the late 1990s has been aimed at the question of habitability, which includes questioning the evolution of past and present climatic conditions.[33] Past climate is being inferred by study of mineral composition and layering in geologic structure; even without the current fascination with global warming on Earth, planetary scientists would probably still be interested in planetary climate and planetary paleoclimate as part of their interest in the living universe.

Climate Science from the Geoscience Perspective

While the Mariner voyages to Venus and Mars were taking place, Earth scientists were making new discoveries about Earth's climate. Prior to the 1960s, geoscientists conceived of the Earth as a relatively stable, unchanging home. Then-current interpretations of glacial deposits indicated four great ice ages millions of years apart. The mathematical speculations of Milankovich that suggested many ice sheet advances and retreats taking place over only few thousand years, and not millions, had been summarily rejected and ridiculed for decades. That started to change in the 1960s.

In a recent article, Mott Greene examines model studies done in the Soviet Union and in the United States in the later 1960s and early 1970s that suggested a radically different view of Earth. These suggested an Earth climate that had two stable modes: an ice-free mode and an iceball mode. Any state in between was unstable in these models. The model climates would change radically with only very small changes in the amount or distribution of incoming sunlight, or with very small changes in atmospheric opacity, or greenhouse gas levels, or albedo (reflectivity). A 1 percent change in incoming sunlight could send ice sheets advancing toward the tropics, or eliminate all ice from the Earth.[34] This was not at all the Earth scientists had expected. Yet observational evidence emerged during the late 1960s that also supported this view of a very changeable Earth.

In the 1950s, the US Army had built a ballistic missile base inside the Greenland Ice Sheet known as "Camp Century." In support of its construction, the Army had a drilling team extract ice cores from which to examine the density and strength of the ice. Then, the Army team turned one core over to a Danish team of scientists led by Willi Dansgaard. He had developed a technique based on oxygen isotopes that could recover past annual average temperatures from air bubbles trapped in the ice. From the Camp Century core, his team was able to create a 100,000-year climate history for Greenland in great detail.[35] It showed frequent, rapid climate shifts, some of which seemed to coincide with Milankovich's orbit cycles. Some of these occurred on time scales of tens of years, not millions. In 1969, they published their detailed reconstruction. It seemed to suggest that the simple climate models were right. Earth's climate wasn't very stable. But, of course, this was only Greenland—not the entire Earth.

In 1973, British scientist Nicholas Shackleton published an analysis of a 1-million-year-long sediment core pulled from the tropical Pacific. It matched Dansgaard's Greenland reconstruction for the first 100,000 years and extended that record through many earlier ice ages.[36] Combined, these two pioneering efforts demonstrated that Milankovich had been largely right (he missed one important orbit cycle), and the rest of the Earth science community largely wrong. Earth had had many ice ages in its recent past, and not the four great classical ice advances. Further, these were driven by tiny changes in the distribution of sunlight on the surface—a percent, sometimes a little more, sometimes a little less. Thus these observational results confirmed both Milankovich and the earliest climate models. Earth's currently cozy climate was not very stable at all.

For Earth's climate to undergo such wide swings from small initial changes, very powerful feedback effects had to exist to reinforce the initial changes. Two were already obvious by 1970. As Greene argues, the climate model effort indicated ice was one factor. Because of its high reflectivity, ice affected planetary albedo substantially, so small changes to ice distribution reinforced initial solar changes. The other obvious factor was greenhouse gases. Carbon dioxide and methane concentrations in the atmosphere (again extracted from ice cores) moved in lockstep with climate. Both had been known since the nineteenth century as greenhouse gases. They must also play a role, though it was not clear in the 1970s how solar irradiance changes produced greenhouse gas feedbacks.

At a big-picture level, at least, by the early 1970s, paleoclimate evidence supported the model results that suggested Earth's climate could be changed by very small changes in climatic boundary conditions—incoming sunlight and greenhouse gas levels among them. This is not to say the two lines of evidence were perfectly in accord; indeed, there were important differences. The climate models did not have topography or oceans, and thus could not address issues of time scale or local patterns of change. Paleoclimate evidence, like nearly all geological evidence, reflected specific locations—generalizing from the local to the global was quite problematic for geologists. Yet each revealed an unexpected sensitivity to change that was not widely expected.

Taking the Planetary View of Climate

If the striking views of Venus and Mars provided by the early space age had forced planetary scientists to begin reinterpreting what they thought they knew about planetary evolution, the demonstration that Earth had had many ice ages, forced by tiny changes in sunlight, caused a painful reexamination of knowledge within the geosciences too. Geologists had to go redate glacial deposits to reconstruct the Earth's most recent past. Geophysicists had to adapt their thinking to a world that suddenly seemed capable of rapid change. Paleoclimate became an important field of study.

The late 1970s, however, saw the winding down of NASA's planetary science program. It was nearly cancelled entirely in the early Reagan administration.[37] This left planetary scientists with little new data, and many of them turned toward the geosciences. In this they were aided by NASA's refocusing on its Space Shuttle, which could only achieve low Earth orbit. NASA had sold the shuttle in part on its potential benefits for Earth remote sensing, and this was the route many planetary scientists took to staying funded. In the process, they brought with them a worldview that favored global data over local, and remote sensing over in situ measurements. They also brought a tendency toward comparative studies, relating the three "Earthlike" planets to each other to tease out similarities and differences.

One line of investigation planetary scientists entered in the 1970s was the role of volcanic eruptions in the short-term alteration of Earth's climate. This was not a new subject. Scientists had looked at the question of whether volcanoes produced measurable cooling trends for more than a century without reaching a conclusion. As historian Matthias Dörries has pointed out recently, the combination of few eruptions and the generally poor quality of temperature records had left the subject moribund by the late 1950s.[38]

A confluence of factors revived interest in the subject during the late 1970s. At the NASA Ames Research Center, James Pollack was interested in quantifying the effects of sulfate aerosols in planetary atmospheres. In 1972, astronomers had determined that the planetary cloud sheet covering Venus was primarily composed of sulfuric acid droplets. Suspended high in the atmosphere, these produced a highly reflective "surface" that reduced the amount of sunlight reaching the planet's surface by about 80 percent. On Earth, volcanoes produce these aerosols, as do many industrial processes, especially burning coal. Pollack, who was a modeler of planetary atmospheres, sought to understand their disparate roles in the two atmospheres so that he could improve his models.

At the same time, at the Goddard Institute for Space Studies in New York, James E. Hansen and Andrew Lacis, both recent graduates of the University of Iowa, had been working on a radiative transfer model of the Venusian atmosphere. Radiative transfer models are essential parts of general circulation and climate models, calculating the details of the movement of photons of various wavelengths through an atmosphere.[39] The sulfate aerosols Pollack was interested in on the other side of the continent were an essential part of this, so Hansen's team was also interested in the subject of aerosol effects in planetary atmospheres. There was, then, a bit of competition between the two NASA centers.

NASA also started a program at Langley Research Center in Virginia to fund the development of instruments for atmospheric chemistry. Some of these were aimed at various aspects of ozone chemistry in both the stratosphere and troposphere. Others were aerosol instruments, aimed at quantifying the wide range of aerosols in Earth's atmosphere. Some of these were ground-based, many were aircraft- and/or balloon-based, and a few, like the Stratospheric Aerosols and Gas Experiment, were spaceborne.

The promise of volcanoes as climate research resources, though, was inhibited by a lack of eruptions before 1979. The eruptions of Soufriere on the island of St. Vincent in that year, of Mt. St. Helens in the United States the following year, and then El Chichón in Mexico in 1982 allowed the first measurements by the new instruments as well as the first real progress toward understanding the atmospheric effects of volcanoes in a generation.[40] NASA, the National Oceanic and Atmospheric Administration, and the National Center for Atmospheric Research deployed research aircraft to sample the plumes while ground stations and satellites monitored their motion. The three volcanoes allowed the researchers to bound the impacts of eruptions, demonstrating that Plinian eruptions could inject large amounts of material into the stratosphere, confirming that the ash and soot components of the plume did not stay in the stratosphere while the sulfate aerosols did, and showing that the material from mid-latitude eruptions tended to stay in a single hemisphere while equatorial eruptions could circulate ejecta globally.

Further, neither Soufriere nor St. Helens had a measurable impact on surface temperatures. The research teams traced this to unusually low sulfur emissions in the two eruptions. But El Chichón had unusually high sulfur emissions, leading to a very large increase in stratospheric sulfate aerosols. Climate models projected that there should have been a surface temperature impact from this, but this did not seem to happen. An analysis by James K. Angell argued that it had happened, but a simultaneous El Nino masked the cooling signal. Not everyone would accept that argument, so it was not until the 1991 eruption of Mt. Pinatubo that volcanic effects on climate could be ascertained definitively. For that eruption, James E. Hansen at the Goddard Institute for Space Studies used a climate model to forecast a global cooling that did actually happen; perhaps more important from a modeling perspective, the forecast pattern of cooling was also largely (though not perfectly) correct. In the eyes of one reviewer, this fact gave global climate modeling a substantial credibility boost.[41]

There are yet other threads to the story of planetary exploration's role in understanding climate processes on Earth. In the late 1960s, the famously independent scientist James Lovelock and collaborators Dian Hitchcock and Lynn Margulis laid out what's often called the Gaia hypothesis.[42] This posits that the Earth functions as a living organism, with the biosphere and geosphere interacting through chemical means to sustain a relatively stable climate. Early in the decade, Lovelock had been asked by JPL to think about how to look for life on other planets. He gradually convinced himself that atmospheric chemistry was the true signature of a living planet. Biologic activity produces waste products that are thermodynamically unstable (on Earth these are nitrogen, oxygen, and methane); over geologic time scales, these gases would be removed from the atmosphere through chemical reactions if not continually refreshed by living processes.[43] The result would be similar to the atmospheres of Mars and Venus, overwhelmingly carbon dioxide. Thus finding a planet with thermodynamically unstable atmospheres was, to him, proof of life. He and JPL parted company over this; most other scientists wanted to look in soils for microbes. But this made little sense to Lovelock. It was far simpler technologically to study the atmospheres of other planets remotely than to land on another planet and retrieve a soil sample.[44]

This relationship between life and atmospheric chemistry led him to conceive of biological regulation of climate through manipulation of atmospheric chemistry. Lovelock came out of the early Space Age believing that living and evolutionary processes were so powerful that they could reform a planet's climate to one more suitable to life—if life had existed at all on a planet.

Gaia got Lovelock in considerable trouble with his colleagues over the obvious religious implications, but no one reputable in scientific circles rejects the idea that biology, atmospheric chemistry, and climate are intimately related any more, although they still argue extensively over relative importance.[45] As I argue elsewhere, Gaia was stripped of its goddess and rebranded "Earth System Science" in the mid-1980s to eliminate the undesirable religious baggage that came with Lovelock's name.[46] Also sometimes called "biogeochemistry," this science is a study of interlinked processes. In a very important sense, it represents the study of how Earth has diverged so strongly from its planetary neighbors. Living processes made the Earth the way we experience it; conversely, no such science as "biogeochemistry" is possible on the lifeless rocks that are our neighbors.

Finally, a last thread is the famous, and infamous, nuclear winter hypothesis. In 1983, Pollack's group at Ames Research Center, working with Carl Sagan, proposed what's popularly called "Nuclear Winter." Badash has examined this in the context of anti-nuclear politics, but not in its scientific context.[47] This posited that a large nuclear exchange would throw so much dust into the atmosphere that the surface would rapidly cool, freezing crops, destabilizing natural ecosystems, and producing giant, unseasonable storms. Widespread starvation would follow.

The idea had more than one source. Scandinavian scientists had looked at the local effects of severe smoke plumes and wondered what global smoke and dust layers might do.[48] Toon's dissertation research had been on the radiative and thermal effects the vast dust storm Mariner 9 had encountered at Mars. James Pollack had spent much of the 1970s examining sulfate aerosols in the Venus and Earth atmospheres.[49] Finally, the group had previously examined the global climate effects of the dinosaur-killing asteroid impact postulated by the Alvarezs.[50] A principal tool in all this work was a climate model they had developed at Ames Research Center for examining the operation of planetary atmospheres. So the planetary view had quite a bit to do with the formation of the hypothesis.

The planetary view opened lines of research into the complex roles played by smoke and aerosols within the Earth's climate system. These are still very active lines of research, with a number of instruments aboard Earth satellites aimed specifically at aerosols.[51] The Intergovernmental Panel on Climate Change lists global aerosol-cloud interactions as one of the largest areas of scientific uncertainty in climate science and has since the early 1990s.

Conclusion

There are several threads to the "big story" of global warming. There is a paleoclimatology thread, a radiative transfer thread, a modeling thread, and, as I hope I have demonstrated, a planetary science thread. In my own view, none of these is the "most important." Instead, the reweaving of Earth's recent history, and its placement within its planetary context, was a complex effort involving many different fields of science. To me the striking thing is that all these threads occur roughly in parallel. As James R. Fleming has pointed out, there were researchers interested in the carbon dioxide-climate intersection even prior to World War II.[52] But significant community interest follows the International Geophysical Year, with all these threads developing during the 1960s.

By the mid-1970s, they all converged in a general understanding that the Earth was a far different place than thought 20 years before. It was, and is, a dynamic, changeable, place with life (not just humans, but many other forms of life) capable of fundamentally altering it. And the changes can happen quickly, on human timescales. This was a hotly contested idea, and political actors like Crichton routinely attack it. There are, unfortunately, also senior scientists who cannot accept it either, and their public denials help sustain the political attacks.[53] Yet this dynamic view of Earth is now dominant in the Earth and planetary sciences.

Notes

* Portions of this essay are adapted from Erik M. Conway, *Atmospheric Science at NASA: A History* (Baltimore, MD: Johns Hopkins University Press, 2008).

1. Michael Crichton, "Aliens Cause Global Warming," Michelin Lecture, California Institute of Technology, January 17, 2003, www.michaelcrichton.com/speech-alienscauseglobalwarming.html, accessed October 8, 2007.
2. For example: J. T. Houghton, Y. Ding, D. J. Griggs, M. Noguer, P. J. van der Linden, X. Dai, K. Maskell, and C. A. Johnson, eds, *Climate Change 2001: The Scientific Basis: Contribution of Working Group I to the Third Assessment Report of the Intergovernmental Panel on Climate Change* (New York: Cambridge University Press, 2001); at the request of the George W. Bush administration, this study was reviewed and reaffirmed by the US National Academy of Sciences: National Research Council, Committee on the Science of Climate Change, "Climate Change Science: An Analysis of Some Key Questions," Washington, DC: National Academy Press, 2001. See also US National Academy of Sciences, "Carbon Dioxide and Climate: A Scientific Assessment," 1979.
3. Erik M. Conway, *Atmospheric Science at NASA: A History* (Baltimore, MD: Johns Hopkins University Press, 2008); "Satellites and Security: Space in Service to Humanity," in Steven J. Dick and Roger D. Launius, eds., *Societal Impact of Spaceflight* (Washington, DC: NASA SP2007–4801, 2007), pp. 267–88; Naomi Oreskes and Erik M. Conway, "Challenging Knowledge: How Climate Science Became a Victim of the Cold War," in Robert N. Proctor and Londa Schiebinger, eds., *Agnotology: The Making and Unmaking of Ignorance* (Palo Alto, CA: Stanford University Press, 2008), pp. 55–89; *Merchants of Doubt: How a Handful of Scientists Obscured the Truth on Issues from Tobacco Smoke to Global Warming* (NY: Bloomsbury USA, 2010).
4. Spencer C. Weart, *The Discovery of Global Warming* (Cambridge: Harvard University Press, 2003). Also see the updated online version at: http://www.aip.org/history/climate/.
5. James Rodger Fleming, *The Callendar Effect* (Boston: American Meteorological Society, 2007); *Historical Perspectives on Climate Change* (New York: Oxford University Press, 1998).
6. Steven J. Dick, *The Biological Universe: The Twentieth Century Life Debate and the Limits of Science* (New York: Cambridge University Press, 1996), chapter 3, provides an excellent survey of the canalae controversy.
7. Ibid., pp. 116–23.
8. The term "global warming" appears to date from 1975. In the late 1960s and through most of the 1970s, the phenomenon was referred to as "inadvertent climate modification," and in the 1980s it gained the label "climate change," which removed the fact of human agency. See my essay on this terminology: "What's in a name?" available online at http://climate.nasa.gov/news/?FuseAction=ShowNews&NewsID=33 (accessed August 21, 2012).

9. For JPL's history, see: Clayton Koppes, *JPL and the American Space Program: A History of the Jet Propulsion Laboratory* (New Haven, CT: Yale University Press, 1982); Peter Westwick, *Into the Black: JPL and the American Space Program, 1976–2004* (New Haven, CT: Yale University Press, 2006).
10. William W. Kellogg and Carl Sagan, "The Atmospheres of Mars and Venus: A Report by the Ad Hoc Panel on Planetary Atmospheres of the Space Science Board," Publication no. 944 (Washington, DC: 1961).
11. Keay Davidson, *Carl Sagan: A Life* (New York: John Wiley and Sons, 1999), 102.
12. Carl Sagan, "The Radiation Balance of Venus," JPL Technical Report No. 32–34, September 15, 1960.
13. Kellogg and Sagan, "The Atmospheres of Mars and Venus," pp. 43–46.
14. Ibid., p. 46.
15. F. T. Barath, A. H. Barrett, J. Copeland, D. E. Jones, and A. E. Lilley, "Mariner 2 Microwave Radiometer Experiment and Results," *The Astronomical Journal* 69, 1 (February 1964): 49–58.
16. Mikhail Ya. Marov and David H. Grinspoon, *The Planet Venus* (New Haven: Yale University Press, 1998), p. 40.
17. Andrew P. Ingersoll, "The Runaway Greenhouse: A History of Water on Venus," *Journal of the Atmospheric Sciences* 26 (November 1969): 1191–98.
18. Marov and Grinspoon, *The Planet Venus*, pp. 43–45.
19. Ibid., p. 63.
20. Ibid., pp. 68–76.
21. Arvydas Kliore, Dan L. Cain, Gerald S. Levy, Von R. Eshleman, Gunnar Fjeldbo, and Frank D. Drake,"Occulation Experiment: Results of the First Direct Measurement of Mar's Atmosphere and Ionosphere," *Science* 149 (September 10, 1965): 1243–48.
22. Crofton B. Farmer and Daniel D. LaPorte, "The Detection and Mapping of Water Vapor in the Martian Atmosphere," *Icarus* 16 (1972): 34–46; also Ronald A. Schorn, *Planetary Astronomy: From Ancient Times to the Third Millennium* (College Station: Texas A&M University Press, 1998), pp. 210–11.
23. Edward Clinton Ezell and Linda Neuman Ezell, *On Mars: Exploration of the Red Planet 1958–1978* (Washington, DC: NASA SP 4212, 1984), pp. 346–47.
24. For example, see: Bruce Murray, Michael Malin, and Ronald Greeley, *Earthlike Planets: Surfaces of Mercury, Venus, Earth, Moon, Mars* (San Francisco: S. H. Freeman, 1981), pp. 113–22; Schorn, *Planetary Astronomy*, p. 242.
25. William K. Hartmann and Odell Raper, *The New Mars: The Discoveries of Mariner 9* (Washington, DC: NASA SP-337, 1974), pp. 94–103.
26. Paraphrased from James B. Pollack and Owen B. Toon, "Quasi-Periodic Climate Changes on Mars: A Review," *Icarus* 50 (1982): 259–87.
27. This is no longer true. There is substantial evidence that the Earth either froze completely, or mostly, in the late pre-Cambrian, and there is an ongoing debate over these two possible events (known respectively as "snowball Earth" and "slushball Earth"). See, e.g.: Paul F. Hoffman Alan J. Kaufman, Galen P. Halverson, and Daniel P. Schrag, "A Neoproterozoic Snowball Earth," *Science* 281, 5381 (August 29, 1998): 1342–46.
28. Carl Sagan and George Mullen, "Earth and Mars: Evolution of Atmospheres and Surface Temperatures," *Science* 177 (July 7, 1972): 52–56.
29. Carl Sagan, O. B. Toon, and P. J. Gierasch, "Climatic Change on Mars," *Science* 181 (September 14, 1973), 1045–49; Carl Sagan and George Mullen,

"Earth and Mars: Evolution of Atmospheres and Surface Temperatures," *Science* 177 (July 7, 1972): 52–56.

30. By the mid-1980s, these "missing glaciations" had largely been found, and Milankovich's analysis had been largely accepted. A detailed review of this from a scientific perspective is: A. Berger, "Milankovitch Theory and Climate," *Reviews of Geophysics* 26, 4 (November 1988): 624–57. See also Weart, *Discovery of Global Warming*, pp. 46–50, 76–77.
31. Obliquity is the angle the rotational axis makes with the plane of a planet's orbit. For both Earth and Mars, this changes slowly—the planets "wobble" like tops. For Earth, the period of the wobble is 41,000 years, for Mars, it is about 50,000.
32. Sagan, Toon, and Gierasch, "Climatic Change on Mars," 1045–49.
33. For a comprehensive examination of this strategy, see: Committee on an Astrobiology Strategy for the Exploration of Mars, "An Astrobiology Strategy for the Exploration of Mars," National Academies Press, 2007.
34. Mott Greene, "Arctic Sea Ice, Oceanography, and Climate Models," in Keith R. Benson and Helen M. Rozwadowski, eds., *Extremes: Oceanography's Adventures at the Poles* (Sagamore Beach, MA: Science History Publications, 2007), pp. 305–12.
35. Weart, *Discovery of Global Warming*, 73–75; W. Dansgaard, S. J. Johnsen, J. Møller, and C. C. Langway Jr., "One Thousand Centuries of Climatic Record from Camp Century on the Greenland Ice Sheet," *Science* (October 17, 1969), 377–80.
36. Weart, *Discovery of Global Warming*, pp. 75–77.
37. For an overview of this period, see Amy Paige Snyder, "NASA and Planetary Exploration," in John M. Logsdon, Amy Paige Snyder, Roger D. Launius, Stephen J. Garber, and Regan Anne Newport, eds., *Exploring the Unknown: Selected Documents in the History of the U.S. Civil Space Program, Volume V: Exploring the Cosmos* (Washington, DC: NASA SP-2001–4407, 2001), pp. 280–93.
38. Matthias Dörries, "In the Public Eye: Volcanology and Climate Change Studies in the 20th Century," *Historical Studies of the Physical Sciences* 37,1 (2006): 87–126.
39. In contrast to a model's "dynamical code," which calculates the movement of air masses in an atmosphere, the earliest atmosphere models consisted of only the dynamical code. During the 1960s, very simple radiative transfer codes that assumed all light energy was the same wavelength (so-called gray codes) were developed; during the 1970s, radiative transfer codes were developed that began to account for different wavelengths (called "non-gray" codes).
40. Dörries, "In the Public Eye."
41. James Hansen, Andrew Lacis, Reto Reudy, and Makiko Sato, "Potential Climate Impact of Mount Pinatubo Aerosol," *Geophysical Research Letters* 19, 2 (January 24, 1992): 215–18; D. J. Carson, "Climate Modeling: Achievements and Prospects," *Quarterly Journal of the Royal Meteorological Society* 125 (January 1999 Part A): 1–27, quote p.10; M. Patrick McCormick, Larry W. Thomason, and Charles R. Trepte, "Atmospheric Effects of the Mt. Pinatubo Eruption," *Nature* 373 (February 2, 1995): 399–404; Hansen interview with author, January 16, 2006.

42. Lovelock published several iterations of this hypothesis, starting with D. R. Hitchcock and J. E. Lovelock, "Life Detection by Atmospheric Analysis," *Icarus* 7, 2 (1967): 149+. Fuller examinations include Lynn Margulis and J. E. Lovelock, "Biological Regulation of the Earth's Atmosphere," *Icarus* 21 (1974): 471–89; James E. Lovelock, *Gaia: A New Look at Life on Earth* (New York: Oxford University Press, 1974).
43. In fact there was a long-running controversy between geophysicists and geologists over the origin of oxygen in Earth's atmosphere. Geophysicists tended to believe Earth's oxygen had been produced by photolysis of water in the upper atmosphere by ultraviolet radiation. See Harold Urey, *The Planets: Their Origin and Development* (New Haven: CT: Yale University Press, 1952). Geologists and geochemists believed the oxygen was biogenic in origin. See Victor Goldschmidt, *Geochemistry,* Alex Muir, ed. (London: Oxford at Clarendon Press, 1954), and G. E. Hutchinson, "Biochemistry of the Terrestrial Atmosphere," in Gerald E. Kuiper, ed., *The Solar System* (Chicago: University of Chicago Press, 1954). A review by Berkner and Marshall established the first recognizably modern "synthesis," proposing a history of oxygen integrating both processes, but dominated by biogenesis. It was refined by geologist Preston Cloud, who linked the thesis much more tightly to the known geology of the late Precambrian.
44. All molecules have characteristic spectral lines, allowing identification through examination of their emissions (or absorption). Oxygen radiates in the microwave portion of the spectrum, for example.
45. On this controversy, see: Stephen H. Schneider and Penelope J. Boston, eds., *Scientists on Gaia* (Cambridge, MA: MIT Press, 1993).
46. National Aeronautics and Space Administration Advisory Committee, *Earth Systems Science: A Program for Global Change* (Washington, DC: NASA, 1986). My analysis of this is in Erik M. Conway, *A History of Atmospheric Science in NASA* (Baltimore, MD: Johns Hopkins University Press, 2008).
47. R. P. Turco, O. B. Toon, T. P. Ackerman, J. B. Pollack, and Carl Sagan, "Nuclear Winter: Global Consequences of Multiple Nuclear Explosions," *Science* (December 23, 1983): 1283–92; also Paul R. Ehrlich, Carl Sagan, Donald Kennedy, and Walter Orr Roberts, *The Cold and the Dark: The World after Nuclear War* (New York: W. W. Norton and Co., 1984), pp. 83–85. Also see Lawrence Badash, "Nuclear Winter: Scientists in the Political Arena," *Physics in Perspective* (2001): 76–105.
48. Paul J. Crutzen and John W. Birks, "The Atmosphere after a Nuclear War: Twilight At Noon," *Ambio* 118, 2–3 (1982).
49. James Pollack, E. Erickson, F. Witteborn, C. Chackerian, A. Summers, G. Aguason, and L. Caroff, "Aircraft Observations of Venus' Near-Infrared Reflection Spectrum: Implications for Cloud Composition," *Icarus* 23 (1974): 8–26; James Pollack et al, "A Determination of the Composition of the Venus clouds from Aircraft Observations in the Near Infrared," *Journal of the Atmospheric Sciences* 32 (1975): 376–90; James B. Pollack, Owen B. Toon, Carl Sagan, Audrey Summers, Betty Baldwin, and Warren Van Camp, "Volcanic Explosions and Climatic Change: A Theoretical Assessment," *Journal of Geophysical Research* 81, 6 (February 20, 1976): 1071–83; also O. Brian Toon interview with author, February 13, 2004. The group detailed their sulfate aerosols work in: R. C. Whitten, O. B. Toon, and

R. P. Turco, "The Stratospheric Sulfate Aerosol Layer: Processes, Models, Observations, and Simulations," *Pure and Applied Geophysics* 118 (1980): 86–127.

50. See Sagan's essay in *The Cold and the Dark*, pp. 3–6, for a summary of this work.
51. There is a very large body of literature on this now, but a few key articles are: S. I. Rasool and S. H. Schneider, "Atmospheric Carbon Dioxide and Aerosols: Effects of Large Increases on Global Climate *Science* 173, 3992 (July 9, 1971): 138–41; James E. Hansen and Andrew A. Lacis, "Sun and Dust versus Greenhouse Gases: an Assessment of their Relative Roles in Climate Change," *Nature* 346 (August 23, 1990): 713–19; R. J. Charlson, S. E. Schwartz, J. M. Hales, R. D. Cess, J. A. Coakley Jr., J. E. Hansen, and D. J. Hofmann, "Climate Forcing by Anthropogenic Aerosols," *Science* 255 (January 24, 1992): 423–30; Yoram J. Kaufman, Didier Tanre, and Olivier Boucher, "A Satellite View of Aerosols in the Climate System," *Nature* 419 (September 12, 2002): 215–23.
52. James Rodger Fleming, *Historical Perspectives on Climate Change* (New York: Oxford University Press, 1998); *The Callendar Effect: The Life and Work of Guy Stewart Callendar (1896–1964), the Scientist Who Established the Carbon Dioxide Theory of Climate Change* (Boston: American Meteorological Society, 2007).
53. For one example, see Chris Mooney, *Storm World* (New York: Harcourt, 2007), p. 77. I give a political interpretation of this in Erik M. Conway, "Satellites and Security," in Dick and Launius, eds., *Societal Impact of Spaceflight*, pp. 267–88; Conway, *A History of Atmospheric Science in NASA*; and Oreskes and Conway, *Merchants of Doubt.*

Chapter 9

Exploring Planet Earth: The Development of Satellite Remote Sensing for Earth Science

Andrew K. Johnston

Advancing the understanding of our home planet has always been one of the central goals of space exploration. Much has been learned about the Earth's biophysical systems using satellite-based remote sensing, and online access to remote sensing data has revolutionized the way scientists, specialists, and technicians use the data. Moreover, millions of people have become familiar through the Internet with Earth images from satellites. Just about everyone with Internet access is able to view and download information that shows their backyard or their entire nation. At the same time, the relative lack of processed data remains a challenge for the scientific community.

This essay outlines the development of satellite remote sensing and its continued impact within the Earth science community. It begins by briefly describing the three broad technological tracks that have led to the availability of satellite imagery, pointing out different types of orbits utilized by remote sensing satellites. The second section provides an overview of management of US remote sensing programs and the challenge of understanding the Earth's climate systems. The final section outlines how remote sensing capabilities and applications continue to develop to address those challenges. Included is a review of technical and administrative changes being explored to allow Earth science to best take advantage of technological advances.

Three Technological Pillars of Satellite Remote Sensing

Satellite remote sensing technology has developed over several decades along three broadly defined tracks. Aerospace technology provided access to space. Sensor technology developed to allow images to be captured, including film and digital systems. Finally, information technology advanced so that large amounts of data could be processed on inexpensive platforms. Advances proceeded rapidly in the first two fields due to significant government investments. Major advances in the third area came last. The arrival of Internet-based "earth information systems" made it possible for a wide range of users to access new sources of raw remote sensing data.

Access to Space

Aerospace technology, including the methods of building spacecraft and getting them into orbit, developed rapidly during the Cold War competition between the United States and the Soviet Union. Reports from Douglas Aircraft Engineering Division (the future RAND Corporation) were among the early attempts to outline technical details for applications of orbital spacecraft.[1] The first such report in 1946 described possible design features and applications.[2] It included descriptions of weather observations, imaging of the land surface, and types of orbits that could be utilized. Several different types of orbits may be used by an artificial satellite, each utilized for a different type of mission. What follows is a review of the main types of orbits used by artificial satellites.

Low earth orbit (LEO), the simplest orbit and the first to be reached, is used for manned spaceflight and several types of unmanned missions. These satellites orbit in altitudes less than approximately two thousand kilometers.

Spacecraft can also be placed into a polar orbit that passes over or close to the Earth's north and south poles. In this orbit, the planet continues to rotate while the satellite moves in a circular orbit. The advantages of using a polar orbit were recognized in early studies to determine how a highly included or polar orbit would be useful. At first, attention was given to weather applications. A RAND report from 1960 demonstrated how a satellite at an altitude of 350–450 miles with an orbit inclined by 56 degrees could scan the majority of the Earth north of 45 degrees latitude.[3]

Most land remote sensing satellites are placed into a specific type of near-polar *sun synchronous* orbit. In these orbits, a satellite passes near the Earth's north and south poles but not directly over them. The orbital plane precesses from west to east, thereby maintaining a constant angle with the Sun. In this way the satellite always observes the Earth's surface from a relatively constant solar illumination angle. This allows images from different years to be compared, because the solar angle determines lighting conditions on the surface.

The advantages of a sun synchronous orbit were recognized early. The 1946 Douglas Aircraft report described this type of orbit: "A satellite on a North-South orbit could observe the whole surface of the world once a day, and entirely in daylight."[4] The first attempt to reach polar orbit was the first Corona/Discoverer launch in February 1959. This was first in a long series of reconnaissance platforms. After several attempts, Discoverer 13 was successfully recovered, making it the first object to be recovered from orbit. Today it is displayed at the National Air and Space Museum. A little more than one year later, polar orbit was used by a civilian program. TIROS 1, the first weather observation satellite, was launched in April 1960.

The ESSA series of weather monitoring satellites were placed into sun synchronous orbits beginning in 1965. Following these flights, the earth resources technology satellite (ERTS, later renamed Landsat) also used sun synchronous orbits. Sun synchronous orbits were also used for NASA's large earth observing system (EOS) satellites to establish "morning" and "afternoon" satellites.[5] The use of sun synchronous orbits allows for the time of day to be selected for observation. The EOS-AM satellite, eventually named Terra, was placed into an obit allowing observations at mid-morning. The EOS-PM satellites (Aqua and Aura) are in afternoon orbits.

Another prime destination for artificial satellites is geostationary orbit. The basic description of the geostationary orbit was outlined by Herman Potočnik (Hermann Noordung) in 1928 and Arthur C. Clarke in 1945.[6] At a distance of 35,794 kilometers (42,164 miles) a satellite orbits the Earth within a period of 24 hours. This is known as a *geosynchronous* orbit. If the satellite's orbit is aligned with the equatorial plane, the satellite remains over one spot on the Earth's surface as it rotates. This is known as a *geostationary* orbit. The first satellite placed into that orbit was the communications experiment Syncom 2, launched in July 1963. Most satellites in geostationary orbit today are used for communication. A geostationary orbit allows a communications satellite to be easily targeted from the ground because it appears stationary in the sky. A small percentage of geostationary satellites are used for remote sensing of weather patterns, such as the geostationary operations environmental satellites (GOESs).[7]

Sensor Development

Vast arrays of technologies are available to allow the systematic capture and storage of images. The ability to record images goes back to the origins of photography in the 1820s. Astronauts began taking photographs on film since the first flights. Handheld Hasselblad cameras were carried on the four orbital Mercury missions. The cameras were only slightly modified to allow for recording images out the small window of their capsules. More than thirteen hundred images were returned during the Gemini program. These photographs were later important for building support for the Earth Resources Observation Satellite (EROS) program, created in 1966.[8] This would eventually develop into the Landsat program.

Astronaut photography began to return significantly greater volumes of images during the Apollo program. The Apollo 7 flight alone returned about five hundred images with Hasselblad cameras. Some Apollo images have left a significant lasting impact. Notable images include the "Earthrise" photograph taken during Apollo 8, and the famous "Blue Marble" image of the Earth taken during Apollo 17. After the Apollo lunar flights, astronauts on Skylab returned tens of thousands more images. Some cameras on Skylab were handheld, others were designed to fit into a bracket mounted on the side of the window facing the Earth's surface.

During the Space Shuttle program vastly more images were collected by astronauts. In the early stages of the Space Shuttle program, almost all of these images were stored as film products. Through the 1990s the 70-mm Hasselblad remained the most common camera.

Two large film cameras intended for photogrammetric applications were flown in the Space Shuttle. In December 1983 the Metric Camera flew as part of the European Spacelab module on board Space Shuttle Columbia. The largest and most capable film camera ever carried on US human flights was the large format camera (LFC), carried in the Space Shuttle payload pay. It flew on Space Shuttle Challenger in October 1984. The film was 23 centimeters wide, more than three times the size of a handheld Hasselblad.[9]

Recording images on film from space has also been performed on robotic spacecraft. The most extensive of these efforts was the US Corona program, which operated from 1960 to 1972 with more than 140 launches, returning a total of more than eight hundred thousand photographs. CORONA, along with the Argon, Lanyard, and other related reconnaissance programs all used film cameras. Exposed film was returned to earth in reentry capsules.[10]

Electronic image capture and transmission can be accomplished using analog or digital methods. Analog methods included the first generation of television cameras (figure 9.1). In digital methods such as modern digital cameras, images are collected and stored as digital data. Before digital imaging technology was refined for remote sensing, analog imaging methods such as television cameras were deployed. During the late 1960s it was not clear if digital or analog technology would be the way forward for satellite remote sensing. Analog sensors were better tested at the time and offered advantages in simplicity of interpretation. However, digital sensors held the

Figure 9.1 Global view produced using MODIS data acquired in 2001. (Credit: NASA Goddard Space Flight Center.)

promise of better performance if they could be perfected. There were proponents of both types of sensors for satellite remote sensing. Within the US government, the Interior Department advocated the use of an analog vidicon system, while the US Department of Agriculture advocated for the digital Multispectral Scanner. Eventually ERTS would carry both sensors. The first three satellites in the Landsat series carried the analog Return Beam Vidicon (RBV) and the digital Multispectral Scanner (MSS). The RBV was a simple imaging sensor similar to a television camera.

Analog image sensors, however, lacked the ability to detect narrow sections of the electromagnetic spectrum, making them less useful for certain applications. During the late 1950s significant skepticism existed that satellite observations using analog sensors could yield anything useful for weather forecasting. A 1960 RAND report suggested that satellites could provide overhead images of clouds, but not the vertical data required by weather forecasters.[11] The solution would eventually be provided by digital atmospheric sounding instruments that utilize infrared wavelengths to provide data at different atmospheric levels.

The development of the charged-coupled device (CCD) was the primary advance responsible for allowing digital technology to replace analog remote sensing. The CCD was first developed in 1969 at Bell Labs as an outgrowth of work on semiconductor memory storage devices. Today CCDs can be found in every digital camera as well as in remote sensing and astronomical instruments.

ERTS (Landsat 1) satellite was launched in July 1972. The MSS it carried demonstrated the strength of using a digital imaging sensor. A new sensor with greater capability was developed for Landsat 4 and 5. This new sensor, the "Thematic Mapper," was designed to return higher spatial and spectral resolution and greater dynamic range than the MSS. Both Landsat 4 and 5 also carried an MSS sensor to provide data continuity. The TM acquired imagery at higher spatial resolution and with greater dynamic range, two issues that seriously limited use of MSS data.

Since the 1990s a new generation of commercial endeavors have made use of high-resolution digital sensors on orbital platforms. Space Imaging, Inc. launched the IKONOS satellite, DigitalGlobe the QuickBird satellite, and ORBIMAGE distributed data from the OrbView satellites. ImageSat International, an Israeli company, operated two EROS satellites to provide high-resolution data. These satellites shared a technological heritage with Russian high-resolution sensors on reconnaissance satellites. The market has not been kind to all these ventures. Space Imaging and ORBIMAGE merged in 2006 to form GeoEye, leaving two major US companies in the market. The two compete for government contracts, while also competing to provide data to Internet-based mapping services.

Information Technology

Information technology, including software, hardware, and storage, is necessary to store, process, and distribute remote sensing image data. This was the last of the technologies to advance to make satellite data widely available.

Internet sites that deliver satellite imagery became so pervasive so quickly that it is easy to forget they did not exist only a decade ago. In the recent past, access to remote sensing data required that orders be placed over the phone. Data would later be shipped to the user on digital storage media. Today only archival data are delivered in this fashion, instead of being transmitted digitally.

Systems of archiving images can be evaluated by the length of time those images can be maintained. Digital media can lose storage ability after decades. A key advantage for hardcopy storage of images is the lifelong ability to read the "format." After a thousand years, a properly maintained photograph can still be readable.

The storage of digital data provides a different set of challenges. Magnetic tapes were the media of choice for most of the history of remote sensing. Various formats of magnetic tapes still provide archival storage for the majority of satellite remote sensing data. Hard disk drives, introduced in 1956 by IBM, were later used for quick-access storage of remote sensing data. Today hard disks remain the storage methods for the vast majority of desktop computer systems, and in recent years have been used widely for some remote sensing archives.

It was the arrival of Internet access that played the most central role in spreading the use of satellite remote sensing data. Although the development of electronic communication was not an integral part of remote sensing development, as in many fields the Internet changed forever the way people had access to satellite imagery. At first, Internet connectivity was used as a command-line method of moving data files between distant computers. Gradually this changed graphical interfaces including web pages. The sheer size of satellite image databases was a serious challenge to allowing Internet access.

Since about 2000 a new type of Internet site has become widely available: "earth information systems." An earth information system can be described as a linked computer system that provides storage, searchable access, and distribution to Earth observation data. Examples include the Earth Observations System Data and Information System (EOSDIS), a NASA portal for data from the Earth Observing System. Data are collected and distributed through servers at NASA's Goddard Space Flight Center (figure 9.2). The US Geological Survey developed websites to provide access to data from Landsat, airphotos, and other systems. These have included GloVis, followed by the improved USGS Earth Explorer. Outside the US government, a primary example of an earth information system is the Global Land Cover Facility at the University of Maryland, which provides data from Landsat satellites and other sources.

The development of earth information systems was a response to perceived needs to make vast amounts of satellite data more accessible. An internal NASA assessment in 1989 recommended that NASA make use of existing distribution points at research labs and universities.[12] The result were the Distributed Active Archive Centers (DAACs). Each DAAC specialized in distribution of a certain type of Earth science data. A 1994 National

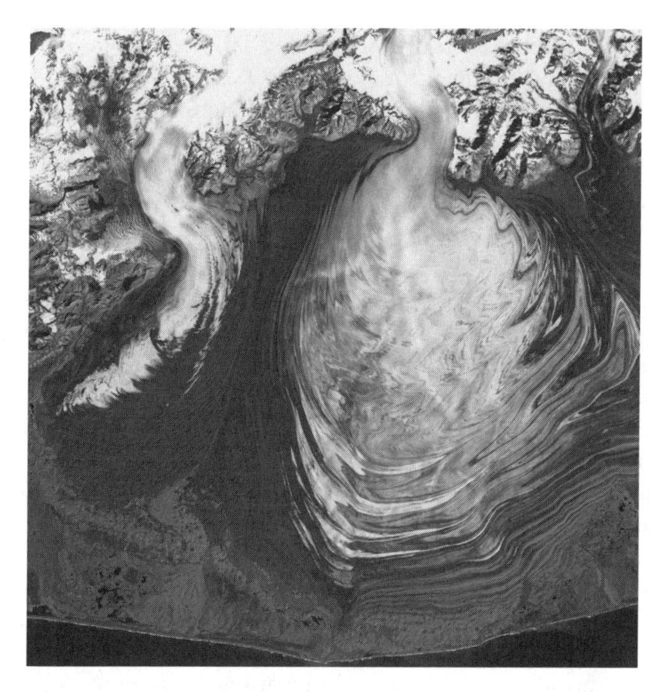

Figure 9.2 The tongue of the Malaspina Glacier. One of the largest glaciers in Alaska, it lies west of Yakutat Bay and covers roughly 1,500 square miles (3,880 square kilometers). This image was acquired by Landsat 7's Enhanced Thematic Mapper plus (ETM+) sensor on August 31, 2000. This is a false-color composite image made using infrared, near infrared, and green wavelengths. The image has also been sharpened using the sensor's panchromatic band. (Credit: NASA Goddard Space Flight Center.)

Research Council report called on NASA to get more usage from EOSDIS by making it more open to the wider research community.[13] Today a wide range of scientific disciplines make use of satellite remote sensing data.

Supporting Earth Science: Program Management

The management of a remote sensing program can have significant impact on the distribution of image data. Before land remote sensing satellites were even in orbit, significant disputes arose between US federal government agencies. NASA, Department of Agriculture, Department of Interior, and others attempted to play a role. Interior attempted to move the satellite remote sensing effort forward by issuing a statement that it was proceeding with development.[14] In fact, the Department of Interior lacked a satellite program or any method of launching one. After years of budgetary disagreements, the first ERTS satellite was eventually launched by NASA in 1972.

In the late 1970s it was proposed to commercialize the Landsat program, as reflected in a NOAA Satellite Task Force report.[15] In 1984 a new entity, known as EOSAT, took over operations of the Landsat 5 satellite and marketed the data as a for-profit corporation. The embrace of a corporate structure was almost universally viewed as a failure. Despite resources and a sophisticated data management system, EOSAT was given an impossible task: to turn a profit from selling moderate resolution satellite imagery. The market simply did not exist to support for-profit operation. After years of losses, the problems were outlined in a 1995 congressional report.[16] The Landsat program returned to government control in 1998, with the USGS distributing Landsat data from its EORS Data Center in South Dakota. The cost of Landsat scenes dropped to about 10 percent of the EOSAT pricing, allowing the majority of science users to make use of the data once again.

During the 1980s NASA determined to build a range of Earth observation satellites. In 1987 NASA named the effort "Mission to Planet Earth" and called for large platforms to be launched on the Space Shuttle.[17] This created serious problems as the Space Shuttle program was scaled back. Fewer launches and the loss of polar orbits meant that the Space Shuttle was not useful for launching Earth observation satellites. This led to significant delays and vastly increased costs while satellites were redesigned to fit on small launch vehicles.[18]

The use of data buys was an innovation in the late 1980s to provide data at lower cost to NASA. In this scenario, NASA requested bids to provide satellite data. The contract would list the required characteristics of the image data, while leaving details of launching and operating the satellite to the contractor. Unfortunately this required new management and operations systems to be created, leading to delays and costs.[19] A prime example is the experience with the sea-viewing wide field-of-view sensor (SeaWiFS). This was originally intended to fly on polar orbiting weather satellites as a follow-on of the Coastal Zone Color Scanner (CZCS) on Nimbus 7. A NASA contract was signed with Orbital Sciences Corp. in 1991, but the satellite did not launch until 1997.

The data buy model was initially followed for the Landsat Data Continuity Mission (LDCM), the successor to the Landsat series that began in 1972. The request for proposals was released in June 1999, but it failed to attract proposals that met the demands of the program. In 2003 the data buy was cancelled. Eventually NASA and USGS agreed to an operational plan for LDCM, but the data buy experiment delayed the program for many years.

Studying the Earth

The Earth's climate system can be conceptualized as an interconnected system of systems. Earth system science has developed into the primary method of organizing research into climate dynamics. Early descriptions of ERTS satellite data included speculation on how images from space could be applied to understanding complex climate systems.[20]

It was not long before it was recognized that satellite data would be especially useful in this type of scientific endeavor. A 1982 report requested by NASA detailed the importance of satellite-based observations to understanding the continued habitability of the planet.[21] This report also introduced the term "Global Change" to a wide audience. "Global Change" was a widely used term in the geoscience community through the 1980s. The focus on "habitability" for human society was maintained within NASA by an internal report the next year.[22] The outline of a plan to use orbital platforms to support an Earth system science program was formalized, also in 1983, in a report commissioned from the National Research Council.[23] This provided the basic outline for what had become described as Earth system science: a holistic look at many interconnected systems. Specific orbital missions were recommended in the 1986 report by the NASA Advisory Council to create an "Earth Observing System," which would include several orbiting platforms as well as data distribution systems.[24]

The Challenge of Understanding Earth's Climate System

Satellite observation of the Earth provides a wealth of data to understand environmental changes (figure 9.3). However, the complexity of the Earth's physical and biological systems presents significant challenges to understanding the behavior of the Earth's environment. Models conceptualized as simplifications of biophysical systems are valuable for describing Earth system processes, but are less effective at making predictions. For instance, most global climate models use cell sizes as large as hundreds of kilometers across. Processes that operate over smaller scales play important roles

Figure 9.3 Oblique view of Hurricane Erin in the Atlantic Ocean off the coast of North America acquired in 2001 by the SeaWiFS sensor on the OrbView-2 satellite. (Credit: NASA Goddard Space Flight Center and GeoEye.)

in determining weather and climate changes. A lack of appropriate remote sensing data presents significant challenges.

Many modeling studies have focused on weather forecasting and climate change, where laws of fluid dynamics can be parameterized into models. Even so, it has proven difficult to predict atmospheric dynamics. Interactions between the Earth's atmosphere and surface are even more complex, making it even less likely that these models can predict future changes. The use of satellite remote sensing allows for observations to be made of smaller parts of the climate system. Although complete understanding of biophysical dynamics may not be possible, these observations provide insight into individual subsystems.

The El Niño Southern Oscillation (ENSO) is the periodic movement of warm water across the southern Pacific operating over shorter time periods (two–seven years). These large-scale oscillations of sea temperature have been extensively modeled but are difficult or impossible to predict. In the Pacific Ocean, decadal and inter-decadal oscillations are responsible for about half of the global variation in surface temperature.[25] A period of warmer ocean temperatures in the late 1970s coincided with a period of more intense ENSO events.[26] However, the data are lacking to understand the links between ENSO and global temperature dynamics. Warmer temperatures may drive a more intense ENSO cycle, but ENSO itself has significant impact in global temperature. A warming ocean may play a significant role in encouraging ENSO events but data and models are not yet of high enough precision to determine if this will continue into the future.

Understanding the dynamics of atmospheric carbon dioxide is an important focus of Earth system science because of its role in climate change. Models have predicted CO_2 concentrations ranging from 540 to 970 ppm by 2100, increasing from about 365 ppm in 2000. Various levels of sensitivity to responses in the oceans and terrestrial biosphere lead to at least 40 percent uncertainty in most of these projections.[27] Despite the quantity of modeling studies, the size of the global terrestrial carbon sink remains highly uncertain. There are a wide range of estimates of carbon exchange between terrestrial ecosystems and the atmosphere. Looking 50 years into the future, models disagree by as much as six Pg C annually, an uncertainty that is likely far greater than the entire terrestrial carbon sink.[28] The failure of the Orbiting Carbon Observatory (OCO) in February 2009 may prove to be a significant setback in understanding carbon dioxide dynamics. The OCO spectrometer instrument was designed to detect carbon dioxide while observing the atmosphere from different viewing angles. It would have provided much more detailed observations of atmospheric carbon dioxide than existing instruments.

Improvements in Observational Capabilities to Benefit Earth Science

The challenges outlined in the previous section present obstacles for better understanding of the Earth's climate system. Although remote sensing has made great strides in the understanding of how biophysical systems change through time, remote sensing tools require improvements to provide data of higher quality so that climate modeling can continue to improve. Advances in technology can provide remote sensing data better suited for Earth system science applications.

Experimental remote sensing programs at NASA include new orbiting sensors. NASA's Earth System Science Pathfinder (ESSP) program has funded five experimental Earth observation satellites including CALIPSO (described elsewhere in this chapter), the Orbiting Carbon Observatory (a 2009 launch failure), and Aquarius, launched in 2011 to globally measure the salinity of seawater. The NPOESS Preparatory Project (NPP) was planned to bridge the gap between the Aqua satellite now in orbit and the launch of the first NPOESS satellite, which has been delayed until 2013.

Operational remote sensing programs under NOAA will continue to evolve. The GOES spacecraft track global weather patterns from geostationary orbit. The first in the next generation of GOES was launched in 2006. GOES-14, launched in 2009, provides data at finer spatial resolution. Like its predecessors, it carries a visible and infrared imager and sounder sensors.

The Need for Improved Remote Sensing Data

The processes connecting climate and global change are an important focus of Earth system science. Agriculture and forestry applications need more precise data on vegetation status and growth. The use of remote sensing for management of water and energy resources is limited, but better data will allow more imagery to be applied in these fields. Disaster relief and homeland security are also areas where remote sensing data at high spatial resolution, if disseminated in a timely fashion, can be widely applied. Weather forecasting makes extensive use of satellite data but the current state of remote sensing limits the accuracy of forecasts. The ability to make seasonal climate predictions may be enhanced with more precise remote sensing data.

The application of remote sensing to support international environmental conventions is of particular importance. For instance, the Kyoto Protocol would have required nations to account for their emissions of atmospheric carbon and the absorption of carbon though vegetation growth and other means.[29] The remote sensing tools currently available do not provide the precise data needed to track these changes. The United States was never party to the Kyoto Protocol, but negotiations for future phases of the treaty continued through 2007. Several nations including the United States began exploring the possibility of "cap and trade" systems that would harness market forces to discourage greenhouse emissions. Among the proposed components was the implementation of credits for avoided deforestation or reforestation. These and other measurements of land surface dynamics will require better remote sensing data.

Integration of Remote Sensing Data

The potential of remote sensing data cannot be fully realized unless they are assimilated into biophysical models. The assimilation of remote sensing data into physical models must be assessed before spacecraft are launched, so the data they return can be used more easily in modeling applications. Increasing numbers of remote sensing products have been generated for application in biophysical modeling in recent years. Products being generated from the MODIS sensors on the Terra and Aqua satellites include global maps of vegetation indices, sea temperature, and atmospheric composition.

More powerful computers will make it possible to create atmospheric circulation models at higher spatial and temporal resolution. The precision of models has been rising in recent years. Vertical analysis models under development are capable of integrating continuous data at high spatial and temporal resolution. The resolution of these models often exceeds that of remote sensing data, therefore requiring higher resolution remote sensing data so that complex physical models can be calibrated and validated.

The integration of land remote sensing data into biophysical models is an area where future improvements can be made. While the importance of remote sensing capabilities for understanding weather, climate, and oceans is widely recognized, the role of land surfaces often does not receive as much attention. A bureaucratic division may be partially responsible. The National Oceanographic and Atmospheric Administration is the US agency responsible for providing data on weather, atmosphere, and oceans. Although NOAA also provides data on land surface processes, this has traditionally been a role for USGS or NASA.

Remote sensing holds the promise for measuring environmental variables that impact human health and well-being. The application of remote sensing to ensuring human health and food security are preliminary. The integration of derived products from remote sensing with socioeconomic data would foster better understanding for phenomena such as disease outbreaks.[30] There is a need to archive long-term data to study food security over time.

The most prolific source of remote sensing data for the land surface has been the Landsat program. While the Landsat program has provided a significant range of data, it had never been given "operational" status. A future land surface sensor was planned for the National Polar Orbiting Environmental Satellite System (NPOESS) initiated in 1994 by NOAA, NASA, and the Department of Defense. At the time this appeared to be an important breakthrough, because for the first time a land surface sensor will fly on an operational NOAA spacecraft. However, significant engineering hurdles were identified; so in 2005 it was decided by NASA and NOAA that a Landsat-type sensor would fly on its own platform in polar orbit. This was planned for a 2011 launch, which then slipped to late 2012. Along with the delays incurred during the ill-fated Landsat data buy process during 1999–2003, this pushed the launch beyond the operational lifetime of the Landsat satellites that remained in orbit. Landsat 5 experienced a failure of its Thematic Mapper sensor in November 2011. Landsat 7 continued to provide only partial data after a previous malfunction in one of its scanning mirrors.

Advances in Data Processing

Advances in data processing capabilities have allowed remote sensing techniques to become widely applied. As recently as a decade ago, it was common for remote sensing data to be processed on large mainframe computers. Now most image processing can be performed on desktop computers. However, software technology has not kept pace with hardware development. The software used to process remote sensing data remains primitive compared to most types of commercial software. While most software has become more capable and work flows have become simpler, remote sensing packages remain difficult to use.

Georeferencing and calibrating image data, for instance, requires many complicated steps in most software packages. This is true despite the growing use of remote sensing data and the presence of healthy competition in the field. The commercial market for remote sensing software remains relatively small, limiting innovation and keeping costs high. The cost of most popular remote sensing software packages is higher than the cost of the hardware on which they operate.

New standards for data distribution and user interfaces, if widely applied, would enable greater numbers of researchers in a wide range of fields to make use of satellite data. For instance, the geoscience community can move toward defining land vegetation variables important to understand climate change that can be incorporated into NASA or NOAA data distribution systems.

Improvements in Dissemination of Remote Sensing Data

Dissemination of data could be enhanced by advances in information systems over the next decade. Earth information systems, accessible from anywhere, may provide global georeferenced quantitative descriptions of the Earth at high resolution. Advances in remote sensing and information technologies make this a real possibility in the coming decade. The future development of more comprehensive and reliable earth information systems will allow more accurate predictions of the behavior of Earth systems.[31]

Older archival satellite data are commonly maintained in long-term storage using magnetic tapes. However, the costs of maintaining an active archive on hard disk are falling. Transitioning to an active archive, as recommended by the National Research Council, would significantly improve access to historical data and understanding of climate dynamics.[32]

The cost of satellite data is often prohibitive, even for government sources. After the failed experiment with commercializing the Landsat program, the US government has distributed civilian satellite data close to the cost of dissemination. This was outlined explicitly in NASA's 1999 Earth Science Enterprise data management strategy.[33] The cost from some sources declined again in the next decade. Data from sensors on NASA EOS satellites were distributed by NASA at nominal or no cost. New sources such as the University of Maryland's Global Land Cover Facility freely distributed Landsat data. In 2009 the US Geological Survey began to provide archived Landsat data at no cost.

Advances in Remote Sensing Technology

Technological advances have provided new opportunities for Earth observations. The use of GPS occultation can provide high-resolution thermodynamic soundings of the atmosphere. In 2006 the Constellation Observation System for Metrology, Ionosphere, and Climate (COSMIC) began observing GPS occultations using six satellites. A GPS occultation sensor is also planned for the NPOESS satellites.

Light detection and ranging (LIDAR) has diverse remote sensing applications. Differential Absorption LIDAR (DIAL) can be used to obtain vertical profiles of atmospheric temperature and moisture. LIDAR data should be particularity useful for improving the accuracy of estimating atmospheric water vapor. CALIPSO (a joint NASA/CNES mission) uses LIDAR to observe the atmosphere. ICESat, launched in 2003, uses LIDAR to measure topography and ice thickness. The ALADIN instrument, designed to use LIDAR to study tropospheric winds, launched on March 27, 2009, on the European ADM-Aeolus spacecraft. LIDAR can also be used from aircraft to observe important land surface characteristics, including height and density of forest canopy.[34] The orbital Vegetation Canopy Lidar was selected as NASA's first Earth System Science Pathfinder mission, but the mission was not able to solve technical problems and did not proceed to launch. NASA's ICESat-2 satellite planned for launch in 2016 will carry a LIDAR sensor intended for observing ice sheet dynamics and vegetation canopy.

Sensors such as MODIS and the Enhanced Thematic Mapper Plus on Landsat 7 use scanning mirrors to observe the Earth. These "whisk-broom" sensors are simpler to calibrate because they have fewer CCD detectors, but their moving mirrors are prone to mechanical failure. The instrument on Landsat 7 experienced such a failure in 2003, and its image data are now useable only with extensive postprocessing. Improvements in technology have made CCDs more sensitive and consistent, allowing a large number of detectors to be used reliably. Future sensors will be constructed with little or no moving parts to achieve greater reliability.

The future of land surface remote sensors may lie in the use of instruments that can adjust their spectral resolution. Using hyperspectral sensors, broadband remote sensing data can be synthesized in orbit without transmitting the entire detected spectra. This would allow the bands to be broadened, narrowed, or shifted as needed.

The useful spatial resolution of a sensor is determined by its pixel size and its modulation transfer function (MTF). Although raster image data are composed of pixels, only a portion of the radiance detected for each pixel is reflected by the land surface within that pixel. The remainder originates in neighboring pixels. Aggregating small pixels into large ones produces a more square-shaped response curve and reduces the effect of the MTF. Remote sensing satellites could incorporate this digital image processing technique onboard by using a high-resolution sensor and aggregating pixels before transmitting data. This would also allow adjustment of the spatial resolution of the data.

Advances in Space Technology

New space technologies are being explored for remote sensing missions. Constellation flying allows satellites to semiautonomously maneuver to maintain a constant formation. This can allow more than one spacecraft to observe diverse phenomena. Constellation flying is being used in the GRACE and COSMIC missions to observe the Earth's gravity field and ionosphere, respectively. Formations can also allow satellites to make simultaneous observations. For instance, the EO-1 satellite was placed into orbit in formation with Landsat 7.

New data acquisition techniques can allow more data to be downloaded for analysis. Higher data capacity will allow future satellites to store much more data. Better compression techniques will be developed so more data can be received in shorter time periods. Future remote sensing platforms in orbit will be capable of returning several terabytes of data daily. A space-based communication system, similar but more capable than the TDRS system deployed by NASA in the 1980s, may be required to transmit large amounts of data.

The use of modular spacecraft has been proposed for solar system exploration. The Mariner Mark II series was intended for missions to the outer planets. The same spacecraft bus could be outfitted with instruments, power equipment, and thrusters for missions to different planets.[35] However, the program was cancelled in the face of rising costs in 1993. Only the Cassini mission to Saturn survived budget cuts. Despite the end of the program, the use of modular spacecraft may be a useful approach for terrestrial remote sensing. Modular spacecraft may be more appropriate for the Earth because, unlike the outer solar system, requirements of fuel and power are similar for all missions.

Large remote sensing satellites have been designed to observe with many sensors to understand multiple biophysical systems. However, the importance of making simultaneous observations in different wavelengths is not usually necessary, except when observing cloud systems. More numerous and smaller satellites may provide cost savings.

A NASA report in 1991 recommended significant changes that would lead to smaller satellites being launched on Titan vehicles.[36] While budgets shrank for the earth observing system during the 1990s, NASA developed the Earth System Science Pathfinder program to fund smaller and more flexible satellites.[37] Smaller missions such as QuickSCAT required the use of sensor technology that was developed outside mission funding. The future success of small inexpensive satellites will require dedicated funding for development of new small sensors.

Very small satellites may provide cheaper access to space during the next decade. Satellites weighing just over 100 kilograms are being placed into orbit for remote sensing applications. These "microsatellites" carry remote sensing sensors capable of returning three–four channel multispectral data with 50–100 meter spatial resolution. Other remote sensing satellites are under development that have a mass of less than 50 kilograms. The small size and cost of these satellites has allowed several developing nations to participate in remote sensing programs for the first time.

Several remote sensing microsatellites are planned or under construction. For instance, Topsat was launched in 2005 to provide 2.5 meter panchromatic images and 5 meter multispectral images. RapidEye, a provider of geospatial information, operates its own five satellite constellation that acquires 6.5 meter multispectral data. Larger optical systems are still required for many remote sensing applications, so microsatellites will not completely replace large satellites. However, over the next decade microsatellites should allow technologies to be tested at much lower cost.

Application of New Technologies to Operational Remote Sensing

The operation of orbital remote sensing programs shifts through time. To make future advances in remote sensing a reality, technologies must move from research to operational uses. While NASA conducts research into development of measurement technologies and analysis techniques, NOAA operates US civil observing systems. The TIROS program of the 1970s provides one example of a successful transition from research to operational status. However, a comprehensive strategy is lacking to smooth the transition from NASA research to operations under NOAA for contemporary remote sensing programs.

A 1993 National Performance Review stated that NOAA polar orbiting satellite program should be merged with the Department of Defense program, along with data collection and distribution from planned sensors.[38] The resulting national polar-orbiting operational environmental satellite system, with an office located at NOAA headquarters, also housed DoD and NASA staff.

For land remote sensing, oversight of the Landsat program has changed frequently. After the experiment with commercializing the program failed, responsibility was split between the Department of Defense and NASA.[39] Today Landsat data are distributed by the US Geological Survey, Department of Interior. A joint NASA-NOAA office has been proposed to coordinate the transition process, as reflected in a 2003 National Research Council report. This could lay the foundation for successful transition to operational remote sensing missions. A formal mechanism to allow transition to operations would permit new technologies to be used in Earth observations as quickly as possible. This would allow remote sensing missions and earth system science to make the best use of rapidly advancing technology.

Conclusion

In recent decades the technology of satellite remote sensing has become widely applied to understanding the Earth's biophysical systems. Development of these capabilities has moved forward slowly and unevenly. It has usually been seen that it is easier to collect images than distribute them across the world. Lack of useable software, limited storage, and difficulty in data access and sharing often prevented the scientific community from making use of remote sensing data. Advances in software technology will allow remote sensing data to be transformed into physical units more easily. The incorporation of features from widely available desktop software, such as simple data structures and useful graphical interfaces, will allow future software developers to produce more widely used programs. Common data formats and standards would enhance the ability of the wider scientific community to use remote sensing data.

Better-quality remote sensing data in the future can lead to more complete understanding of the Earth system. In recent years sensor technology has been developing faster than spacecraft designs. Using modular designs, sensors can be incorporated into standardized spacecraft designs. It also allows instruments to be replaced or upgraded as new technologies become available.

Management strategy of US Earth observation programs has varied significantly through time. Earth observation programs have often been altered in response to reports commissioned from the National Research Council. These reports have provided important guidance for NASA and NOAA. As the Department of Interior takes a greater operational role in land remote sensing, the same mechanism can provide important guidance for that agency as well.

Notes

1. Merton E. Davies and W. R. Harris, *RAND's Role in the Evolution of Balloon and Satellite Observation Systems and Related U.S. Space Technology* (Santa Monica, CA: RAND Corporation, 1988).
2. Louis Ridenour, "Significance of a Satellite Vehicle," in F. H. Clauser, ed., *Preliminary Design of an Experimental World-Circling Spaceship* (Los Angeles, CA: Douglas Aircraft Engineering Division, 1946, Report SM-11827).
3. S. M. Greenfield and W. W. Kellog, *Inquiry into the Feasibility of Weather Reconnaissance from a Satellite Vehicle* (Santa Monica, CA: RAND Corporation, 1960, Report R-365).
4. Clauser, ed., *Preliminary Design of an Experimental World-Circling Spaceship.*
5. National Research Council, *Strategy for Earth Explorers in Global Earth Sciences* (Washington, DC: National Academy Press, 1988).
6. Herman Noordung, *The Problem of Space Travel: The Rocket Motor* [1929, Ljubljana, Slovenia: Cultural Centre of European Pace Technologies (KSEVT), 2011 rep. ed.]; Arthur C. Clarke, "Extra-Terrestrial Relays," *Wireless World*, October 1945, pp. 305–308.

7. David J. Whalen, *The Origins of Satellite Communications, 1945–1965* (Washington, DC: Smithsonian Institution Press, 2002).
8. P. D. Lowman Jr., "T plus Twenty-Five Years: A Defense of the Apollo Program," *Journal of the British Interplanetary Society* 49 (February 1996): 71–79.
9. B. H. Mollberg and B. B. Schardt, *Mission Report on the Orbiter Camera Payload System (OCPS) Large Format Camera (LFC) and Attitude Reference System (ARS)* (Houston, TX: Johnson Space Center, 1988).
10. For a comprehensive discussion of the CORONA system and its contributions, see Robert A. McDonald, "CORONA: Success for Space Reconnaissance, A Look into the Cold War, and a Revolution for Intelligence," *Photographic Engineering and Remote Sensing* 61 (June 1995): 689–720; Robert A. McDonald, *Corona between the Sun and the Earth: The First NRO Reconnaissance Eye in Space* (Bethesda, MD: ASPRS Publications, 1997); Dwayne A. Day, John M. Logsdon, and Brian Latell, eds., *Eye in the Sky: The Story of the Corona Spy Satellite* (Washington, DC: Smithsonian Institution Press, 1998); Curtis Peebles, *The Corona Project: America's First Spy Satellites* (Annapolis, MD: Naval Institute Press, 1997).
11. Greenfield and Kellog, *Inquiry into the Feasibility of Weather Reconnaissance.*
12. NASA, "Initial Scientific Assessment of the EOS Data and Information System (EOSDIS)," in Science Advisory Panel for EOSDIS, 1989.
13. National Research Council, *Panel to Review EOSDIS Plans, Final Report* (Washington, DC: National Academy Press, 1994).
14. Department of Interior News Release Office of the Secretary, "Earth's Resources to be Studied from Space," September 21,1966.
15. D. S. Johnson, chairman, Satellite Task Force, "Planning for a Civil Operational Land Remote Sensing Satellite System: A Discussion of Issues and Options" (Washington, DC: Department of Commerce, National Oceanic and Atmospheric Administration, 1980).
16. D. P. Radzanowski and S. J. Garber, "An Overview of NASA's Mission to Planet Earth," Congressional Research Service, Report 95–312 SPR, Washington, DC, 1995.
17. J. H. McElory and R. A. Williamson, "The Evolution of Earth Science Research from Space: NASA's Earth Observing System," in John M. Logsdon, gen. ed., *Exploring the Unknown: Selected Documents in the History of the U.S. Civil Space Program, Volume 6, Space and Earth Science* (Washington, DC: NASA Special Publication 4407, 2004).
18. C. Mathews, "ERTS-1: Teaching Us a New Way to See," *Astronautics and Aeronautics* 11 (1972): 3–40.
19. McElory and Williamson, "Evolution of Earth Science Research from Space."
20. Mathews, "ERTS-1."
21. R. Goody, Chairman, *Global Change: Impacts on Habitability, A Scientific Basis for Assessment* (Woods Hole, MA: NASA, Workshop Executive Committee, 1982).
22. B. I. Edelson, Associate Administrator for Space Science and Applications, "Global Habitability," 1983.
23. H. Friedman, chairman, Commission on Physical Sciences, Mathematics, and Resources, *Toward an International Geosphere-Biosphere Program: A Study of Global Change* (Washington, DC: National Research Council, 1983).

24. NASA, "Earth System Science Overview," Earth System Sciences Committee, NASA Advisory Council, 1986.
25. R. W. Higgins , A. Leetmaa, Y. Xue, and A. Barnston, "Dominant Factors Influencing the Seasonal Predictability of US Precipitation and Surface Air Temperature," *Journal of Climate* 13 (2000): 3994–4017.
26. IPCC, Fourth Assessment Report, Intergovernmental Panel on Climate Change, 2007.
27. Ibid.
28. W. A. Cramer, F. I. Bondeau, I. C. Woodward, R. A. Prentice, V. Betts, P. M. Brovkin, V. Cox, J. A. Fisher, A. D. Foley, C. Friend, M. R. Kucharik, N. Lomas, S. Ramankutty, B. Sitch, A. Smith, A. White, and C. Young-Molling, "Global Response of Terrestrial Ecosystem Structure and Function to CO 2 and Climate Change: Results from Six Dynamic Global Vegetation Models," *Global Change Biology* 7 (2001).
29. PCC, Fourth Assessment Report, Intergovernmental Panel on Climate Change, 2007.
30. National Research Council, "Contributions of Land Remote Sensing for Decisions About Food Security and Human Health," Geographical Sciences Committee, Committee on the Earth System Science for Decisions About Human Welfare, 2007.
31. National Research Council, "Satellite Observations of the Earth's Environment: Accelerating the Transition of Research to Operations," Committee on NASA-NOAA Transition from Research to Operations, 2003.
32. National Research Council, "Utilization of Operational Environmental Satellite Data: Ensuring Readiness for 2010 and Beyond," Committee on Environmental Satellite Data Utilization, 2004.
33. NASA, NASA Earth Science Enterprise Statement of Data Management, 1999.
34. A. Lefsky , W. B. Cohen, G. G. Parker, and D. J. Harding, "Lidar Remote Sensing for Ecosystem Studies," *Bioscience* 52 (2002): 19–38.
35. SAIC, "Low-Cost Outer Planet Mission Definitions: Report to NASA Headquarters," Science Applications International Corporation, 1995.
36. E. Frieman, chairman, Earth Observing System Engineering Review Committee, "Report of the Earth Observing System Engineering Review Committee," NASA, 1991.
37. NASA, "Mission to Planet Earth Biennial Review," Office of Mission to Planet Earth, 1997.
38. Department of Commerce, "Establish a Single Civilian Operational Environmental Polar Satellite Program," Department of Commerce, National Performance Review, 1993.
39. DoD/NASA, "Management Plan for the Landsat Program," Department of Defense, National Aeronautics and Space Administration, March 10, 1992.

Chapter 10

Venus-Earth-Mars: Comparative Climatology and the Search for Life in the Solar System

Roger D. Launius

Both Venus and Mars have captured the human imagination during the twentieth century as possible abodes of life. Venus had long enchanted humans—all the more so after astronomers realized it was shrouded in a mysterious cloak of clouds permanently hiding the surface from view. It was also the closest planet to Earth, with nearly the same size and surface gravity. These attributes brought myriad speculations about the nature of Venus, its climate, and the possibility of life existing there in some form. Mars also harbored interest as a place where life had existed or might still exist. Seasonal changes on Mars were interpreted as due to the possible spread and retreat of ice caps and lichen-like vegetation.

A core element of this belief rested with the climatology of these two planets, as observed by astronomers, but these ideas were significantly altered, if not dashed during the space age. Missions to Venus and Mars revealed strikingly different worlds. The high temperatures and pressures found on Venus supported a "runaway greenhouse theory," and Mars harbored an apparently lifeless landscape similar to the surface of the Moon. While hopes for Venus as an abode of life curtailed without ending, the search for evidence of past life on Mars, possibly microbial, remains a central theme in space exploration.[1]

This essay explores the evolution of thinking about the climates of Venus and Mars as life-support systems, in comparison to Earth. At some level this belief rests on little more than high hopes and exceptionally thin air, at least on Mars, as well as thick air on Venus. It also suggests the central role of post–World War II scientific priorities, often dictated by national security concerns in the United States vis-à-vis the Soviet Union. The influence of Cold War institutions on postwar science, technology, and medicine was pervasive, and the search for life on Venus and Mars, while it had a legitimate scientific element, also engaged the attention of sponsors because of Cold

War priorities. Those involved in the pursuit, as Audra Wolfe has shown, helped to define "the American space program as one with 'scientifically valid' goals" and at the same time demonstrating American capabilities.[2]

Venerean Visions

Planetary exploration began in the early 1960s in a race between the United States and the Soviet Union to see who would be the first to place some sort of spacecraft near Venus. This was not just an opportunity to best the rival in the Cold War; scientists in both the United States and the Soviet Union recognized the attraction of Venus as a near twin to this planet in terms of size, mass, and gravitation. As Earth's "sister planet," a near twin, scientists and the public speculated about the nature of Venus and the possibility of life existing there in some form. Through much of the nineteenth century observers harbored hopes that Venus might be a place teeming with life. As R. A. Proctor wrote in 1870, because of its similarity to Earth "on the whole, the evidence we have points very strongly to Venus as the abode of living creatures not unlike the inhabitants of earth."[3]

In the latter part of the nineteenth century, however, a series of astronomical observations suggested Venus may be much less conducive to life similar to that seen on Earth than previously expected (figure 10.1). These observations found the evidence to ask whether or not Venus might perpetually face the

Figure 10.1 A family portrait showing (from left to right) Pioneers 6–9 (far left), 10, and 11 (second from left) and the Pioneer Venus Orbiter (third from left) and Multiprobe series (far left). These were the stalwarts of the missions to Venus and Mars until the 1990s. (Credit: NASA.)

Sun with one side, perhaps leaving one side too baked and the other too frozen for life.[4] Because of the thick clouds covering the planet's surface, however, the observations were tentative and many scientists refused to accept this idea, in no small part because of the desire to believe that life lived at the bottom of that dense cloud cover. R. G. Aitken, an astronomer at Lick Observatory, argued that if Venus did rotate similar to the Earth "we have reason to believe that it is habitable, for the conditions we named as essential to life—air, water in its liquid form and a moderate temperature—are undoubtedly realized." Absent that rotation, however, Aitken admitted that the possibility of life there "must be utterly desolate."[5]

Even so, and perhaps surprisingly, in the first half of the twentieth century a popular theory held that the Sun had gradually been cooling for millennia and that as it did so, each planet in the solar system had a turn as a haven for life of various types. Although it was now Earth's turn to harbor life, the theory suggested that Mars had once been habitable and that life on Venus was now just beginning to evolve. Beneath the clouds of the planet, the theory offered, was a warm, watery world and the possibility of aquatic and amphibious life. "It was reasoned that if the oceans of Venus still exist, then the Venusian clouds may be composed of water droplets," noted JPL researchers as late as 1963; "if Venus were covered by water, it was suggested that it might be inhabited by Venusian equivalents of Earth's Cambrian period of 500 million years ago, and the same steamy atmosphere could be a possibility."[6]

This theory was popularized by Svante Arrhenius, a Nobel Prize–winning chemist who took it to millions with popular lectures and publications. Arguing for a tropical environment of more than 100 degrees Fahrenheit Arrhenius posited a strikingly wet atmosphere on Venus, one conducive to the rise of aquatic and amphibian life. He wrote:

> We must therefore conclude that everything on Venus is dripping wet... A very great part of the surface of Venus is no doubt covered with swamps, corresponding to those on the Earth in which the coal deposits were formed... The constantly uniform climactic conditions which exist everywhere result in an entire absence of adaptation to changing exterior conditions. Only low forms of life are therefore represented, mostly no doubt belonging to the vegetable kingdom; and the organisms are nearly of the same kind all over the planet. The vegetative processes are greatly accelerated by the high temperatures.

Arrhenius speculated that more complex life forms might have evolved at the Venusian poles since the temperatures would not be quite as hot there, and with that "progress and culture... will gradually spread from the poles toward the equator."[7]

The director of Smithsonian observatory, Charles Greeley Abbot, took these ideas even further. He argued concerning Venus, a twin of Earth, in the 1920s that its "high reflecting power seems to show that Venus is largely covered by clouds indicative of abundant moisture, probably at almost identical temperatures to ours." He then concluded that it "appears lacking in no essential to habitability."[8] For Abbot, Venus was even more attractive an abode of life than Mars and he made the case in several studies that followed. He even speculated that Earth might make contact with Venereans, evincing his excitement at coming "into fluent communication by wireless with a race brought up completely separate, having their own systems of government, social usages, religions, and surrounded by vegetation and animals entirely unrelated to any here on earth. It would be a revelation far beyond the opening of Japan, or the discoveries of Egyptologists, or the adventures of travelers in the dark continent."[9] Of course, spectroscopic investigation, and the failure by this means to find oxygen in the Venerean cloud system, made many scientists question Venus as an abode of life.

Thus the debate over the climate of Venus portended a larger debate over the possibilities of life in the solar system. As historian Erik M. Conway has theorized, "The question of life, at least life like Earth's, is fundamentally a question of climate. Astrophysicists prior to the space age sought measurements of climatic factors in order to inform speculations about the presence or absence of Earth-like life."[10] Venus's atmosphere, the pressures it had, the presence or absence of atmospheric oxygen, water, and carbon dioxide fundamentally informed this debate. It led to a succession of planetary theories concerning Venus. Measurement of these climate characteristics constrained theoretical models of planetary evolution while also restraining some of the more exotic speculations about Martian and Venusian life.

By the 1930s the detection of carbon dioxide in its thick atmosphere forced scientists grudgingly to abandon the idea that Venus contained a carboniferous swamp. The scientists investigating Venus replaced the pre-Cambrian environment, as Carl Sagan noted in 1961, with "an arid planetary desert, overlain by clouds of dust from the wind-swept surface."[11] They continued to search for water vapor, but failed to find it. What scientists found was carbon dioxide, a lot of it; a layer of gas roughly equivalent to a two-mile-deep ocean at a pressure similar to that of Earth.[12] In 1939 astronomer Rupert Wildt postulated a "greenhouse effect" with temperatures far above what was present on Earth. As astronomer Ronald A. Schorn concluded, "By 1940 there was good reason to believe that conditions on Venus were harsh and life impossible."[13] Charles Greeley Abbot, for one, refused to change his perspective. He offered concerning Venus as late as 1946 that "the conditions may possibly be as favorable for life there as on our earth."[14]

Few scientists wanted to accept that conclusion, however, at least as yet. Astronomer/cosmographer/iconoclast Fred Hoyle tried to explain away the lack of water. According to Carl Sagan, "Hoyle explained the lack of water by assuming a great excess of hydrocarbons over water on primitive Venus, and subsequent oxidation of the hydrocarbons to carbon dioxide, until all the water was depleted. He suggested that the surface is now covered with the remainder of the hydrocarbons, and that the cloud layer is composed of smog."[15] In such an environment, the planet Venus might be a global petroleum field awaiting human exploitation. Some advocates even suggested mining it for energy.

Even so, as late as the middle part of the 1950s Donald H. Menzel and Fred L. Whipple found another explanation for the data that allowed the possibility, however narrow, that life might have emerged on Venus by replacing both the wind-swept desert and planetary oil field theories with speculation that Venus was covered with a seltzer ocean. By arguing for a global seltzer ocean, Menzel and Whipple suggested that the chemical processes of the transformation of carbon dioxide into silicates would be hampered. Stranger things had happened, but many other scientists refused to accept this theory; Otto Struve even called it unworthy of serious discussion.[16]

In fits of wishful thinking, a few others continued to claim that they saw signs of life through their telescopes. Soviet astronomer G. A. Tikhoff in 1955 emphasized:

> Now already we can say a few things about the vegetation of Venus. Owing to the high temperatures on this planet, the plants must reflect all the heat rays, of which those visible to the eye are the rays from red to green inclusive. This gives the plants a yellow hue. In addition, the plants must radiate red rays. With the yellow, this gives them an orange color. Our conclusions concerning the color of vegetation on Venus find certain confirmation in the observation...that in those pats of Venus where the Sun's rays possibly penetrate the clouds to be reflected by the planet's surface, there is a surplus of yellow and red rays.[17]

Subsequent measurements largely subverted the idea of Venus as a planet teeming with life even before the dawn of interplanetary travel.

So bleak did the situation appear by 1961, at the point when the first spacecraft were being dispatched to Venus, that even Carl Sagan thought it unlikely that the planet has ever harbored life. As he concluded:

> At such high temperatures, and in the absence of liquid water, it appears very unlikely that there are indigenous surface organisms at the present time. If life based upon carbon-hydrogen-oxygen-nitrogen chemistry ever developed in the early history of Venus, it must subsequently have evolved to sub-surface or atmospheric ecological niches. However, since, as has been mentioned, there can have been no appreciable periods of time when Venus had both extensive bodies of water and surface temperatures below the boiling point of water, it is unlikely that life ever arose on Venus.[18]

Sagan's conclusion squared with a Space Science Board's analysis completed near the same time.[19]

This would have been a difficult conclusion for Sagan to accept judging from a career dedicated to finding life beyond Earth, and even here he held out for the prospect of reengineering the planet for biological habitation. He concludes:

> Ideally, we can envisage the seeding of the upper Cytherean atmosphere with appropriate strains of Nostocaceae after exhaustive studies have been performed on the existing environment of Venus. As the carbon dioxide content

> of the atmosphere fails, the greenhouse effect is rendered less efficient and the surface temperature fails. After the atmospheric temperatures decline sufficiently, the decreasing rate of algal decomposition will reduce the water abundance slightly and permit the surface to cool below the boiling point of water...At somewhat lower temperatures, rain will reach the surface, and the Urey equilibrium will be initiated, further reducing the atmospheric content of carbon dioxide to terrestrial values. With a few centimeters of perceptible water in the air, surface temperatures somewhere near room temperature, a breathable atmosphere, and terrestrial microfiora awaiting the next ecological succession, Venus will have become a much less forbidding environment than it appears to be at present.[20]

Such a faith statement in the face of the wealth of countervailing evidence is breathtaking for scientists dedicated to the analysis of objective data.

After carrying out ground-based efforts in 1961 to view the planet using radar, which could "see" through the clouds, and learning among other things that Venus rotated in a retrograde motion opposite from the direction of orbital motion, both the Soviet Union and the United States began a race to the planet with robotic spacecraft to learn the truth about the planet and its prospects for life. The Soviets tried first, launching Venera 1 on February 12, 1961. Unlike lunar exploration, however, the Soviets did not win the race to Venus; their spacecraft broke down on the way. The United States claimed the first success in planetary exploration during the summer of 1962 when Mariner 1 and Mariner 2 were launched toward Venus. Although Mariner 1 was lost during a launch failure, Mariner 2 flew by Venus on December 14, 1962, at a distance of 21,641 miles. It probed the clouds, estimated planetary temperatures, measured the charged particle environment, and looked for a magnetic field similar to Earth's magnetosphere (but found none). Most important, it confirmed that the planet's surface was an inferno:

> Earth-based measurements of microwave emissions from Venus had indicated a temperature of about 600° F., but researchers did not—and could not—know whether the emissions came from the surface, from cloud layers in the atmosphere or from a dense ionosphere high overhead. The question was answered by a microwave radiometer aboard Mariner 2, which revealed "limb-darkening" (weaker emissions at the edge of the planet's disk than at the center). The conclusion was not only that the surface was the hot part, but that, at about 800° F., it was even hotter than the earth-based data had implied. An infrared radiometer, meanwhile, took temperatures high in the atmosphere, revealing, to the scientists' disappointment, no breaks in the clouds.

Certainly, such an environment made unlikely the theory that life—at least as humans understood it—had ever existed on Venus.[21]

Although the Soviet Union made several more attempts to reach Venus, only in 1965 was it successful in reaching the surface when Venera 3 crashed there without returning any scientific data. In 1967 the Soviets sent Venera 4, which successfully deployed a probe into the atmosphere and returned further information about the makeup of the planet's surface. In the same year the Americans sent Mariner 5 to Venus to investigate its atmosphere. Both spacecraft demonstrated that Venus was a very inhospitable place for life to exist. Collectively, these and subsequent planetary probes revealed that Venus was superheated because of the greenhouse effect of the cloud layer and that the pressure on the surface was about 90 atmospheres, far greater than even in the depths of the oceans on Earth. Add to this the observations of James Pollack and others using aircraft-based near-infrared spectroscopy in 1974 that found on Venus a cloud sheet made predominantly of sulfuric acid and the possibilities of life on the planet appeared as remote as they had ever been.[22]

Even with what was already known about Venus, the planet remained a priority for exploration, and because of Venus's thick cloud cover, scientists early on advocated sending a probe with radar to map Venus. Pioneer 12 had made a start toward realizing this goal, orbiting the planet for more than a decade to complete a low-resolution radar topographic map. Likewise, the Soviets' Venera 15 and 16 missions in 1983 provided high-resolution coverage over the northern reaches of the planet. The best opportunity to learn the features of the Venerean surface came in the early 1990s with what turned out to be a highly successful Magellan mission to Venus. Launched in 1989, Magellan mapped 95 percent of the surface at high resolution, parts of it in stereo. This data provided some surprises; among them the discoveries that plate tectonics was at work on Venus and that lava flows showed clear evidence of volcanic activity. For five years Magellan yielded outstanding scientific results, showing volcanoes, faults, impact craters, and lava flows. It failed to deliver any data that suggested possibilities for life on the planet.[23]

Although one would think that evidence from the probes sent to Venus would be conclusive, overwhelmingly altering most of the beliefs held as recently as a generation ago about Venus as an abode of life, this was only partially true. For example, data from the Pioneer Venus probe suggested that in the distant past Venus had an ocean that may have existed for as long as a billion years, certainly enough to spawn primitive life. Planetary scientist Thomas M. Donahue, University of Michigan in Ann Arbor, reported that he and his team of researchers had found traces of water molecules in the atmosphere of Venus. As *Science News* reported in 1993:

> This chemical signature comes from the abundance of two atoms—hydrogen and its less abundant isotope deuterium, which has twice hydrogen's mass...The craft's early measurements revealed that this deuterium-to-hydrogen ratio is at least 150 times greater on Venus than in any other known place in the solar system...And since hydrogen readily bonds with oxygen to produce water, this suggests that Venus once had a minimum of 150 times as much water as it does now.

Donahue concluded, "The data indicate that Venus was a pretty wet planet." If this is proven correct, it may signal another shift in thought about Venus as a place where life might once have resided, even if conditions are no longer conducive to its survival.[24]

With the same evidence from Venus being interpreted in different ways the scientific community has long been divided into two large groups when considering prospects for life on Earth's "twin." Essentially, some insisted that large amounts of water vapor existed on Venus, as much as one hundred microns of perceptible vapor, while others insisted that results showed a lack of water. There has been no resolution of the debate, although the preponderance of evidence is in favor of a lifeless Venus. But some hope remains, because the community fundamentally wants to hope, and the 1993 announcement about water vapor in the Venerean atmosphere is one shred that enabled some hopefuls to cling to this belief despite the plethora of countervailing evidence. For most, however, beliefs held about Venus as a tropical, proto-organic planet had proven a bust.

The Lure of the Red Planet

But what of Mars? No planet has been more consistently held up as a possible location for life than the red planet. As with Venus, there had long been speculation of intelligent life there, and astronomical observations had lent credence to the idea, at least in the public mind. For instance, Percival Lowell became interested in Mars during the latter part of the nineteenth century. Using personal funds and grants from other sources, he built what became the Lowell Observatory near Flagstaff, Arizona, to study the planets. His research led him to argue that Mars had once been a watery planet and that the topographical features known as canals had been built by intelligent beings. Over the course of the first 40 years of the twentieth century others used Lowell's observations of Mars as a foundation for their arguments. The idea of intelligent life on Mars stayed in the popular imagination for a long time, and it was only with the scientific data returned from probes to the planet since the beginning of the Space Age that this began to change.[25]

The United States had reached Mars by July 15, 1965, when Mariner 4 flew within 6,118 miles of the planet taking 21 close-up pictures. These photographs dashed the hopes of many that life might be present on Mars, for these first close-up images of Mars showed a cratered, lunar-like surface. They depicted a Mars without artificial structures and canals, nothing that even remotely resembled a pattern that intelligent life might produce. *U.S. News and World Report* announced that "Mars is dead."[26] Even President Lyndon Johnson pronounced that "life as we know it with its humanity is more unique than many have thought."[27] Since the vast distance to the nearest star precluded visits to another solar system in the foreseeable future, life scientists gave up their hope of direct contact with other intelligent life on Mars and focused on the dual search for evidence that it had once existed there and that non-intelligent or microbial life existed there.[28]

Mariners 6 and 7, launched in February and March 1969, each passed Mars in August 1969, studying its atmosphere and surface to lay the groundwork for an eventual landing on the planet. Their pictures verified the Moon-like appearance of Mars, but they also found that volcanoes had once been active on the planet, that the frost observed seasonally on the poles was made of carbon dioxide, and that huge plates indicated considerable tectonic activity in the planet's past. Mariner 9 entered Martian orbit in November 1971, and its pictures showed the remains of giant extinct volcanoes dwarfing anything on the Earth. Later pictures showed a canyon, "Valles Marineris," 2,500 miles long and 3.5 miles deep, and meandering "rivers" appeared indicating that at some time in the past fluid had flowed on Mars. Suddenly, Mars fascinated scientists, reporters, and the public once again, largely because of the possibility of past life that might have emerged because of the evidence of flowing water.[29]

To find the answer about the Martian past, and perhaps its present as well, NASA developed what became Project Viking, a soft landing mission to Mars, but it also included what turned out to be two significant Mars orbiters that mapped the surface.[30] Very clearly, the search for signs of life prompted this emphasis on the exploration of Mars. NASA administrator James C. Fletcher, for example, supported the Viking mission because of his belief that life was present in the universe and that greater knowledge of this might probably be found on Mars:

> Although the discoveries we shall make on our neighboring worlds will revolutionize our knowledge of the Universe, and probably transform human society, it is unlikely that we will find intelligent life on the other planets of our Sun. Yet, it is likely we would find it among the stars of the galaxy, and that is reason enough to initiate the quest...We should begin to listen to other civilizations in the galaxy. It must be full of voices, calling from star to star in a myriad of tongues. Though we are separate from this cosmic conversation by light years, we can certainly listen ten million times further than we can travel.

Fletcher supported the Viking Mars lander in part because of its biological research on the red planet.[31]

The Viking mission consisted of two identical spacecraft, each consisting of a lander and an orbiter. Launched in 1975 from the Kennedy Space Center, Florida, Viking 1 spent nearly a year cruising to Mars, placed an orbiter in operation around the planet, and landed on July 20, 1976, on the Chryse Planitia (Golden Plains), with Viking 2 following in September 1976. While one of the most important scientific activities of this project involved an attempt to determine whether there was life on Mars, the scientific data returned mitigated against the possibility. The two landers continuously monitored weather at the landing sites and found both exciting cyclical variations and an exceptionally harsh climate that mitigated the possibility of life. Atmospheric temperatures at the more southern Viking 1 landing site, for instance, were only as high as +7 degrees Fahrenheit at midday, but the predawn summer temperature was -107 degrees Fahrenheit. And the lowest predawn temperature was -184 degrees Fahrenheit, about the frost point of carbon dioxide.[32]

Although the three biology experiments discovered unexpected and enigmatic chemical activity in the Martian soil, they provided no clear evidence for the presence of living microorganisms in soil near the landing sites. According to mission biologists, Mars was self-sterilizing. They concluded that the combination of solar ultraviolet radiation that saturates the surface, the extreme dryness of the soil, and the oxidizing nature of the soil chemistry had prevented the formation of living organisms in the Martian soil.[33] The uncertainty of the conclusions from Viking haunted the program's chief scientist Gerald Soffen ever after. He was known to second guess his judgment; perhaps he should have installed a microscope on the lander. But, he also believed he did the best he could. "I think what we did was ahead of our time. We were young enough not to know that it couldn't be done," Soffen recalled.[34]

The failure to find evidence of life on these two planets in the solar system devastated the optimism present for astrobiology in the era of great expectations. Collectively, these missions led to the development of two essential reactions. The first was abandonment by most scientists that life might exist elsewhere in the solar system, but that did not mean that it was not present throughout the universe. JPL director Bruce Murray believed that the legacy of failure to detect life, despite the billions spent and a succession of overoptimistic statements, would spark public disappointment and perhaps a public outrage.[35] Murray was right. The immediate result was that NASA did not return to Mars for two decades. As Soffen commented in 1992: "If somebody back then had given me 100 to 1 odds that we wouldn't go back to Mars for 17 years, I would've said, 'You're crazy.'"[36]

The second reaction was never accepted by the majority of scientists, but it found a powerful public and it also accepted that extraterrestrial life did exist. Essentially, this position argued that a powerful and corrupt Federal government, and its mandarins of science, had found evidence of life beyond Earth but was keeping it from the public for reasons ranging from stupidity to diabolical plots to enslave the planet.[37] This theme really fed into the second era of pessimism and distrust of the outcome of the search for life.

At least by the Viking landings in 1976 it began to be seen that the prospects for discovering extraterrestrial life on Mars had been oversold. Planetary scientist and JPL director Bruce Murray complained at the time of Viking about the lander being ballyhooed as a definite means of ascertaining whether or not life existed on Mars. The public expected to find it, and probably so did many of the scientists, and what would happen when hopes were dashed? Murray argued that "the extraordinarily hostile environment revealed by the Mariner flybys made life there so unlikely that public expectations should not be raised." Carl Sagan (figure 10.2), who fully expected to find something there, accused Murray of pessimism. Murray accused Sagan of far too much optimism. And the two publicly jousted over how to treat the Viking mission. Murray, as well as other politically savvy scientists and public intellectuals, argued that the legacy of failure to detect life, despite billions spent on research since the beginning of the Space Age and overoptimistic

Figure 10.2 Carl Sagan with the Viking lander mock-up in Death Valley, California, on October 26, 1980. (Credit: Jet Propulsion Laboratory.)

statements that a breakthrough was just around the corner, would spark public disappointment and perhaps an outrage manifested in reduced public funding for the effort.[38]

Such activities tended to oversell the possibilities for extraterrestrial life and its discovery. The result was overarching disinterest, and in some cases distrust of those making the claims, on the part of the public. "Something wonderful is about to happen," says the voice of Dave Bowman in the motion picture *2010: Odyssey Two*, and like Bowman scientists interested in this subject said it far too often and with too much certitude. Political scientist Howard E. McCurdy makes the case in space policy that there must be a close linkage between the belief system of the public, the possibilities claimed, and the returns on public investment. With the issue of extraterrestrial life, the public expected to find something. This expectation was fueled by enthusiasm for discovery offered by scientists and NASA officials. Failure to deliver on those expectations, fueled as they had been by official pronouncements, fostered disillusionment and ultimately distrust.[39]

Couple this with an overarching suspicion of government officials that began to emerge in the late 1960s and early 1970s and the result was genuine difficulty for the effort to discover extraterrestrial life. For example, in broader terms, public confidence in the capability of government declined after 1964. Responding to a public opinion survey in 1964, 76 percent of the Americans polled expressed confidence in the ability of government "to do what is right" most or all of the time. This was an all-time high in the history of polling, and it allowed the government to charge ahead with large initiatives during the 1960s, including all types of reforms. This consensus collapsed in the post-Vietnam and post-Watergate era of the 1970s, to a low of less than 25 percent of Americans believing that the government would seek to do right all or even a majority of the time by the early 1990s.[40]

For space exploration, too often the culture of the "X-Files" had replaced the culture of "Star Trek" as the dominant one by the 1990s. With the sense of conspiracy and evil manifest in such a transition, there has been a decided belief by a segment of the American population that NASA is not fully truthful about what it has found concerning extraterrestrial life. While there is a whole range of issues that could be explored here, nothing points this up more effectively than an incident that has been in the news of late, the so-called face on Mars. In 1976 Viking spacecraft photographed the Cydonia region of Mars from orbit and returned a single photograph showing what appeared to be a sculpted image of a human face. No one quite knew what to do with this admittedly poor-quality image. Was it a sunlight-induced optical illusion of a natural geological formation, or an artificial structure built on the surface by some perhaps long-extinct civilization? The reality was that no one really knew. Investigations into it yielded results that seemed to reflect more preconceived notions than most scientific findings and certainly more smoke than light on both sides of the debate.

Those who believed that the face was artificial explained the fact that it did not receive official status as such by means of a conspiracy on the part of those in power, especially those at NASA with vested interests in the status quo.[41] One of these investigators, Stanley V. McDaniel, summarized the controversy up in this way:

> Possibly the best place to start is the virtual censorship that has been imposed on the publication of serious commentary on the subject in the United States. This began long ago, when the first paper on the subject...was inexplicably expunged from the published papers of the first Case For Mars Conference in 1984. Subsequent attempts to publish papers on the topic, by scientists with impeccable credentials and a long list of published scientific papers, were uniformly refused consideration by the primary American journals of planetary science. These scientists were forced by this censorship to turn to publishing their work in books for the general public, whereupon NASA characterized them as seeking personal gain and running "cottage industries."
>
> Over the course of time, as individual citizens, having read such publications, began to ask questions of NASA, a long string of spurious arguments were put forward against the idea that the Face on Mars might be artificial. The services of that powerful propagandist, Carl Sagan, were evidently engaged in this task...Then, in 1985, Sagan published an article in *Parade* Magazine debunking the Face, characterizing anyone who took it seriously as a kind of a "zealot," and including a doctored version of one of the Viking Frames that used false color to make it look as though the Face is actually not there...
>
> This incessant stream of negative propaganda put out by NASA has had a damping effect on the American Press.[42]

A conspiracy has defeated the effort thus far, McDaniel charged, not evidence and not plausibility. But the really interesting question is, and there is no reliable evidence concerning this, what has been the effect of such controversies on the public's willingness to support research into extraterrestrial life?

Gerald Soffen, who had led the science team on the Viking program, had to respond to these charges repeatedly throughout the remainder of his life. This issue first arose on July 25, 1976, when the Viking 1 orbiter took the image of the Cydonia region of Mars that looked like a human face. All evidence suggests that this was the result of shadows on the hills, and Gerry Soffen said so at a press conference, but some refused to accept this position. The "face" remains a sore point to the present, with Soffen being asked about it many times over the years. Always, he stated it was not the remnant of some ancient civilization but was a natural feature lit oddly in this one image but not in any others. As NASA stated officially in 2001,

> The "Face on Mars" has since become a pop icon. It has starred in a Hollywood film, appeared in books, magazines, radio talk shows—even haunted grocery store checkout lines for 25 years! Some people think the Face is *bona fide* evidence of life on Mars—evidence that NASA would rather hide, say conspiracy theorists. Meanwhile, defenders of the NASA budget wish there *was* an ancient civilization on Mars.[43]

Equally important, no spacecraft went to Mars for more than 20 years after Viking, adding fuel to the contention that NASA did not want to learn the truth. Not until 1988 did the Soviet Union, just a year away from collapse and the end of the Cold War, sent Phobos 1 and 2 to Mars, but both failed en route to Mars. The Mars Observer launched by the United States on September 25, 1992, fared little better. Intended to provide the most detailed data available about Mars as it orbited the planet since what had been collected by the Viking probes of the mid-1970s, the mission was progressing smoothly until August 21, 1993, three days before the spacecraft's capture in orbit around Mars. Suddenly and without warning, controllers lost contact with it. The engineering team working on the project at the Jet Propulsion Laboratory responded with a series of commands to turn on the spacecraft's transmitter and to point the spacecraft's antennas toward Earth. No signal came from the spacecraft, however, and the Mars Observer was not heard from again. The loss of the nearly $1 billion Mars Observer probably came as a result of an explosion in the propulsion system's tanks as they were pressurized.[44]

One wit offered an alternative explanation, suggesting that after the landing by the Vikings in 1976 the Martians had developed a planetary defense system and it was now knocking out everything aimed at the red planet. Some conspiracy theorists might have agreed, but many thought that NASA had made contact with extraterrestrials and had concocted the loss of Mars Observer as a cover story. Richard C. Hoagland, for instance, held a press conference alleging that "a 'rogue group in NASA' many have sabotaged the Mars Observer mission to suppress information about the Cydonia Plain structures."[45] Not until 1998, when the Mars Global Surveyor photographed the Cydonia region again, were these charges laid to rest for all but the most stalwart conspiracy theorists as the images returned showed what appeared to be natural features.

Even though all of the elements of disillusionment remained present in the mid-1990s, those who had created heightened expectations seem to have learned from their errors and conspiracy theorists seem to have been increasingly discredited in popular conception. But most important, activities in several related fields of astrobiology seem to be offering renewed hope for discovering extraterrestrial life. Earthly studies have helped to show additional possibilities for extraterrestrial life, as marine biologists have discovered life—even though very far removed from what we have known before—in the depths of the ocean and under polar caps and in strata previously thought uninhabitable. Might creatures of such type be able to survive in what we have usually thought of as barren planets such as Mars and Venus?

A change to the beliefs in life on Mars took place in August 1996 when a team of NASA and Stanford University scientists announced that a Mars meteorite found in Antarctica contained possible evidence of ancient Martian life. When the 4.2-pound, potato-sized rock, identified as ALH84001, formed as an igneous rock about 4.5 billion years ago, Mars was much warmer and probably contained oceans hospitable to life. Then, about 15 million years ago, a large asteroid hit the red planet and jettisoned the rock into space where it remained until it crashed into Antarctica about 11,000 BC. The scientists presented three intriguing, but not conclusive, pieces of evidence that suggest that fossil-like remains of Martian microorganisms, which date back 3.6 billion years, are present in ALH84001. While there is no consensus on the truth of these findings, they did lead to added support for an aggressive set of missions to Mars to help discover the truth. Even as scientists modestly called for more research, the findings electrified the public and set in motion popular support for an aggressive set of missions to Mars by the year 2010 to help discover the truth of these theories.[46]

Thereafter the strategy for much of Mars exploration has been built upon the motto "follow the water." In essence, this approach noted that life on Earth is built upon liquid water and that any life elsewhere would probably have chemistries built upon these same elements. Accordingly, to search for life on Mars, past or present, NASA's strategy would be to follow the water. If scientists could find any liquid water on Mars, probably only deep beneath the surface, the potential for life to exist was also present.[47]

The Mars of today, without any evidence of water whatsoever on the surface, probably had water flowing freely in its ancient history. One scientist concluded as early as 1988:

> Even though the Martian epoch of liquid water was short, it apparently coincided with the period of Earth history when life originated. Gross similarities in the early geophysical history of the two planets hold open the possibility

> that life arose on Mars as well...(A discussion of the biological potential of Venus, whose early geophysical history bears resemblance to that of Mars, would lead to the same conclusion.) Therefore, the search for evidence of extinct life on Mars should be among the highest scientific priorities in future explorations of the planet.[48]

Evidence of changes to the planet's surface from fast-flowing water has been collected by many space probes orbiting the planet since the latter 1990s. The spacecraft to open this possibility was Mars Global Surveyor, reaching the planet in 1998 and a new and exciting era of scientific missions to study the red planet. Its recent discoveries offer titillating hints for learning about the possibility of life on Mars, at least in the distant past. In an exciting press conference in June 2000, astronomer Michael Malin discussed his analysis of imagery from Mars Global Surveyor, a stunningly successful NASA probe. He showed more than 150 geographic features all over Mars probably created by fast-flowing water. He suggested that there might actually be water in the substrata of Mars, and our experience on Earth has indicated that where water exists life as we understand it exists as well.[49]

Operating for several years, Mars Global Surveyor continued to send back views of the Martian surface that seemed to show evidence of dry riverbeds, flood plains, gullies on Martian cliffs and crater walls, and sedimentary deposits that suggested the presence of water flowing on the surface at some point in the history of Mars. This led scientists to theorize that billions of years ago, Earth and Mars might have been very similar places. Of course, Mars lost its water and the question of why that might have been the case has also motivated many Mars missions to the present. At that point, a consensus emerged that on any mission to Mars we should "follow the water" and seeking the answer to the ultimate question: "Are we alone in the universe?"

At present, most scientists believe the odds are almost nonexistent that complex life forms could have evolved on Mars because of its extremely hostile environment. The stories of "advanced civilizations," as proposed by Percival Lowell, or "little green men" are just that, stories. But many scientists believe there is sufficient evidence to think that microscopic organisms might once have evolved on the planet when it was much warmer and wetter billions of years ago. There are even a few scientists who would go somewhat further and theorize that perhaps some water is still present deep inside the planet. In that case simple life forms might still be living beneath Mars's polar caps or in subterranean hot springs warmed by vents from the Martian core. These might be Martian equivalents of single-celled microbes that dwell in Earth's bedrock. Scientists are quick to add, however, that these are unproven theories for which evidence has not yet been discovered.[50] This strategy of "follow the water" has dominated all planning for Mars science missions for more than a decade and results thus far have encouraged many to believe that the discovery of incontrovertible evidence supporting the contention that life will be found on Mars is just a matter of time.

The Desire to Belief and the Search for Life on Venus and Mars

When the twentieth century began, the idea of a solar system brimming with life had a foothold in the collective consciousness of the American people. Moreover, many in the scientific community accepted the "fact" of life on Venus and Mars. While a completely unproven assertion, and unprovable using the scientific tools available at the time, the belief persisted. It also enjoyed a rather variegated reputation as a province of cranks, philosophers, and almost anyone other than scientists. At the end of the twentieth century what has changed? The belief in extraterrestrial life is more firmly ensconced in the collective consciousness of the public and has gained legitimacy in the scientific world. There is still no proof acceptable to all, but there have been signs of possibility—notably the evidence from Mars—and the tools, methods, and scientific wherewithal to determine with some accuracy exists. In the more than 50 years since the beginning of the Space Age the best attempts of science to answer this age-old question have yielded contested results in the solar system, enticing possibilities for life among the stars, and the emergence of a cosmology that gives hope for a life-filled universe.

In the First Annual Carl Sagan Memorial Lecture in December 1997 before the American Astronautical Society Bruce Murray summarized many of the possibilities for the future of "the search for life elsewhere." He described that search as a "unifying idea" of what space science has before it, and he was guardedly optimistic. He offered a fivefold program for the future:

1. Complete the search for life in our solar system, especially beneath the surface of Mars and on Jupiter's moon Europa;
2. complete the search for preserved evidence of past life on the surface of Mars;
3. inventory planetary bodies orbiting other stars;
4. identify Earth-like planets around those stars, and search for spectral indications of life in their atmospheres; and
5. expand SETI to Earth orbit for submillimeter and shorter wavelength searches.

Humanity has come far in understanding about this issue since the beginning of the Space Age, Murray declared, and should be ready for exciting new efforts in the future.[51]

During this 40-year effort, the separate disciplines of Solar System Astrobiology, Origins and Evolution of Life, Planetary Systems, and SETI have merged into a powerful force for scientific research and study. Equally important, the necessary ingredients for any exploration—political will, economic stability and growth, enabling technologies, scientific sophistication necessary to conceptualize and incorporate new findings into the scientific paradigm, and public perception that these efforts are both achievable and desirable—have coalesced. At century's end a modest, optimistic, and rational approach seems to offer a bright future for a discipline and for the discovery of extraterrestrial life, perhaps even on Mars.

But one must ask why that may be the case. With a succession of inconclusive results at every point that space probes reached other worlds, especially Venus and Mars, why the persistence in the belief of life beyond Earth? As a truly unusual aspect of the studies of Venus and Mars, despite no clear evidence to support this belief after more than 50 years of scrutiny, a perception that life probably existed in the solar system at some point in the past, if not presently, remains a powerful ideal among many planetary scientists. In essence, they have followed a classic cognitive dissonance model, defined by Leon Festinger in his seminal 1956 book, *When Prophecy Fails.* Festinger asked the question, what happens when a prediction to which a social group subscribes fails completely and without ambiguity? What happens to all its faithful supporters? Reason would suggest that the members of the group would abandon the commitments that proved faulty, adjusting their perspectives to reflect mainstream ideals. But true believers do not automatically abandon their cause when reality intrudes in discomforting ways. They rarely admit that they were wrong or change their behavior, especially those who remain close to the original group. As Festinger found, groups holding firm beliefs, however disconfirmed they might be by fact, were nonetheless committed at considerable expense to maintain it. He wrote, "If more and more people can be persuaded that the system of belief is correct, then clearly it must after all be correct." Instead they increase their level of proselytizing, working hard to spread the underlying belief. They seek out new evidence to validate their old behavior and new explanations as to why initial expectations failed. Sometimes they even deny that their prophecies in fact did fail.[52]

Festinger's model works well with the scientific community using comparative climatology to postulate life on Venus and Mars. Most of them have made little attempt to deny their failure but they rationalized what had taken place and restructured their perspectives without abandoning them. They did not give up their faith in life on Mars and Venus, rather they changed their ideas on what it would consist of and the structure it would take. This is a fascinating development.

The case of Mars is the most straightforward. Numerous distinguished scientists argued for civilizations on Mars during the prespace age, in no small part because of the fabled canals advocated by Percival Lowell and others. As scientific instruments became more capable they proved that those canals were nonexistent, but scientists did not abandon their belief in life on the planet, suggesting as late as the 1950s that vegetation changed colors through planetary seasons on Mars. Gerard P. Kuiper, using a powerful ground-based telescope at the McDonald Observatory, claimed seeing a "touch of moss green" on Mars, sending fellow astronomers searching for a chlorophyll signature using spectroscopic analysis.[53] Some claimed to find it and for several years scientists reported that vegetation on the plant, probably lichens or some other type of plant life, changed with the seasons. These vivid colors finally proved illusory, and the greens and blues reported turned out to be visual tricks when neutral-toned areas were surrounded by yellow-orange fields.[54]

When missions such as Mariner 4 reached Mars in the mid-1960s and revealed a cratered, moon-like planet, scientists had to dial back their expectations yet again. But more did not abandon their beliefs. What about the potential for microbial life? The Viking landing mission was predicated on the belief that microbial life would be found in the Martian surface. Failure to discover any microbes proved devastating to the cause of life on Mars, but still some clung to the hope that they had looked in the wrong locations, or had designed the experiments in a way that did not yield useful results. But the idea did not completely die. By the 1990s scientists had begun to approach the issue of life on Mars in another way, modifying but not abandoning earlier conceptions of a solar system containing life beyond Earth. They admitted that liquid water on the surface of Mars would either freeze or evaporate almost immediately, and that the atmosphere was also almost waterless. Even so, they asserted that features seen from space looked like they had been carved by rivers and fast-flowing floods. The last decade of the twentieth century brought new possibilities as data from the ALH84001 meteorite and the mission of the Mars Pathfinder spacecraft showed the potential of past liquid water flowing freely on the planet. With water as the fundamental building block of life the search for life on Mars entered a new arena, a scaling back to past life now extinct on a dead world. Scientists may yet find it; certainly the evidence of past water on the planet is compelling. If fossils of prehistoric Martian creatures are found it will hold important—even profound—implications for humanity. But the important point is that with every disconfirming piece of evidence about the lack of life on the planet many scientists do not abandon hope; instead they modify the desire just enough to continue their hope for life beyond Earth.

Recent disconfirming evidence of life elsewhere led scientists to modify their expectations yet again, asking the simple question—so simple that one must ponder why it has not been asked before—about whether or not humanity would recognize extraterrestrial life were they to encounter it. As a 2007 National Academies study noted concerning the search for life in the solar system:

> The tacit assumption that alien life would utilize the same biochemical architecture as life on Earth does means that scientists have artificially limited the scope of their thinking as to where extraterrestrial life might be found, the report says. The assumption that life requires water, for example, has limited thinking about likely habitats on Mars to those places where liquid water is thought to be present or have once flowed, such as the deep subsurface. However, according to the committee, liquids such as ammonia or formamide could also work as biosolvents—liquids that dissolve substances within an organism—albeit through a different biochemistry.

John Baross, professor of oceanography at the University of Washington, commented on this shift: "The search so far has focused on Earth-like life because that's all we know, but life that may have originated elsewhere could be unrecognizable compared with life here. Advances throughout the last decade in biology and biochemistry show that the basic requirements for life might not be as concrete as we thought." Broadening the search might yield positive results, according to this report.[55]

Interestingly, the possibility that life might not be out there never really entered into the discussion. Mars and Venus revealed themselves to be far forbidding places than anticipated when the space age began. The Earth, by comparison, appears to be a far more unique and precious sphere. Few want to embrace this as a possibility, although a few scientists do. In their book *Rare Earth*, Peter Ward and Donald Brownlee suggest that complex life forms like those found on Earth require local conditions that rarely occur. Simple life forms might appear under a variety of circumstances, but complex life is extraordinarily atypical.[56] This would account for the apparent absence of evidence of extraterrestrial life anywhere in the solar system and beyond. But even Ward and Brownlee believe that microorganisms probably emerged in many environments beyond Earth; they have just not been discovered yet.

This begs the fundamental question, why do those engaged in the scientific pursuit of knowledge persist in a belief that life is present, or at least was once present, on planets such as Mars and perhaps Venus. Sometimes this is a straightforward belief predicated on nonscientific reasoning. For instance, James C. Fletcher, NASA administrator between 1971 and 1976 and again in 1986–1989, was enthusiastic about this search for life beyond and supported the Viking mission to Mars because of it. His Mormon faith explicitly asserted the existence of life on other planets. In a June 1830 revelation, Joseph Smith, Jr., the founding prophet of Mormonism, described an encounter between Moses and God in which Moses was told about the enormity and populousness of the universe. In this supposed revelation God told Moses:

> And worlds without number have I created; and I also created them for mine own purpose; and by the Son I created them, which is mine Only Begotten.
>
> And the first man of all men have I called Adam, which is many.
>
> But only an account of this Earth, and the inhabitants thereof, give I unto you. For behold there are many worlds which have passed away by the word of my power. And there are many also which now stand, and innumerable are they unto man; but all things are numbered unto me, for they are mine and I know them.[57]

Established early on as a part of the Mormon faith, therefore, was the idea of a plurality of worlds inhabited by other beings.

Fletcher was committed to the ideas expressed in Mormon religion and emphasized the belief that humans were not alone as intelligent beings in the universe. He was interested in the probability of finding other civilizations in space and commented on it repeatedly. For example: "It is hard to imagine anything more important then making contact with another intelligent race. It could be the most significant achievement of this millennium, perhaps the key to our survival as a species."[58] More recently Fletcher remarked that "intelligent life on other planets around other suns is a likely possibility" and he saw it as critical that NASA look for it. He added, "The public's misconception is that they don't realize how real a possibility such intelligence is." He calculated at one point that the universe probably has 5 billion worlds capable of supporting life as humans understand it, and that NASA had a responsibility to seek them out. On another occasion he expressed his belief that NASA should prosecute one of the "more adventuresome undertakings, such as communicating with other intelligent life in the universe and establishing colonies in space—perhaps before the end of this century."[59] Accordingly, while at NASA Fletcher supported as important agenda items efforts to search for life beyond Earth. For instance, he emphasized the astrobiological aspects of Viking.[60]

Most scientists, however, do not possess the well-understood religious convictions of a James Fletcher. What they do possess is a belief predicated on faith that one might characterize as a form of civil religion.[61] The search addresses deep-seated needs that strike to the confluence of the scientific pursuit of knowledge and the philosophical understanding of humanity's place in the universe. According to planetary scientist Michael Caplinger:

> The question of whether life is common or rare in the universe has deep philosophical implications. It is uncertain exactly how life arose on Earth, so it is difficult to determine how common such mechanisms are. But if life also arose on Mars, this would show that those mechanisms operated not just once, but twice, arguing that life may well be common elsewhere. However, the search for life on Mars thus far has been unsuccessful. Some portion of the scientific community feels that further searches are a waste of time, while another portion remains neutral or guardedly optimistic. In principle, it's simple to prove that there *is* life on Mars—all one need do is find an example. Proving there isn't life on Mars is much harder. Even a prolonged negative search can be countered with the suggestion of yet another, more inaccessible place in which to look. In the case of Mars, the issue has been complicated by the emotional belief in an Earthlike Mars, which has largely been shown to have been a myth. Mars is a spectacular place, and will remain so even if it is finally proved to be lifeless. Today, we don't know for sure if there is or ever was life on Mars. But one thing is certain—one day, there will be.[62]

This faith in positive results is repeated in many settings with similar expectations. Perhaps it is a part of the human condition. Perhaps it is somewhat like the tagline from the "X-Files," the 1990s television series concerning the search for extraterrestrial visitation of Earth, "I Want to Believe." We all collectively want to believe and are willingly expending millions of dollars each year seeking confirmation of that desire. Perhaps we will find someday, in another tagline from the "X-Files" that "The Truth is Out There." Meantime, we investigate and with every disconfirming piece of evidence modify the nature of the search in an act of cognitive dissonance as great as any chronicled in Leon Festinger's work.

Notes

1. Roger D. Launius, "'Not Too Wild a Dream': NASA and the Quest for Life in the Solar System," *Quest: The History of Spaceflight Quarterly* 6 (Fall 1998): 17–27.
2. Audra J. Wolfe, "Germs in Space: Joshua Lederberg, Exobiology, and the Public Imagination, 1958–1964," *Isis* 93 (June 2002): 183–205, quote on 185.
3. Richard A. Proctor, *Other Worlds Than Ours: The Plurality of Worlds Studied Under the Light of Recent Scientific Researches* (New York: J.A. Hill and Co., 1870), p. 94; Dale P. Cruikshank, "The Development of Venus Studies," in D. M. Hunten, L. Colin, T. M. Donahue, and V. I. Moroz, eds., *Venus* (Tucson: University of Arizona Press, 1983), pp. 1–9.
4. J. L. E. Dreyer, "Schiaparelli's Researches on the Rotation of Venus and Mercury," *Monthly Notices of the Royal Astronomical Society* 51 (February 1891): 246–49; Patrick Moore, *The Planet Venus* (New York: Macmillan, 1960), pp. 80–89.
5. R. G. Aitken, "Life on Other Worlds," *Journal of the Royal Astronomical Society of Canada* 5 (September–October 1911): 291–308, especially 300–303.
6. Jet Propulsion Laboratory, *Mariner: Mission to Venus* (New York: McGraw-Hill, 1963), p. 5. This became an enormously popular conception in science fiction literature. See the 1949 short story by Arthur H. Clark, "History Lesson," in *Expedition to Earth* (New York: Ballantine Books, 1953), pp. 73–82, for an explanation of the theory.
7. Svante Arrhenius, *The Destinies of the Stars* (New York: G.P. Putnam's Sons, 1918), pp. 250–53.
8. Charles Greeley Abbot, "The Habitability of Venus, Mars, and Other Worlds," *Annual Report of the Board of Regents of the Smithsonian Institution...for 1920* (Washington, DC: Government Printing Office, 1922), pp. 165–71, quote from p. 170.
9. Charles Greeley Abbot, *The Earth and the Stars* (New York: D. Van Nostrand, 1946), pp. 74–75.
10. Communication to author, October 13, 2008, copy in possession of author.
11. Carl Sagan, "The Planet Venus," *Science*, New Series, 133 (March 24, 1961): 849–58, quote from 849.
12. Charles E. St. John and Seth B. Nicholson, "The Absence of Oxygen and Water-Vapor Lines in the Spectrum of Venus," *The Astrophysical Journal* 56 (1922): 380; W. S. Adams and T. Dunham Jr., "Absorption Bands in the Infra-Red Spectrum of Venus," *Publications of the Astronomical Society of the Pacific* 44 (1932): 243; Harold C. Urey, *The Planets: Their Origin and Development* (New Haven: Yale University Press, 1952).
13. Ronald A. Schorn, *Planetary Astronomy: From Ancient Times to the Third Millennium* (College Station: Texas A&M University Press, 1998), p. 120.
14. Abbot, *Earth and Stars*, p. 109.

15. Sagan, "The Planet Venus," 849.
16. Donald H. Menzel and Fred L. Whipple, "The Case for H2O Clouds on Venus," *Publications of the Astronomical Society of the Pacific* 67 (June 1955): 161–68; Steven J. Dick, *The Biological Universe: The Twentieth-Century Extraterrestrial Life Debate and the Limits of Science* (New York: Cambridge University Press, 1996), pp. 133–34.
17. Quoted in David Harry Grinspoon, *Venus Revealed: A New Look Below the Clouds of our Mysterious Twin Planet* (Reading, MA: Addison-Wesley, 1996), p. 51.
18. Sagan, "The Planet Venus," 857.
19. Space Science Board, Panel on Planetary Atmospheres, *The Atmospheres of Mars and Venus* (Washington, DC: National Academy of Sciences Publication 944, 1961), pp. 33, 37–50.
20. Sagan, "The Planet Venus," 857.
21. Jonathan Eberhart, "The Hard Ride of Mariner 2," *Science News* 122 (December 11, 1982): 382–83.
22. Ladislav E. Roth and Stephen D. Wall, *The Face of Venus: The Magellan Radar-Mapping Mission* (Washington, DC: NASA SP-520, 1995), pp. 1–9; Robert Reeves, *The Superpower Space Race: An Explosive Rivalry through the Solar System* (New York: Plenum Press, 1994), pp. 195–278; Richard O. Fimmel, Lawrence Colin, and Eric Burgess, *Pioneer Venus* (Washington, DC: NASA SP-461, 1983); R. Stephen Saunders and Michael H. Carr, "Venus," in Michael H. Carr, ed., *The Geology of the Terrestrial Planets* (Washington, DC: NASA SP-469, 1984), pp. 57–77; James Pollack, E. Erickson, F. Witteborn, C. Chackerian, A. Summers, G. Aguason, and L. Caroff, "Aircraft Observations of Venus' Near-Infrared Reflection Spectrum: Implications for Cloud Composition," *Icarus* 23 (1974): 8–26; "A Determination of the Composition of the Venus Clouds from Aircraft Observations in the Near Infrared," *Journal of the Atmospheric Sciences* 32 (1975): 376–90.
23. Space Science Board, *Venus: A Strategy for Exploration* (Washington, DC: National Academy of Sciences, 1970); Roth and Wall, *The Face of Venus*; Grinspoon, *Venus Revealed*; Carolynn Young, ed., *The Magellan Venus Explorer's Guide* (Pasadena, CA: Jet Propulsion Laboratory, 1990); Peter Cattermole and Patrick Moore, *Atlas of Venus* (New York: Cambridge University Press, 1997); Earle K. Huckins III, Charles Elachi, and Dan V. Woods, "Exploring the Solar System—A Current Overview," IAF-99-Q.2.01, paper presented at the 50th International Astronautical Congress, October 4–8, 1999, Amsterdam, The Netherlands; S. B. Calcutt and F. W. Taylor, "The Deep Atmosphere of Venus," *Philosophical Transactions: Physical Sciences and Engineering* 349 (November 15, 1994): 273–83.
24. R. Cowen, "New Evidence of Ancient Sea on Venus," *Science News* 143 (April 3, 1993): 212.
25. See Percival Lowell, *Mars* (Boston: Houghton Mifflin Co., 1895), pp. 201–12; William Sheehan and Richard McKim, "The Myth of Earth-based Martian Crater Sightings," *British Astronomical Association Journal* 104 (December 1994): 281–86; William Graves Hoyt, *Lowell and Mars* (Tucson: University of Arizona Press, 1976); Martin Caiden and Jay Barbree, *Destination Mars: In Art, Myth, and Science* (New York: Penguin Studio, 1997), pp. 83–95.
26. "An End to the Myths about Men on Mars," *U.S. News and World Report*, August 9, 1965, p. 4.

27. Lyndon B. Johnson, “Remarks Upon Viewing New Mariner 4 Pictures from Mars,” July 29, 1965, *Public Papers of the Presidents of the United States, 1965* (Washington, DC: Government Printing Office, 1965), p. 806.
28. Robert B. Leighton, Bruce C. Murray, Robert P. Sharp, J. Denton Allen, and Richard K. Sloan, “Mariner IV Photography of Mars: Initial Results,” *Science* 149 (August 6, 1965): 627–30.
29. See William K. Hartman and Odell Raper, *The New Mars: The Discoveries of Mariner 9* (Washington, DC: NASA SP–337, 1974); Carl Sagan and George Mullen, “Earth and Mars: Evolution of Atmospheres and Surface Temperatures,” *Science*, New Series, 177 (July 7, 1972): 52–56.
30. For a complete treatment of this subject, see Edward C. Ezell and Linda Neuman Ezell, *On Mars: Exploration of the Red Planet, 1958–1978* (Washington, DC: NASA SP-4212, 1984), pp. 83–120. The average percentage of the total NASA appropriation allotted for space science in 1959–1968 was 17.6 percent; it was 17 percent in 1969–1978. By 1996 it was only 10 percent.
31. James C. Fletcher, *NASA and the “Now” Syndrome* (Washington, DC: National Aeronautics and Space Administration, 1975), p. 7; Roger D. Launius, “A Western Mormon in Washington, D.C.: James C. Fletcher, NASA, and the Final Frontier,” *Pacific Historical Review* 64 (May 1995): 217–41.
32. G. E. Hunt, “A New Look to the Martian Atmosphere,” *Proceedings of the Royal Society of London. Series A, Mathematical and Physical Sciences* 341 (December 10, 1974): 317–30.
33. An entire issue on the “Scientific Results of Viking” appeared in the *Journal of Geophysical Research* 82 (1977): 3959–4680. See also Ezell and Ezell, *On Mars*, pp. 363–420; Reeves, *Superpower Space Race*, pp. 392–98.
34. Billy Cox, “Mars Pioneer Remembered,” *Florida Today* (Orlando), December 6, 2000; “From Mars to Earth: A Conversation with Gerald Soffen,” *Space World*, July 1986, pp. 17–18.
35. Bruce Murray, *Journey into Space* (New York: W.W. Norton and Co., 1989), p. 74.
36. Cox, “Mars Pioneer Remembered.”
37. The literature on this subject is both exhaustive and exhausting. Useful overviews of the issues may be found in Paul Devereux and Peter Brookesmith, *UFOs and Ufology: The First 50 Years* (New York: Facts on File, 1997); Carl Sagan and Thornton Page, eds., *UFO’s: A Scientific Debate* (Ithaca, NY: Cornell University Press, 1972); Jodi Dean, *Aliens in America: Conspiracy Cultures from Outerspace to Cyberspace* (Ithaca, NY: Cornell University Press, 1998); Thomas M. Disch, *The Dreams Our Stuff is Made Of: How Science Fiction Conquered the World* (New York: The Free Press, 1998); Curtis Peebles, *Watch the Skies! A Chronicle of the Flying Saucer Myth* (Washington, DC: Smithsonian Institution Press, 1994).
38. Murray, *Journey into Space*, pp. 61, 68–69, 74, 77.
39. Howard E. McCurdy, *Space and the American Imagination* (Washington, DC: Smithsonian Institution Press, 1997).
40. Paul R. Abramson, *Political Attitudes in America* (San Francisco: W. H. Freeman, 1983), p. 12. See also Seymour Martin Lipset and William Schneider, *The Confidence Gap* (New York: Free Press, 1983).
41. The classic statement of the conspiracy thesis about the “face on Mars” may be found in Richard C. Hoagland, *Heritage of Mars* (New York: North Atlantic Books, 1998).

42. Stanley V. McDaniel, "The Harpendon Lecture," unpublished commentary at Harpenden, England, on Saturday, September 27, 1997.
43. "Unmasking the Face on Mars," May 24, 2001, Science@NASA, available at http://science.nasa.gov/headlines/y2001/ast24may_1.htm (accessed October 16, 2008, 6:35:51 A.M.). See also James B. Garvin, "The Emerging Face of Mars: A Synthesis from Viking to Mars Global Surveyor," *Astrobiology* 1, 4 (2001): 513–21.
44. Reeves, *Superpower Space Race*, pp. 410–16.
45. "Mars Watchers See Extraterrestrial Cover-Up," *Los Angeles Times*, August 25, 1993, p. 6.
46. David H. Onkst, "Life on Mars and Europa? NASA Reveals Possible Evidence of Extraterrestrial Existence," *Space Times: Magazine of the American Astronautical Society* 35 (September–October 1996): 4–7.
47. Orlando Figuroa, NASA Mars program director, "Following the Water: The Mars Exploration Program," briefing available online at http://www.hq.nasa.gov/mars/presentations/FTW/index.html (accessed October 15, 2008, 3:57:27 P.M.)
48. S. Chang, "Planetary Environments and the Conditions of Life," *Philosophical Transactions of the Royal Society of London. Series A, Mathematical and Physical Sciences* 325 (July 29, 1988): 601–10, quote from 609.
49. Stephen M. Clifford, D. Crisp, D. A. Fisher, K. E. Herkenhoff, S. E. Smrekar, P. C. Thomas, D. D. Wynn-Williams, R. W. Zurek, J. R. Barnes, B. G. Bills, E. W. Blake, W. M. Calvin, J. M. Cameron, M. H. Carr, P. R. Christensen, B. C. Clark, G. D. Clow, J. A. Cutts, D. Dahl-Jensen, W. B. Durham, F. P. Fanale, J. D. Farmer, F. Forget, K. Gotto-Azuma, and H. J. Zwally, "The State and Future of Mars Polar Science & Exploration," *Icarus* 144 (April 2000): 210–42; James E. Graf, Richard W. Zurek, Howard J. Eisen, Benhan Jai, and M. D. Johnston, "The Mars Reconnaissance Orbiter Mission," IAC-05-A.3.3 paper, 2005; Michael C. Malin and Kenneth S. Edgett, "Evidence for Recent Groundwater Seepage and Surface Runoff on Mars," *Science* 288 (June 30, 2000): 2330–35.
50. NASA Press Release, "Opportunity Rover Finds Strong Evidence Meridiani Planum Was Wet," March 2, 2004, available online at http://marsrovers.jpl.nasa.gov/newsroom/pressreleases/20040302a.html (accessed October 15, 2008, 4:16:17 P.M.); N. J. Tosca, A. H. Knoll, and S. M. McLennan, "Water Activity and the Challenge for Life on Early Mars," *Science* 320 (2008): 1204; V. A. Krasnopolskya, J. P. Maillard, and T. C. Owen, "Detection of Methane in the Martian Atmosphere: Evidence for Life?" *Icarus* 172, 2 (2004): 537–47; NASA Press Release, "Mars Rover Spirit Unearths Surprise Evidence of Wetter Past," May 21, 2007, available online at http://www.nasa.gov/mission_pages/mer/mer-20070521.html (accessed October 15, 2008, 4:21:37 P.M.).
51. Bruce Murray, "The Search for Life Elsewhere," First Annual Carl Sagan Memorial Lecture before the American Astronautical Society, December 3, 1997, Pasadena, CA.
52. Leon Festinger, with Henry W. Riecken and Stanley Schachter, *When Prophecy Fails: A Social and Psychological Study* (Minneapolis: University of Minnesota Press, 1956), p. 28.
53. Gerald P. Kuiper, "Visual Observations of Mars, 1956," *Astrophysical Journal* 125 (1957): 307–17.

54. D. G. Rea, B. T. O'Leary, and W. M. Sinton, "The Origin of the 3.58- and 3.69-Micron Minima in the Infrared Spectra," *Science* 147 (1965): 1286–88; Dean B. McLaughlin, "Interpretation of Some Martian Features," *Publications of the Astronomical Society of the Pacific* 66 (1954): 161–70; L. D. Kaplan, G. Münch, and H. Spinrad, "An Analysis of the Spectrum of Mars," *Astrophysical Journal* 139 (1964): 1–15.
55. National Academies Press Release, "Life Elsewhere in Solar System Could Be Different From Life as We Know It," July 6, 2007, available online at http://www8.nationalacademies.org/onpinews/newsitem.aspx?RecordID=11919 (accessed October 16, 2008 9:03:56 A.M.).
56. Peter D. Ward and Donald Brownlee, *Rare Earth: Why Complex Life is Uncommon in the Universe* (New York: Copernicus, 2000).
57. *Pearl of Great Price* (Salt Lake City, UT: Church of Jesus Christ of Latter-day Saints, 1968 ed.), Moses 1:33–35; Robert Paul, "Joseph Smith and the Plurality of Worlds Idea," *Dialogue: A Journal of Mormon Thought* 19 (Spring 1986): 13–36; Robert Paul, *Science, Religion and Mormon Cosmology* (Urbana: University of Illinois Press, 1992).
58. Fletcher, *NASA and the "Now" Syndrome*, p. 7.
59. "Interview, James Fletcher," *Omni*, December 1987, p. 22; James C. Fletcher to Jonathan Eberhart, August 23, 1974; James C. Fletcher, "Space: 30 Years into the Future," *Acta Astronautica* 19 (1989): 855–57.
60. Fletcher, *NASA and the "Now" Syndrome*, p. 22; Ezell and Ezell, *On Mars*, pp. 1–4, 51–82, 235–36, 404–14; *Life Beyond Earth & the Mind of Man* (Washington, DC: Government Printing Office, 1973); Philip Morrison, John Billingham, and John Wolfe, *The Search for Extraterrestrial Intelligence: SETI* (Washington, DC: NASA SP-419, 1977); Fletcher, "Space," pp. 855–57.
61. This concept was first characterized in Robert N. Bellah, "Civil Religion in America," *Dædalus: Journal of the American Academy of Arts and Sciences* 96 (Winter 1967): 1–21. See also Robert N. Bellah, *Broken Covenant: American Civil Religion in a Time of Trial* (Chicago: University of Chicago Press, 1992).
62. Michael Caplinger, "Life on Mars," April 1995, available online at http://www.msss.com/http/ps/life/life.html (accessed October 16, 2008 9:35:09 A.M.); emphasis in the original.

Chapter 11

Missions to Mars: Reimagining the Red Planet in the Age of Spaceflight

Robert Markley

Mars has been the object of both popular fascination and scientific inquiry since the seventeenth century. Comparatively close to Earth and offering launch windows for spacecraft every 26 months, after the Moon the red planet has become the most visited and studied object in the solar system during the age of planetary exploration. Because the planet's surface has been reshaped by floods, volcanism, and glaciation, the history of Mars exploration has been entwined with changing perceptions of Earth and of humankind's place in the cosmos, and, more speculatively, with dreams of our species' potential to become a spacefaring civilization. Since the late 1990s with the arrival of the Mars Global Surveyor, NASA has maintained a continual robotic presence orbiting the planet or on its surface. Landers and orbiters have sent back terabytes of data that have revolutionized our understanding of the fourth planet; hundreds of scientific studies published since 1997 have fueled ongoing debates about the planet's past and present conditions, the possibility that it harbored (or still harbors) life, and its potential as a destination for human explorers and colonists. Given the wealth of data and the increasingly sophisticated technologies deployed to study Mars, writing about Martian exploration comes with the guarantee that whatever one publishes (this chapter included) will need to be updated almost as soon as it appears in print.[1] Nonetheless, the last half-century of missions to Mars reveals the extent to which the planet remains crucial to answering fundamental questions about the history of the solar system, the evolution of planetary atmospheres, and the origins of life.

Early Missions

Even before the launch of Sputnik in 1957, scientists were speculating about what it would take to send humans to Mars and what future astronauts might find. While President Kennedy made landing humans on the Moon a national priority, the new National Aeronautics and Space Administration in the early 1960s was brainstorming ambitious plans to reach Mars within a decade, justifying its efforts by emphasizing "the similarity of Mars to earth" and the "likelihood of finding life" on its surface.[2] Like the Apollo program, NASA's vision for Mars was motivated, in part, by competition with the Soviet Union: the Russians tried repeatedly and unsuccessfully to land spacecraft on the planet during the 1960s and 1970s. With the advantage of 20–20 hindsight, NASA's early plans seem more like science fiction, or the kind of visionary speculation that Werner von Braun offered in his futuristic *The Mars Project* (1955), rather than a sober assessment of the limitations of the mid-century technologies that would prove essential for spaceflight: rocketry, guidance systems, life support, and computer science.[3] The Mars Mariner program was conceived as an ambitious prelude to a heroic but overhyped strategy to colonize Mars before 2000.

NASA's plans to land on Mars in 1964 had to be scrapped because aerospace engineers, lacking basic design parameters, envisioned a craft "too ambitious for its time, representing too large a technological jump."[4] Redesigns and cost overruns forced NASA to scale back the Mariner project to two planetary flybys and an orbiter mission to map the planet's surface. Since launch windows occurred more than two years apart, problems in design, construction, and testing meant that mission delays tended to pile up, particularly in the mid and late 1960s as NASA struggled with tight budgets during the escalation of the Vietnam War.

In 1964, the United States and the Soviet Union launched a total of four spacecraft bound for Mars. The Russian probes were lost on their way to the planet, and NASA's Mariner 3 failed in Earth orbit. Its twin craft, Mariner 4, negotiated the difficult, eight-month journey, and passed within seventy-four hundred miles of Mars in July 1965 on what was effectively a brief reconnaissance mission. As the spacecraft neared its goal, Bruce Murray, then a mission geologist and later director of the Jet Propulsion Laboratory (1976–1982), wrote that "the expectation of an earthlike Mars was still very high."[5] The 19 grainy photographs that Mariner 4 returned, with each pixel covering three miles, took days to transmit and dramatically dashed those expectations. The photographs covered only shreds and patches of the planet's surface, but revealed a cratered surface that seemed more lunar than earthlike: no signs of life, no bodies of water, and no legendary Martian canals. Seeing no evidence of erosion, weathering, or volcanic activity in the photographs, scientists concluded that the planet's surface had changed little in the eons since its surface features were formed.[6] Mars looked far less hospitable and significantly less interesting than they had imagined.

Even as NASA's mission planning went forward, the rationale for exploring Mars was much less compelling after Mariner 4. The original Mariner '71 mission to land instruments on Mars was cancelled, and the Mariner '69 mission scaled back. The new Viking lander program also ran into budgetary and technical difficulties, and its launch was postponed to 1975. NASA settled for two sets of missions for launch windows in 1969 and 1971: another flyby to photograph part of the southern hemisphere and study its atmosphere, and an orbiter to map the planet and measure its temperature and ultraviolet radiation. Mariners 6 and 7 carried more sophisticated instruments than Mariner 4, and, in addition to returning 59 close encounter photographs that covered one-fifth of the planet's surface, sent back enough data about the planet's atmospheric and surface chemistry, to confirm the presence of ice.[7] But surface temperatures were so low that scientists concluded the polar caps were largely frozen carbon dioxide, and therefore Mars lacked the water necessary to sustain anything remotely like an earthlike ecology. Although both craft came within two thousand miles of Mars, they passed over the southern highlands and therefore missed photographing the massive volcanoes of the northern hemisphere and the three-thousand-mile-long Valles Marineris. These results reinforced the view of Mars that had emerged from Mariner 4. A NASA press release on September 11, 1969, described the planet in terms that seem reminiscent of what Neil Armstrong had found on the moon that July: a surface "heavily cratered, bleak, cold, dry, nearly airless and generally hostile to any Earth-style life forms."[8] Haunting this description are the blasted, pre-Mariner dreams of exploring an earthlike Mars.

Nonetheless, the search for life continued to drive NASA's plans to explore Mars. Mariners 8 and 9 were designed to orbit the planet for four months and photograph most of the surface so that NASA could identify sites for the Viking landers. Twice the size of Mariners 6 and 7, each craft included upgraded scientific instruments to study the planet's surface and atmosphere, and an improved all-digital camera.[9] Although Mariner 8 failed shortly after launch, Mariner 9 entered its orbit around Mars during a massive dust storm that obscured the planet's surface. A Russian lander, preprogrammed for its descent, disappeared into the dust clouds. When the dust storm finally died down early in 1972, Mariner 9 began returning spectacular photographs—seven thousand of them—that forced a wholesale reconsideration of what scientists thought they had known about Mars. The massive shield volcano Olympus Mons covered an area larger than Arizona, Valles Marineris dwarfed the Grand Canyon, and the layered polar caps revealed a complex record of periodic changes in the planet's climate. Studying a dynamic record of meteorite bombardments, volcanic eruptions, and massive floods, planetary astronomers recognized after Mariner 9 that Mars had a fascinating climatological as well as geological history. In this respect, this mission gave rise to new directions in comparative planetology as scientists started to piece together a partial mosaic of best-guess assumptions about Mars's history. One fundamental fact stood out: for water to flow across the Martian surface, the planet's atmosphere would have had to have been much thicker and the surface much warmer in the remote past. The volcanoes, outflow channels from massive floods, and teardrop-shaped islands characteristic of prolonged water flow offered striking evidence that comparisons to Earth, and not the Moon, provided the best framework for understanding the mechanisms that had shaped and reshaped the surface of Mars.

Mariner 9 revived a long-standing tradition of scientific analogies between Mars and Earth that had structured understanding of the planet since the eighteenth century.[10] The belief in the early 1970s that Mars might preserve a four-billion-year record of chemical, geological, and temperature change both reinforced and was reinforced by NASA's commitment to make the search for life a priority for the Viking mission. Although some geologists complained that the "hugely expensive search for Martian microbes" was "scientifically unjustified" and came at the expense of a basic understanding of surface chemistry and mineralogy, most of the media's attention was focused on the three biology packages that were designed to search for microorganisms and perhaps answer fundamental questions about the origin and evolution of life.[11] Operating without much of a geochemical context for its high-profile investment in life-detection, NASA hedged its bets by including three separate experiments to hunt for Martian microbes. The controversies that surrounded these experiments and their results had important consequences during and long after the mission.

From the start, the biology experiments faced design and conceptual problems. The pyrolytic release (PR), the labeled release (LR), and the gas exchange (GEX) experiments, as well as the gas chromatograph mass spectrometer (GCMS) designed to search for organic molecules, were the most sophisticated instruments ever sent into space. But for these instruments to work, biologists had to assume that the chemistry of alien life would be similar to what scientists knew about microorganisms on Earth: Martian organisms would have to be carbon-based, metabolize nutrients in a recognizable manner, and require water to survive. The principal investigators on the three experiments, Vince Oyama (GEX), Norman Horowitz (PR), and Gilbert Levin (LR), tailored their experiments to conform to what they knew or suspected about conditions on the surface of Mars. While all their experiments involved heating soil samples and then measuring chemical by-products to look for familiar signatures of metabolic processes, each test reflected a different view about how microscopic Martians would react to water, heat, light, and nutrients. The debates that emerged during the Viking mission, in this regard, reflected the three scientists' competing views about the water content, energy budget, and soil chemistry of Mars; in turn, these assumptions ultimately led to incommensurate interpretations of the data.

The GEX experiment measured the production or absorption of carbon dioxide, nitrogen, methane, hydrogen, and oxygen in either water or in a water-nutrient mix. The LR experiment was designed to detect metabolic activity in a soil sample moistened with a water-based solution of simple organic compounds. Micro-martians, Levin believed, would break down these compounds into carbon dioxide, and this breakdown could be detected by the release of gases from a mixture of radioactive compounds (carbon 14) supplied during the experiment.[12] He suggested that the cold and low atmospheric pressure on Mars meant that the tenuous air a few inches above the surface could retain enough water vapor to allow the relative humidity to reach 100 percent, and Martian organisms might have evolved to exploit this resource. Horowitz's PR experiment differed fundamentally from the other two. It added no nutrients or water to the soil sample because Horowitz wanted to preserve what he understood as the arid conditions of the planet's surface. He maintained, even while the months of testing were going on, that "the absence of liquid water on the Martian surface excludes any possibility of terrestrial types of organisms."[13] Unlike the GEX and LR experiments, PR looked for metabolic processes identified with terrestrial plant life. By heating the soil at high temperatures to "crack" organic compounds (pyrolysis), detectors would register whether any microorganisms had metabolized the radioactive carbon that they had been fed.[14] Horowitz's differences with Levin, then, existed long before their disputes about interpreting the Viking data. Given their radically different perceptions of the Martian environment, it would have been surprising if the two men did *not* disagree about the experimental data.

In addition to the life-detection experiments, the molecular analyses conducted by the GCMS experiment was designed to identify long chains of carbon atoms, characteristic of the complex organic compounds that, in turn, are needed to form the long chains of nucleotides. If microorganisms existed, their presence would be marked by an environment rich in organic compounds.[15] But because organic compounds had to be identified from the spectra of their components, the GCMS experiment depended on finding the by-products of millions of dead microorganisms or concentrations of living microbes of at least a million per gram of soil. GCMS, in short, was designed to detect life in a robust biosphere, and not comparatively sparse populations of alien extremophiles.

Both Viking 1 and 2 succeeded in reaching Mars, but the initial orbital photographs from the first craft revealed much rougher terrain at the favored landing sites than the NASA landing team had expected. After much debate, the eventual target areas for touchdowns on Chryse and Utopia Planitia were selected because they posed fewer hazards than the scientifically more interesting sites in the polar regions or ancient, water-carved channels.[16] Both craft landed safely—then, and in retrospect, a stunning engineering achievement. Viking 1's initial photographs from the surface of Mars on July 20, 1976, were humankind's first eye-level view of the surface of another planet. Mars became an identifiable place, often eerily earthlike, rather than the stuff of dreams (figure 11.1).

The life-detection experiments surprisingly—and repeatedly—produced strong positive responses. After several trials by test equipment on both landers, Levin suggested publicly that the experiments probably had found Martian microorganisms. But the intensity of the reactions led Oyama and Klaus Biemann, the leader of the molecular analysis team, to conclude that GEX was registering chemical reactions involving either peroxides or superoxides, unstable compounds that contain excess oxygen atoms that can be released through contact with water. For months, scientists struggled to explain the puzzling results. Additional trials of all three experiments registered readings that met the standards for biology that had been defined before the mission, but the GCMS experiment found no organic molecules. Biemann, Horowitz, and Oyama argued that these compounds had been

(a)

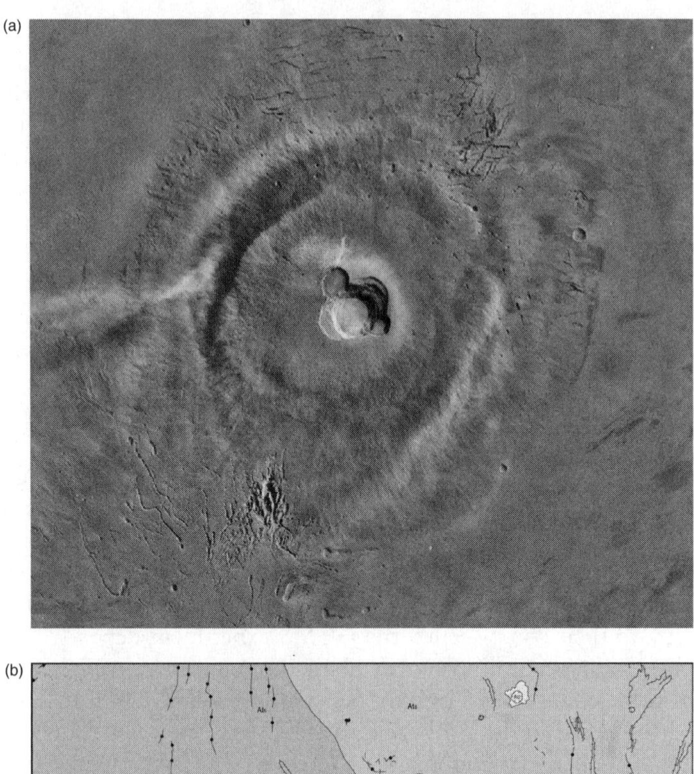

(b)

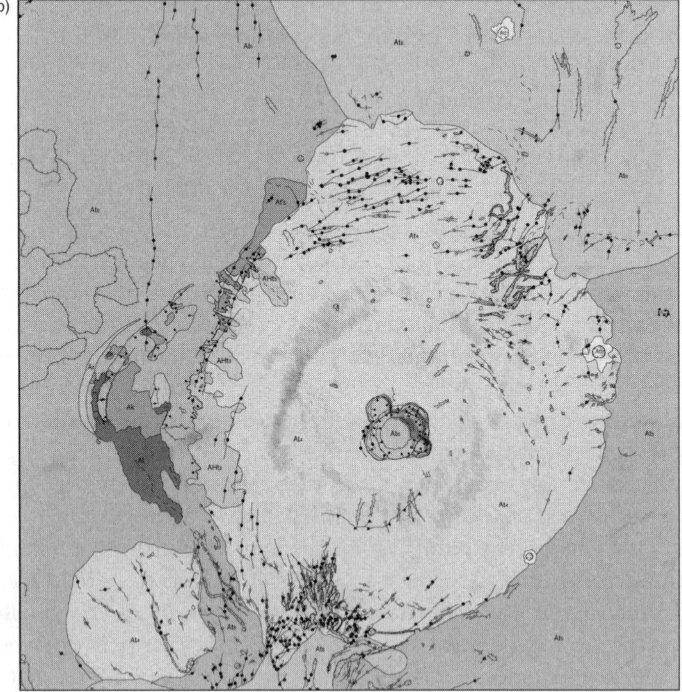

Figure 11.1 (a) and (b) Imagery from the Viking orbiters. Planetary geologists attempt to reconstruct the history of an alien world. This is done by using images from various orbiting spacecraft. Different sections of the surface are interpreted to determine their likely origin. Map colors indicate different geological units, or ages of material. Structures such as faults and volcanic flows are indicated with black symbols. (Credit: NASA/JPL.)

broken down by the ultraviolet radiation that penetrated the thin Martian atmosphere and therefore the biology experiments were registering exotic modes of chemical oxidation, not Martians. Levin countered that the results were atypical of chemical reactions: a strong oxidant would have continued to react with the carbon and not leveled off as his LR data had done. While other scientists on the biology team argued that chemical reactions were far more likely than exotic biological processes, Levin maintained that nonbiological explanations of the data relied on untested suppositions about the chemistry of the Martians surface and ignored the agreed-upon experimental protocols: positive reactions indicated biological activity. Thirty years later, Levin, Biemann, and proponents of their rival interpretations were still debating the Viking results.[17]

By 1977 most scientists were convinced by the GCMS results and concluded that the lack of organic compounds on Mars ruled out the presence of life. Although no superoxides had been identified, an oxidizing, self-sterilizing soil seemed a more elegant and robust explanation than alien life forms capable of withstanding the planet's tenuous atmosphere, bitter cold, and ultraviolet radiation. Even scientists who speculated that early Mars might have harbored microbial life conceded that the possibilities of its continued survival were minimal. Without liquid water, Mars lacked the means of transport and recruitment of new genetic material necessary to sustain microbial colonies.[18]

Martian Extremophiles?

The Viking experiments date from an era before microbiology, and therefore the question of what constitutes life, underwent complex redefinitions.[19] During the 1990s developments in extremophile microbiology changed scientists' understanding of both cellular evolution and the limits of habitable ecologies. Extremophile microorganisms, all but unknown and unsuspected in 1976, thrive in environments previously considered sterile: deep oceanic thermal vents that provide energy sources and microenvironments for bacteria that can tolerate temperatures above the boiling point of water; the alkaline, oxygen-less mud at the bottom of California's Lake Mono; lakes deep beneath the ice of Antarctica; and basalt deposits, more than a mile below the surface of the Columbia River Gorge, that harbor microorganisms that survive by metabolizing rock.[20] Analogies between extremophiles on Earth and the kinds of organisms that might exist on Mars became common in scientific literature after 1990. After the controversy over fossilized nanobacteria in the Martian meteorite, ALH84001, erupted in 1996, the study of extremophile microorganisms took on new implications for the budding field of exobiology.

Extremophile microbiology, though, was just one factor that helped rekindle the debates about life on Mars that Horowitz and many other scientists thought they had put to rest a decade earlier. Post-Viking studies indicated that Mars had enough subsurface water (or water-ice) to support microbial life, even if that life had died out in the remote past.[21] The Pathfinder mission and the Mars Global Surveyor in the late 1990s gave new impetus to scientific speculation about life on Mars because the more data scientists had, the more robust its hydrology seemed to be.[22] Finally, the advent of digital technologies and the publication of the original Viking data online allowed researchers to reassess old data in new contexts and to expand their earlier arguments. Levin and others marshaled data from the Pathfinder and Mars Global Surveyor missions that supported their view that extremophile organisms could survive on Mars or just below its surface. After 2000, results from the Mars Odyssey and Phoenix missions indicating abundant subsurface water-ice suggested that conditions were closer to Levin's understanding of the Martian environment than to Horowitz's; the Mars Phoenix mission in 2008 added considerably to the still developing scientific knowledge of surface chemistry, hydrology, and thermal disequilibria.[23] By the first decade of the twenty-first century, exobiologists emphasized three major points: Martian microorganisms could secure energy from chemical reactions in the atmosphere; enough water existed in the soil for these organisms to exploit; and the thermal differences between the surface and the atmosphere within a few centimeters of the ground could drive robust processes capable of supporting 15 million bacterial cells per cubic gram of regolith—about the same concentration as terrestrial soil in arid climates.[24]

As NASA prepared for the launch window in 1996 to return spacecraft to Mars, a team of scientists headed by David McKay argued in a controversial article that they had identified possible nanofossils of ancient bacteria in a Martian meteorite.[25] In some respects, the ensuing controversies about ALH84001 were reminiscent of the controversies that surrounded the Viking life-detection experiments: both debates were characterized by fundamental disagreements about conditions on Mars, past and present, about the definition of life, and about different understandings of the standards of evidence that should be applied to exobiology.

ALH84001 (the designation identifies the meteorite as having been the first one found in the Allan Hills of Antarctica in 1984), like other Martian meteorites, contains gas globules trapped within the rock that are identical in composition to the atmosphere of Mars and distinct from other sources of gas in the solar system.[26] ALH84001 therefore offered a kind of time-dilated sample return mission—a four-billion-year-old rock from Mars that had been ejected into space by a meteorite impact on the surface 16 million years ago (dated by the rock's exposure to cosmic rays) and then fell to Earth and remained on the frozen surface of Antarctica until it was discovered. The paper by McKay and his collaborators offered four lines of evidence—none in and of itself conclusive, they conceded, but, taken together, fairly persuasive evidence that they had identified fossilized nanobacteria from Mars. First, the meteorite contained polycyclic aromatic hydrocarbons (PAHs). Although PAHs can be formed by inorganic processes, they are often associated on Earth with decayed organic matter. Second, the sections of the meteorite they examined had perfectly regular crystals of magnetite (an iron oxide), seemingly identical to crystals that are made only by terrestrial bacteria. In addition, there was evidence of incompatible minerals existing in close proximity that, on Earth, would indicate organic action. Finally, photographs of evocative, bacteria-shaped formations in the meteorite were eye-catching, and, to some observers, persuasive.

Each of these conclusions was challenged, and many of these challenges rebutted; the controversy continues to simmer more than a decade later. Skeptics contended that the carbonates could have formed by evaporation; that the bacteria-shaped objects in photos—smaller than the smallest known terrestrial bacteria known—were too miniscule to contain the DNA and RNA necessary for life; and that any fossils were the result of terrestrial contamination. To counter these criticisms, members of the original team and additional collaborators drew on work on extremophile microbiology to conduct extensive tests on the disputed nanofossils. Acknowledging that definitive answers will have to wait for a sample return mission or even human geologists on Mars, the scientists defending the biology thesis pointed out that in the absence of such firsthand knowledge, the possible magnetofossils in the ALH84001 have been far more extensively studied than any such structures on Earth.[27]

In brief, these scientists argued that the evidence they documented about ALH84001 dovetailed with widely accepted criteria for determining life, including the biomarkers that have been identified for terrestrial magnetofossils: crystal size and shape, magnetite chains, elongated crystal structures, anisotopic growth of crystal faces, and chemical purity of the crystals.[28] Several teams extended the original research by comparing the crystal structures in the meteorite to biomarkers that indicate the presence of terrestrial magnetobacteria and countering critics who produced magnetite crystals by chemical means.[29] On Earth, magnetobacteria that orient themselves by means of a global magnetic field are ubiquitous in aquatic environments, and apparently use the magnetic properties of magnetite in order to navigate.[30] Magnetobacteria produce only one type of magnetite, identical (or at least quite similar) to the structures found in ALH84001. Biological crystals are a uniform size; so are those in the meteorite. Magnetic crystals in linear chains occur only in biological configurations on Earth; outside cells, such energetically unfavorable configurations would collapse. The crystals observed in ALH84001 are defect-free; in terrestrial bacteria, such chemically pure magnetite crystals (containing only iron and oxygen) facilitate the uptake of iron. Moreover, the crystals in ALH84001 are separated by dark areas of inorganic substances, again a characteristic feature of terrestrial bacteria. Biological membranes hold such crystals together in flexible chains; inorganic chains are not flexible because there is no elastic material between structures. Finally, the elongated crystals in ALH84001 are oriented along an axis, again in a manner identical to magnetobacteria on Earth; outside the laboratory, inorganic crystals are elongated and lie parallel to each other, forming characteristic bands. The best—that is, the simplest—way to account for the existence of these features in ALH84001, McKay and his supporters maintain, is to attribute them to evolutionary principles: magnetobacteria on ancient Mars evolved in similar ways to magnetobacteria on Earth, developing the same energy-efficient configurations to navigate in watery environments. In contrast, the arguments against Martian nanofossils present obstacles for *any* biological interpretation of the data because skeptics assume that any biogenetic system is too complex to qualify as a simple explanation for structures that could have been produced inorganically.[31]

The debates about ALH84001 pose a fundamental question about the usefulness of biological analogies between Earth and Mars: Is a catalogue of similarities between the structures in the meteorite and known magnetotactic bacteria preferable to invoking inorganic, biomimetic processes that have no analogues outside of specialized laboratories on Earth? Because no comprehensive database exists of terrestrial microorganisms and biomarkers that could provide a firm basis for comparisons between terrestrial microorganisms and suspected exobacteria, this question resists a hard and fast answer. The belief that extraordinary claims require extraordinary proof—that is, life requires extraordinary proof but inorganic chemistry does not—applies to the ALH84001 debate as much as it did to the Viking controversy. Different theoretical approaches, different methodological assumptions, and different values lead to radically different conclusions. These conclusions, moreover, are never distinct from the philosophical and even religious questions that pervade our culture's complicated, even contradictory, views about exobiology.

Return to Mars

For two decades, until the Mars Global Surveyor (MGS) and Pathfinder missions in 1997, the Viking results structured views of the red planet as NASA concentrated on the Space Shuttle program, the International Space Station, and painful internal and external reviews after the Challenger disaster. Since Mariner 9, climate change on Mars had fascinated scientists trying to decode the planet's history. The recognition that a warmer Mars once had been shaped by vast floods raised three fundamental questions: What had happened to the water? How and when had the planet lost much of its atmosphere and consequently its ability to retain solar and geothermal heat? And did this warmer and wetter world give rise to life?[32] Answering these questions, however, required a better understanding of the outlines of the planet's history. After the Viking orbital photographs allowed scientists to map the planet, a consensus timeline, based on the density of craters on the surface, was constructed that divided Martian history into distinct geological periods: the Noachian Era when massive floods scoured the surface; the Noachian-Hesperian boundary (the intense meteorite bombardment of Mars after the floods) that occurred 3.8–3.5 billion years ago; the Hesperian era; the Amazonian period (the flooding and reforming of the northern lowlands, the Vastitas Borealis, that began 1.8 billion years ago); and the resurfacing of the comparatively uncratered regions of Olympus Mons within the last 200 million years. Although Mars had been locked into a deep freeze for at least 3 billion years, the Viking data led some scientists to theorize that water existed in underground aquifers covered by subsurface ice and porous basalt rocks. If these aquifers were still heated by a molten core, they could remain in a liquid state, protected from the frigid atmosphere by subsurface ice. This possibility drove NASA's planning for future missions and helped to renew debates about Martian life.[33]

Amid ongoing budgetary problems that delayed launches and forced a series of redesigns of spacecraft and mission profiles in the 1980s and early 1990s, NASA's plans for a spectacular return to the red planet were dealt major blows by three mission failures in the 1990s: the Mars Observer went dead in 1993 as it approached the planet; the Mars Climate Orbiter missed its orbital insertion and flew past the planet in 1999; and the Mars Polar Lander crashed to the surface a few months later.[34] The loss of Mars Observer, in particular, forced a lot of soul-searching about NASA's approach to planetary exploration. A big-ticket, high-risk mission like Viking but without a backup vehicle, Observer may have been doomed by its reliance on obsolete onboard computers and the long development time that left engineers with a spacecraft that failed to capitalize on advances in computer science, telecommunications, and photography.

In the wake of the Mars Observer failure, NASA's future missions were designed to be comparatively inexpensive and to forego hunting for microbes to concentrate on providing incrementally more detailed and expansive views of Martian geology and hydrology.[35] All of the spacecraft—the Mars Pathfinder (1997), Mars Global Surveyor (1997–2004), Mars Odyssey (2002–2004), the European Space Agency Mars Express Orbiter (2004), the Mars Reconnaissance Orbiter (2006–2012), and the Spirit and Opportunity rovers (2004–2012)—bristled with instruments to search for water, inventory the minerals and chemical make-up of the planet, and photograph the surface at much higher resolutions than the Viking orbiters had done. After 2000, the prospect of global warming and NASA's significant investments in studying climate change on Earth fostered a renewed interest in comparative climatology on an interplanetary scale, and science teams for the Mars missions had as one of their primary goals piecing together a climatological history of the planet.

In 1997 Pathfinder became the first spacecraft to land on Mars since Viking. The mission employed a bare-bones approach to planetary exploration that NASA promoted with the slogan "faster, better, cheaper." Pathfinder, however, had to be engineered almost from scratch because few engineers and scientists who worked on the Viking missions were still at NASA 20 years later and a range of new technologies had to be incorporated into the mission design.[36] Operating on a tight budget, engineers developed new strategies for getting scientific instruments to the surface of Mars. Rather than costly engines to slow the lander's descent for a soft touchdown, Pathfinder's payload was cushioned in a cocoon of air bags. Parachutes slowed the lander's descent, allowing the payload to drop to the surface: the cushioned payload bounced several times, rolled to a stop, the airbags deflated, and the Sojourner rover rolled down a ramp to the planet's surface.[37]

In the emerging era of extremophile microbiology and the ALH84001 controversy, Sojourner was instrumented to do basic geological and chemical analyses of rocks and regolith. Maneuvering across a tiny patch of an ancient flood plain, Sojourner returned more than a gigabyte of data, including ten thousand photographs from the planet's surface.[38] The rocks at the site included some that had been modified by water, confirming the view that in the remote past a warmer, earthlike Mars had been shaped by water and volcanism. As it crawled from rock to rock, Sojourner became a sensation on the still-young World Wide Web. If the photographs of the Martian surface recalled the excitement generated by the Viking landers, their availability online gave NASA a new way to popularize the mission and foster public support for planetary exploration. By the time Sojourner's batteries died after a month on the surface, NASA's web site had received a half a billion hits—a significant milestone in the early years of the web.

The other missions that NASA launched in the late 1990s were more ambitious and long-lived than Pathfinder. The Mars Global Surveyor included updated versions of experiments originally designed for the failed Mars Observer; it entered orbit in 1997 and continued returning data to Earth until November 2006, four times longer than its original mission profile. The spacecraft included a thermal emission spectrometer (TES) to study the mineralogy of the surface; the Mars orbital camera (MOC); the Mars orbiter laser altimeter (MOLA) to measure topographical variations; and a magnetometer to search for traces of magnetic fields on the planet. The data returned by this suite of instruments dwarfed the returns from previous missions and quickly proved as revolutionary as the Mariner 9 photographs had been in 1972.[39]

In its first years of operation, MGS returned paradigm-shattering results; a remnant magnetic field indicated that, early in its history, Mars had enough geothermal activity and sufficient water to reshape its surface. The Tharsis volcanoes (Arsia Mons, Pavonis Mons, Ascraeus Mons) and Valles Marineris seemed, to some scientists, geological evidence of Martian tectonics. More spectacularly, the spacecraft returned over 240,000 high-resolution photographs, providing striking indications of water flow across portions of the Martian surface. Each of the MOC photographs offered a small piece of a planet-wide mosaic, a dynamic and heuristic map of the changing climatic and geological forces that have reshaped Mars. In several series of dramatic images released between 2000 and 2003, the principal investigators on the camera, Michael Malin and Kenneth Edgett, provided evidence of recent channels cut by water; gullies on the crater walls, indicating that water had modified the surface in geologically recent times, even cutting across transient sand dunes; and the delta of an ancient riverbed.[40] While mineralogical inventories of the surface suggested that the planet has been arid for more than a billion years, the MOC photographs showed planet-wide evidence of the layered terrains characteristic of the valley systems and cratered areas. Rather than a rapid descent into planetary senescence, Mars apparently had experienced complex climatological changes that were preserved in the geological record.

The extended MGS mission also identified new craters that appeared between 1999 and 2006. Rather a uniform, clock-like measure of planetary bombardment, Malin and his collaborators concluded that the cratered surface revealed an episodic history of both bombardment from solar system collisions and the covering and uncovering of ancient craters by erosion.[41] In the same article, they demonstrated that the gullies on crater rims in the mid-latitudes are "geologically young features," consistent with subsurface water flows. The article concludes with the kinds of questions that have driven Martian exploration since Mariner 9: "Where is the water coming from? How is it being maintained in liquid form . . . ? How widespread is the water? Can it be used as a resource in further Mars exploration? Finally, has it acted as an agent to promote or sustain a martian biosphere?"[42] For Malin and for many other scientists, questions about the planet's water resources led naturally to the kind of speculation that has motivated visions of human exploration of Mars since the beginning of the age of spaceflight. Since the 1970s, the dream of terraforming Mars—artificially warming the planet by creating a runaway greenhouse effect to release water at the poles and below the surface into the atmosphere—has been a staple of science fiction and serious scientific speculation.[43] Scientific articles that conclude by posing questions about "a Martian biosphere" gesture implicitly to a speculative future as well as to a Noachian past.

Questions about water and its significance for exobiology were given added urgency by results from the Mars Odyssey spacecraft that arrived in 2001 and delivered data for more than ten years. This mission dramatically expanded knowledge of the surface mineralogy, and quickly confirmed the presence of subsurface ice and water. Using nine spectral bands, the Thermal Emission Imaging System (THEMIS) provided data that allowed scientists to map the distribution of minerals on the surface, providing a geological context for suppositions about the planet's history. These results were complemented by a suite of three instruments: the gamma ray spectrometer (GRS) determined the abundance of 20 elements on the Martian surface and revealed the presence of hydrogen bound as water or water-ice below the surface.[44] In some areas, ice comprised as much as 60 percent by volume of the subsurface soil, significantly more than the most optimistic of previous estimates. The volume of water on Mars, in short, led many scientists to argue that early in its history oceans may have existed across the northern lowlands, and advanced geological mapping techniques have modeled an extensive network of valleys that were likely fed by precipitation.[45] Analyses of a few of these regions indicate a complex history that register multiple periods when liquid was present and affecting surface chemistry.[46] Other studies confirm that the Martian hydrological cycle produces frost and even snowfall.[47] The more sophisticated the instruments studying Mars and the more robust the data they return, the more fascinating the planet becomes (figure 11.2).

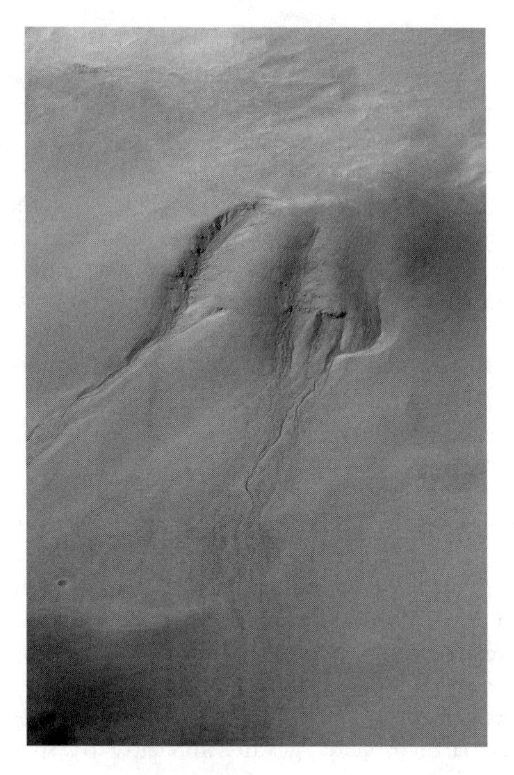

Figure 11.2 Image from the Mars Global Surveyor shows what appears to be seepage of water during a wet part of the Martian past. Such imagery sparked debate on whether there might still be water under the surface in the remains of a pond or lake. (Credit: NASA/JPL.)

Yet as always in the history of Mars exploration, the data posed more questions than it answered. While the lack of weathering of olivine, a common green mineral that is easily transformed by water, suggested that much of the Martian surface has been dry in the geologically recent past, the photographs from the MOC and the recent images from the HiRISE (High-Resolution Imaging Science Experiment that continues returning data in 2012) on the Mars Reconnaissance Orbiter, the discovery of abundant subsurface ice, and the data returned over a nine-year period by Spirit and Opportunity indicate that surface water did exist periodically in the past.[48] Because the obliquity of Mars (the angle of its axial rotation from the ecliptic) varies significantly over periods of approximately 120,000 years, Mars periodically enters ice ages when the polar regions warm enough to melt much of the ice caps and some of the subsurface ice. Water vapor and carbon dioxide thicken and hydrate the atmosphere, triggering a robust hydrological cycle, with ice and possibly snow reaching latitudes as low as 30 degrees.[49] Because the Martian climate is also affected by the response of the ice caps to a climatic precession (the wobbling of the planet as it rotates) of 51,000 years, and variations in its orbit around the sun of 95,000–99,000 years, the planet has enjoyed brief intervals when it warms significantly. Beneath the residual north polar ice cap, layered deposits preserve a history of these cyclical variations in the orbit and rotation of Mars.[50] The inventories of water on Mars from the Odyssey, Reconnaissance, and HiRISE missions confirm the possibility that this cycle could be relatively robust, creating glaciers in the mid-latitudes and bodies of liquid water, for short periods, elsewhere.[51]

The overlapping Mars missions of the first 12 years of this century have turned planetary exploration from a singular venture with established protocols to collaborative and complex integrations of science and engineering teams. Because the instruments to study Mars must fit into rigorous engineering specifications for size, weight, and durability, advances in computer hardware and software have meant exponential increases in the amounts and kinds of data that can be collected, stored, and then transmitted to Earth. Sophisticated software has allowed mission scientists to troubleshoot problems, write patches to resolve malfunctions, and reboot systems. The designs of scientific instrumentation and the ability to correct problems has allowed several missions to outlast even the most optimistic predictions of how long and reliably orbiters and rovers could function in the frigid regions of interplanetary space and on the surface of an alien world.

As the missions from Pathfinder and MGS on demonstrate, the exploration of Mars involves specialists from a wide variety of fields and has expanded the size of science and engineering teams for individual experiments. A 2009 article in *Science* describing the detection of ice at recent crater impacts, for example, included 18 coauthors.[52] The investigation of Martian geology, atmospheric dynamics, and hydrology requires a complement of specializations, as well as coordination among NASA science teams, contractors, principal investigators, and university centers. In the case of the new craters appearing on Mars, Malin Space Science Systems, operating the Context Camera (CTX) on the Mars Reconnaissance Orbiter, photographed the same region of Mars on June 4 and August 10, 2008, noting the appearance of a new cluster of small craters in western Arcadia Planitia sometime between those dates. Because CTX has a resolution of six meters per pixel, the principals put in a request to the HiRISE team (located at the University of Arizona and operated in conjunction with JPL) to photograph the same site at a resolution of approximately one foot per pixel. The HiRISE image showed a light material at the impact site that was determined to be water ice; over a period of weeks the ice sublimed into the atmosphere.[53] Examining the sites of other recent impact craters, the authors of the *Science* article suggested that as Mars begin to warm, subsurface ice tends to retreat, possibly inducing a modest greenhouse effect as the atmosphere incrementally warms. Subsequent photographs of the same areas at intervals of several months show dark boulder tracks, for example, slowly disappearing—evidence of dynamic processes that are continually reshaping the planet's surface. As such multidimensional collaborations suggest, Martian science requires a cross-disciplinary approach that tolerates ambiguity, heuristic questions, multiple avenues of investigation, and intellectual as well as interpersonal and interinstitutional negotiation. In an important sense, Mars has become a planet for polymaths.

Martian Geologists

Bridging the last years of the Mars Global Surveyor, the Mars Odyssey mission, the four-month Phoenix lander mission, and the six years (so far) of the Mars Reconnaissance Orbiter, the Mars Exploration Rover mission (MER) kept the rovers Spirit and Opportunity operating on the surface of the planet since January 2004, although each rover was built to last roughly three months and travel only one half mile from its landing site. Targeted for sites that spectrographic analysis indicated had been reshaped by water, Gusev Crater and Meridiani Planum, Spirit and Opportunity adapted the landing techniques that had been developed for Pathfinder. The twin rovers carried no life-detection experiments but instead were designed to perform more extensive and robust geological fieldwork than Sojourner had done in 1997. In addition to a high-resolution panoramic camera, the rovers carried three different spectrometers to determine the mineral composition of rocks and soil samples, a rock abrasion tool to dig small circles to facilitate studying the interior of rocks, and a microscopic imager for detailed mineralogical observations. While the European Space Agency's Mars Express mission centered on the Beagle 2 lander that was instrumented to search for Martian microbes, the lander disappeared in December 2003 after it had separated from its orbiter.

Spirit and Opportunity proved to be among the most successful planetary probes that humankind has launched, lasting years beyond their original mission. The data returned during their first few months on the planet exceeded scientists' expectations, in large measure because Opportunity had come to rest in a small impact crater, quickly named Eagle, on Meridiani Planum; Steve Squyres, the principal investigator for the science packages on the rovers, called it "a 300-million mile interplanetary hole in one."[54] The crater walls revealed rock outcrops that preserved a record of eons of Martian geological history. Before moving on to the larger Endurance and Victoria craters, Opportunity circled the interior of Eagle for two months, studying rock formations. The crater and surrounding terrain were littered with tiny spheres that were dubbed "blueberries," hematite-rich concretations of minerals formed as water diffused through rocks. Opportunity's Microscopic Imager photographed these tiny spheres embedded in and emerging from porous rocks. As Spirit traveled across the floor of Gusev, Opportunity returned photographs that showed evidence of geological layering, including outcroppings marked by the uneven or intersecting sedimentary layers that geologists call crossbedding, indicating that they had formed in gently flowing water. The chemical composition of these layers was marked by high concentrations of salts, left by water as it evaporated. The strata exposed in Endurance recorded changes that were much older than those at Eagle. Differences in magnesium and sulfur content in older sedimentary layers suggested that those elements had been dissolved and removed by water. For significant periods of time—likely millions of years—water had altered the chemical composition of the surface.

Prepared for a short-term mission, the NASA kept day by day—or really sol by sol—logs of the progress of both rovers and regularly updated images on the mission web page.[55] What began as a three-act play stretched into an epic journey as the rovers slowly rolled across the landscape. The early photographs of blueberries, Eagle and Endurance craters, and the first meteorite discovered on Mars became iconic images of the mission, but the terabytes of data returned over the years have led to attempts by the science teams to synthesize their findings and paint a more comprehensive picture of Martian geology. In summing up the first two years of Opportunity's spectrographic analysis, Squyres and his collaborators concluded that cross-laminations in rock outcrops indicate that flowing, highly acidic water as well as windborne processes shaped the surface of Meridiani Planum.[56] Opportunity's subsequent investigations at Victoria Crater, studying layered deposits more than ten meters thick, demonstrated that the processes that had shaped and eroded Eagle and Endurance acted regionally. The rich history of this region showed epochs of different kinds and durations of modifications, including sedimentary deposits by water as well as dune formation and erosion by wind. The interaction of acidic water and basalts had produced sulfate salts; windblown sands had formed dunes that had been cemented into layers by water; outcrops had been weathered under increasingly arid conditions, and wind-blown erosion of sedimentary rocks had continued throughout geological history.[57] Spirit's two-mile trek from its landing site to the Columbia Hills, a range that rises some three hundred feet above the datum, took several months and culminated in a slow and careful climb to several of its ridges, where it found evidence of chemical alterations in rocks, indicating the prolonged action of water. Studies based on evidence from the orbiters indicate that these forces operate across significant portions of the planet, and that the widespread layering provides evidence of climatological cycles and ancient surface conditions.[58]

In 1997, waiting patiently for images to download over their 28K and 56K modems, millions of people had followed Sojourner's frame-by-frame crawl across a tiny patch of Martian ground. Dozens of digital generations later, during the Spirit and Opportunity missions, NASA's websites received billions of hits. The rovers became robotic photojournalists of an alien landscape of weathered rocks, delicate and treacherous dunes, and vistas that appear like rock-strewn desert landscapes on Earth. Spirit and Opportunity have returned hundreds of thousands of photographs archived by sol on the mission website; the planetary scientist Jim Bell, who led the photography team, calls these "postcards from Mars." The then one hundred and fifty thousand photographs culled for Bell's 2006 book of that title allow readers to experience vicariously one of science fiction's most resilient forms—the epic trek across undiscovered territory.[59] In this respect, the rovers' photographic odyssey, seen from a robotic, but eerily eyewitness perspective, mimics the experience of walking across the Martian landscape. The rovers have given rise to a new sort of coffee-table book.

As Spirit and Opportunity approached three years on Mars, the Mars Reconnaissance Orbiter began studying the interconnected problems of the planet's climate, geological history, and water inventory. Its suite of instruments included HiRISE, subsurface radar to detect water, and CRISM (compact reconnaissance imaging spectrometer for Mars) to search for minerals associated with water such as polyhydrated sulfate (sulfates with more than one water molecule incorporated into each molecule of the mineral). Rather than a set of individual experiments, the instrument packages provided complex, multidimensional views of the planet. In December 2008, the CRISM team reported the discovery of carbonate rocks, created when water and carbon dioxide interact with calcium, iron, or magnesium, that dissolve quickly in acidic environments; therefore, the existence of carbonates in the Nili Fossae region of the planet indicates that the area had a watery past with an alkaline or neutral pH balance.[60] These findings suggest that early Mars had more than one kind of aqueous environment, rather than being dominated by highly acidic bodies of water that shaped Meridiani Planum. Based on what scientists know about terrestrial biochemistry, less acidic waters are more hospitable to life, and the presence of carbonates add one more aspect to consider in speculating whether Mars had, or still harbors, life. Mars's past climate is, in an important sense, locked in minerals that have been formed by reacting with volatiles, notably water and carbon dioxide. A wet and warmer climate, coupled with a much denser atmosphere, would facilitate the formation of carbonates from basaltic rocks.

The presence of carbonates in Nili Fossae dovetails with observations by the Phoenix lander after it touched down in the north-polar region in May 2008. Ice was found just centimeters below the surface, the soil showed strong evidence of interaction with water vapor in the atmosphere, and, in the later Martian summer, snowfall and frost blanketed the surface at night. Calcium carbonate in the soil, the mission scientists suggested, could act as a buffering agent for an alkaline pH, creating conditions similar to those of many habitable environments, such as terrestrial seawater. The identification of perchlorates at the Phoenix site provided a chemical mechanism to lower the freezing point of water to about –70°C, making it possible that briny water could exist for a few hours on the surface during the Martian summer. Scientists actually debated during the mission whether globules that appeared on the legs of the lander were liquid water or icy mud. Moreover, although it indicated a higher concentration of salts than terrestrial extremophiles can tolerate, the chemical analysis conducted by the Phoenix lander suggested an aquatic environment that could provide a medium for biology activity or "pre-biotic organic synthesis." Such openness to the possibility of life represents just how far the exploration of surface chemistry has come since Viking.[61]

No smoking gun has emerged from any of the recent Mars mission to settle the 35-year debate about whether the only life-detection experiments ever conducted on the surface found life. Yet in the eyes of many planetologists, Martian microorganisms have gone from crackpot science to a plausible hypothesis worth investigating. In 2003, scientists at NASA and the European Space Agency independently confirmed the presence of methane on Mars, including a strong and localized release.[62] Unless it is continually replenished, methane will disappear from the Martian atmosphere within two or three hundred years because it quickly oxidizes into carbon dioxide and water vapor. On Earth, methane is typically a by-product of biological processes, although it also can be generated by volcanic activity. The existence of methane on Mars, then, indicates the presence of either subsurface microbial colonies or active volcanic processes, with the localized gas concentrations escaping through the kinds of thermal vents that on Earth harbor rich extremophile ecologies. Teams of scientists drilling into the Greenland ice sheet have discovered high concentrations of methane produced by methanogenic organisms at depths of three thousand meters. These organisms may be expending most of their energy to repair damage to DNA and amino acids rather than to grow, and consequently may provide an idea of the mechanisms that Martian organisms might have evolved to survive.[63] Radiolysis could provide the energy in an environment rich in liquid water and carbon dioxide for methanogens, organisms that produce methane, or methanotrophs, microorganisms that are able to metabolize methane.[64]

In 2012 (months after this book goes to press) NASA will land the sedan-size Curiosity rover, with a range of ten miles, on the planet's surface. With more sophisticated instrumentation to study Martian geology and hydrology than Opportunity and Spirit, Curiosity will mark almost a half century of robotic exploration, and 35 years of the study of the surface. It will go much farther than Opportunity or Spirit and is equipped to tackle tougher—and potentially more interesting—terrain. The questions that it will ask about the planet have expanded exponentially since Mariner 4 or even Viking. While the study of Mars has not returned to the rampant speculation about dying civilizations of canal-builders, it continues to grapple with fundamental questions about exobiology and the limits and insights offered to earthlings by comparative planetology.

Notes

1. This chapter was written while the Mars Curiosity Mission was en route to Mars, and scheduled for an August 2012 landing.
2. Joseph N. Tatarewicz, *Space Technology and Planetary Astronomy* (Bloomington: Indiana University Press, 1990), p. 75.
3. Werner von Braun, *The Mars Project* (Urbana: University of Illinois Press, 1992 ed.); Edward Clifton Ezell and Linda Newman Ezell, *On Mars: Exploration of the Red Planet 1958–1978* (Washington, DC: NASA SP-4212, 1982), pp. 25–39.
4. Tatarewicz, *Space Technology and Planetary Astronomy*, pp. 43, 49.
5. Bruce Murray, "From the Eyepiece to the Footpad: The Search for Life on Mars," in Yervant Terzian and Elizabeth Bilson, eds., *Carl Sagan's Universe* (Cambridge, UK: Cambridge University Press, 1997), pp. 35–48, quote from p. 38.

6. Bruce Murray, *Journey into Space: The First Three Decades of Space Exploration* (New York: W.W. Norton and Co., 1989), pp. 37–45; Paul Raeburn, *Uncovering the Secrets of the Red Planet: Mars* (Washington, DC: National Geographic Society, 1998), pp. 57–59.
7. Robert Godwin, comp. and ed., *Mars: The NASA Mission Reports* (Burlington, Ontario: Apogee Books, 2000), 1:56–61; Ezell and Ezell, *On Mars*, pp. 156–59; Tatarewicz, *Space Technology and Planetary Astronomy*, pp. 96–103.
8. Godwin, comp. and ed., *Mars: The NASA Mission Reports*, 1:52.
9. Ibid., 1:81–87.
10. Robert Markley, *Dying Planet: Mars in Science and the Imagination* (Durham, NC: Duke University Press, 2005).
11. Murray, *Journey into Space*, p. 68.
12. Godwin, comp. and ed., *Mars: The NASA Mission Reports*, 1:132–33; Henry S. F. Cooper Jr., *The Search for Life on Mars: Evolution of an Idea* (New York: Holt, Rinehart, and Winston, 1980), pp. 128–29; Markley, *Dying Planet*, pp. 250–66.
13. N. H. Horowitz, G. L. Hobby, and Jerry S. Hubbard, "Viking on Mars: The Carbon Assimilation Experiments," *Journal of Geophysical Research* 82, 28 (1977): 4659–62, quote from 4659.
14. Cooper, *The Search for Life on Mars*, pp. 146–48; Godwin, comp. and ed., *Mars: The NASA Mission Reports*, 1:131–32.
15. Klaus J. Biemann , J. Oro, P. Toulmin III, L. E. Orgel, A. O. Nier, D. M. Anderson, P. G. Simmonds, D. Flory, A. V. Diaz, D. R. Rushneck, and J. A. Biller, "Search for Organic and Volatile Inorganic Compounds in Two Surface Samples from the Chryse Planitia Region of Mars," *Science* 194 (1976): 72–76; Klaus J. Biemann J. Oro, P. Toulmin III, L. E. Orgel, A. O. Nier, D. M. Anderson, P. G. Simmonds, D. Flory, A. V. Diaz, D. R. Rushneck, J. E. Biller, and A. L. Lafleur, "The Search for Organic Substances and Inorganic Volatile Compounds in the Surface of Mars," *Journal of Geophysical Research* 82 (1977): 4641–62.
16. Ezell and Ezell, *On Mars*, pp. 278–86.
17. Gilbert V. Levin, "Analysis of Evidence of Mars Life," Carnegie Institution Geophysical Laboratory Seminar, May 14, 2007; Klaus J. Biemann, "On the Ability of the Viking Gas Chromatograph–Mass Spectrometer to Detect Organic Matter," *Proceedings of the National Academy of Sciences* 10 (2007): 10310–13.
18. Christopher P. McKay, R. L. Mancinelli, C. R. Stoker, and R. A. Wharton Jr., "The Possibility of Life on Mars During a Water-Rich Past," in H. H. Kieffer, B. M. Jakosky, C. W. Snyder, M. S. Mathews, eds., *Mars* (Tucson: University of Arizona Press, 1992), pp. 1234–45; P. J. Boston, M. V. Ivanov, and C. P. McKay, "On the Possibility of Chemosynthetic Ecosystems in Subsurface Habitats on Mars," *Icarus* 95 (1992): 300–308.
19. Laurence Bergreen, *Voyage to Mars: Mankind's Search for Life Beyond Earth* (New York: Penguin Putnam, 2000).
20. Michael Ray Taylor, *Dark Life: Martian Nanobacteria, Rock-Eating Cave Bugs, and Other Extreme Organisms of Inner Earth and Outer Space* (New York: Charles Scribner's Sons, 1999); Peter Doran, R. A. Wharton, D. J. Des Marais, and C. P. McKay, "Antarctic Paleolake Sediments and the Search for Extinct Life on Mars," *Journal of Geophysical Research* 103 (1998): 28–36.

21. Michael H. Carr, *Water on Mars* (New York: Oxford University Press, 1995); Victor Baker, "Water and the Martian Landscape," *Nature* 412 (2001): 228–36.
22. William K. Hartmann, *A Traveller's Guide to Mars* (New York: Workman Publishing, 2003).
23. P. H. Smith, L. K. Tamppari, R. E. Arvidson, D. Bass, D. Blaney, W. V. Boynton, A. Carswell, D. C. Catling, B. C. Clark, T. Duck, E. Dejong, D. Fisher, W. Goetz, H. P. Gunnlaugsson, M. H. Hecht, V. Hipkin, J. Hoffman, S. F. Hviid, H. U. Keller, S. P. Kounaves, C. F. Lange, M. T. Lemmon, M. B. Madsen, W. J. Markiewicz, J. Marshall, C. P. McKay, M. T. Mellon, D.W. Ming, R. V. Morris, W.T . Pike, N. Renno, U. Staufer, C. Stoker, P. Taylor, J. A. Whiteway, and A. P. Zent, "H2O at the Phoenix Landing Site," *Science* 325 (2009): 58–61.
24. D. P. Glavin, M. Schubert, O. Botta, G. Kminek, and J. L. Bada, "Detecting Pyrolysis Products from Bacteria on Mars," *Earth and Planetary Science Letters* 185 (2001): 1–5.
25. David S. McKay, H. K. Gibson Jr, K. L. Thomas-Keprta, H. Vali, C. S. Romanek, S. J. Clemett, X. D. Chillier, C. R. Maechling, and R. N. Zare, "Search for Past Life on Mars: Possible Relic Biogenic Activity in Martian Meteorite ALH84001," *Science* 273 (1996): 924–30
26. Bruce Jakosky, *The Search for Life on Other Planets* (Cambridge, UK: Cambridge University Press, 1998).
27. David Wharton, *Life at the Limits: Organisms in Extreme Environments* (Cambridge, UK: Cambridge University Press, 2002), pp. 232–40.
28. E. Imre Friedman, Jacek Wierzchos, Carmen Ascaso, and Michael Winklhofer, "Chains of Magnetite Crystals in the Meteorite ALH84001: Evidence of Biological Origin," *Proceedings of the National Academy of Sciences* 98 (2001): 2176–81; Kathie L. Thomas-Keprta, Simon J. Clemett, Dennis A. Bazylinski, Joseph L. Kirschvink, David S. McKay, Susan J. Wentworth, Hojatollah Valii, Everett K. Gibson, Jr., Mary Fae McKay, and Christopher S. Romanek, "Truncated Hexa-octahedral Magnetite Crystals in ALH84001: Presumptive Biosignatures," *Proceedings of the National Academy of Sciences* 98 (2001): 2164–69.
29. D. C. Golden, D. W. Ming2, R. V. Morris, A. J. Brearley, H. V. Lauer Jr., A. H. Treiman, M. E. Zolensky, C. S. Schwandt, G. E. Lofgren, and G. A. McKay, "Evidence for Exclusively Inorganic Formation of Magnetite in Martian Meteorite ALH84001," *American Mineralogist* 89 (2004): 681–95; A. P. Taylor and J. C. Barry, "Magnetosomal Matrix: Ultrafine Structure May Template Biomineralization of Magnetosomes," *Journal of Microscopy* 213 (2004): 180–97.
30. Thomas-Keprta et al., "Truncated Hexa-octahedral Magnetite Crystals in ALH84001," pp. 2164–69.
31. Markley, *Dying Mars*, pp. 323–37.
32. Oliver Morton, *Mapping Mars: Science, Imagination, and the Birth of a World* (New York: Picador, 2002); Andrew Chaikin, *A Passion for Mars: Intrepid Explorers of the Red Planet* (New York: Abrams, 2008).
33. Hartmann, *A Traveler's Guide to Mars*, pp. 29–35; Jean-Pierre Bibrin, Yves Langevin, John F. Mustard, François Poulet, Raymond Arvidson, Aline Gendrin, Brigitte Gondet, Nicolas Mangold, P. Pinet, F. Forget, the OMEGA team, Michel Berthé, Jean-Pierre Bibring, Aline Gendrin, Cécile

Gomez, Brigitte Gondet, Denis Jouglet, François Poulet, Alain Soufflot, Mathieu Vincendon, Michel Combes, Pierre Drossart, Thérèse Encrenaz, Thierry Fouchet, Riccardo Merchiorri, GianCarlo Belluci, Francesca Altieri, Vittorio Formisano, Fabricio Capaccioni, Pricilla Cerroni, Angioletta Coradini, Sergio Fonti, Oleg Korablev, Volodia Kottsov, Nikolai Ignatiev, Vassili Moroz, Dimitri Titov, Ludmilla Zasova, Damien Loiseau, Nicolas Mangold, Patrick Pinet, Sylvain Douté, Bernard Schmitt, Christophe Sotin, Ernst Hauber, Harald Hoffmann, Ralf Jaumann, Uwe Keller, Ray Arvidson, John F. Mustard, Tom Duxbury, François Forget, and G. Neukum, "Global Mineralogical and Aqueous Mars History Derived from OMEGA/Mars Express Data," *Science* 312 (2006): 400–404; Carr, *Water on Mars.*

34. Bergreen, *Voyage to Mars*, pp. 304–10.
35. On these low-cost missions, see Howard E. McCurdy, *Faster, Better, Cheaper: Low-Cost Innovation in the U.S. Space Program* (Baltimore, MD: Johns Hopkins University Press, 2001).
36. Brian K. Muirhead and William L. Simon, *High Velocity Leadership: The Mars Pathfinder Approach to Faster, Better, Cheaper* (New York: HarperBusiness, 1999); Judith Reeves-Stevens, Garfield Reeves-Stevens, and Brian Muirhead, *Going to Mars: The Untold Story of Mars Pathfinder and NASA's Bold New Missions for the 21st Century* (New York: Pocket Books, 2000).
37. Andrew Mishkin, *Sojourner: An Insider's View of the Mars Pathfinder Mission* (New York: Berkley, 2003).
38. M. P. Golombek,R. A. Cook, T. Economou, W. M. Folkner, A. F. C. Haldemann, P. H. Kallemeyn, J. M. Knudsen, R. M. Manning, H. J. Moore, T. J. Parker, R. Rieder, J. T. Schofield, P. H. Smith, and R. M. Vaughan, "Overview of the Mars Pathfinder Mission and Assessment of Landing Site Predictions," *Science* 278 (1997): 1743–48.
39. Hartmann, *A Traveler's Guide to Mars*; Morton, *Mapping Mars.*
40. Michael C. Malin and Kenneth S. Edgett, "Evidence for Recent Groundwater Seepage and Surface Runoff on Mars," *Science* 288 (2000): 2330–35; Michael C. Malin and Kenneth S. Edgett, "Sedimentary Rocks of Early Mars." *Science* 290 (2000): 1927–37.
41. Michael C. Malin, Kenneth S. Edgett, Liliya V. Posiolova, Shawn M. McColley, and Eldar Z. Noe Dobrea, "Present-Day Impact Cratering Rate and Contemporary Gully Activity on Mars," *Science* 314 (2006): 1573–77.
42. Ibid., p. 1575.
43. Markley, *Dying Mars*; Christopher McKay, J. Kasting, and O. Toon, "Making Mars Habitable," *Nature* 352 (1991): 489–96.
44. W. V. Boynton, W. C. Feldman, S. W. Squyres, T. H. Prettyman, J. Brückner, L. G. Evans, R. C. Reedy, R. Starr, J. R. Arnold, D. M. Drake, P. A. J. Englert, A. E. Metzger, Igor Mitrofanov, J. I. Trombka, C. d'Uston, H. Wänke, O. Gasnault, D. K. Hamara, D. M. Janes, R. L. Marcialis, S. Maurice, I. Mikheeva, G. J. Taylor, R. Tokar, and C. Shinohara, "Distribution of Hydrogen in the Near Surface of Mars: Evidence for Subsurface Ice Deposits," *Science* 297 (2002): 81–85.
45. Wei Luo and T. F. Stepinski, "Computer-Generated Global Map of Valley Networks on Mars," *Journal of Geophysical Research* 114 (2009): E11010; Brian M. Hynek and Roger J. Phillips, "Evidence for Extensive Denudation of the Martian Highlands," *Geology* 29 (2001): 407–10.

46. Janice L. Bishop, Eldar Z. Noe Dobrea, Nancy K. McKeown, Mario Parente, Bethany L. Ehlmann, Joseph R. Michalski, Ralph E. Milliken, Francois Poulet, Gregg A. Swayze, John F. Mustard, Scott L. Murchie, and Jean-Pierre Bibring, "Phyllosilicate Diversity and Past Aqueous Activity Revealed at Mawrth Vallis, Mars," *Science* 321 (2008): 830–33.
47. J. A. Whiteway, L. Komguem, C. Dickinson, C. Cook, M. Illnicki, J. Seabrook, V. Popovici, T. J. Duck, R. Davy, P. A. Taylor, J. Pathak, D. Fisher, A. I. Carswell, M. Daly, V. Hipkin, A. P. Zent, M. H. Hecht, S. E. Wood, L. K. Tamppari, N. Renno, J. E. Moores, M. T. Lemmon, F. Daerden, and P. H. Smith, "Mars Water-Ice Clouds and Precipitation," *Science* 325 (2009): 68–70.
48. Joshua L. Bandfield, Timothy D. Glotch, and Philip R. Christensen, "Spectroscopic Identification of Carbonate Minerals in the Martian Dust," *Science* 301 (2001): 1084–87.
49. James W. Head, John F. Mustard, Mikhail A. Kreslavsky, Ralph E. Milliken, and David R. Marchant, "Recent Ice Ages on Mars," *Nature* 426 (2003): 797–802.
50. M. A. Chamberlain and W. V. Boynton, "Response of Martian Ground Ice to Orbit-induced Climate Change," *Journal of Geophysical Research* 112 (2007): E06009; Roger J. Phillips, Maria T. Zuber, Suzanne E. Smrekar, Michael T. Mellon, James W. Head, Kenneth L. Tanaka, Nathaniel E. Putzig, Sarah M. Milkovich, Bruce A. Campbell, Jeffrey J. Plaut, Ali Safaeinili, Roberto Seu, Daniela Biccari, Lynn M. Carter, Giovanni Picardi, Roberto Orosei, P. Surdas Mohit, Essam Heggy, Richard W. Zurek, Anthony F. Egan, Emanuele Giacomoni, Federica Russo, Marco Cutigni, Elena Pettinelli, John W. Holt, Carl J. Leuschen, and Lucia Marinangeli, "Mars North Polar Deposits: Stratigraphy, Age, and Geodynamical Response," *Science* 320 (2008): 1182–85.
51. John W. Holt, Ali Safaeinili, Jeffrey J. Plaut, James W. Head, Roger J. Phillips, Roberto Seu, Scott D. Kempf, Prateek Choudhary, Duncan A. Young, Nathaniel E. Putzig, Daniela Biccari, and Yonggyu Gim, "Radar Sounding Evidence for Buried Glaciers in the Southern Mid-latitudes of Mars," *Science* 322 (2008): 1235–38; Maria T. Zuber, Roger J. Phillips, Jeffrey C. Andrews-Hanna, Sami W. Asmar, Alexander S. Konopliv, Frank G. Lemoine, Jeffrey J. Plaut, David E. Smith, and Suzanne E. Smrekar, "Density of Mars' South Polar Layered Deposits," *Science* 317 (2007): 1718–19.
52. Shane Byrne, Colin M. Dundas, Megan R. Kennedy, Michael T. Mellon, Alfred S. McEwen,Selby C. Cull, Ingrid J. Daubar, David E. Shean, Kimberly D. Seelos, Scott L. Murchie, Bruce A. Cantor, Raymond E. Arvidson, Kenneth S. Edgett, Andreas Reufer, Nicolas Thomas, Tanya N. Harrison, Liliya V. Posiolova, and Frank P. Seelos, "Distribution of Mid-Latitude Ground Ice on Mars from New Impact Craters," *Science* 325 (2009): 1674–76.
53. Ibid.
54. Robert Godwin, comp. and ed., *Mars: The NASA Mission Reports*, Volume 2 (Burlington, Ontario: Apogee Books, 2004), 2:321.
55. Each Martian day, or sol, is 24 hours and 37 minutes long, and the science teams adopted a Martian timeline to describe the rovers' activities.
56. S. W. Squyres, A. H. Knoll, R. E. Arvidson, B. C. Clark, J. P. Grotzinger, B. L. Jolliff, S. M. McLennan, N. Tosca, J. F. Bell III, W. M. Calvin, W. H. Farrand, T. D. Glotch, M. P. Golombek, K. E. Herkenhoff, J. R. Johnson,

G. Klingelhöfer, H. Y. McSween, and A. S. Yen, "Two Years at Meridiani Planum: Results from the Opportunity Rover," *Science* 313 (2006): 1403–07.

57. S. W. Squyres, A. H. Knoll, R. E. Arvidson, J. W. Ashley, J. F. Bell III, W. M. Calvin, P. R. Christensen, B. C. Clark, B. A. Cohen, P. A. de Souza Jr., L. Edgar, W. H. Farrand, I. Fleischer, R. Gellert, M. P. Golombek, J. Grant, J. Grotzinger, A. Hayes, K. E. Herkenhoff, J. R. Johnson, B. Jolliff, G. Klingelhöfer, A. Knudson, R. Li, T. J. McCoy, S. M. McLennan, D. W. Ming, D. W. Mittlefehldt, R. V. Morris, J. W. Rice Jr., C. Schröder, R. J. Sullivan, A. Yen, and R. A. Yingst, "Exploration of Victoria Crater by the Mars Rover Opportunity," *Science* 324 (2009): 1058–61.
58. Kevin W. Lewis, Oded Aharonson, John P. Grotzinger, Randolph L. Kirk, Alfred S. McEwen, and Terry-Ann Suer, "Quasi-Periodic Bedding in the Sedimentary Rock Record of Mars," *Science* 322 (2008): 1532–35.
59. Jim Bell, *Postcards from Mars: The First Photographer on the Red Planet* (London, UK: Dutton, 2006); Gregory L. Vogt, *Landscapes of Mars: A Visual Tour* (New York: Springer, 2008).
60. Bethany L. Ehlmann, John F. Mustard, Scott L. Murchie, Francois Poulet, Janice L. Bishop, Adrian J. Brown, Wendy M. Calvin, Roger N. Clark, David J. Des Marais, Ralph E. Milliken, Leah H. Roach, Ted L. Roush, Gregg A. Swayze, and James J. Wray, "Orbital Identification of Carbonate-Bearing Rocks on Mars," *Science* 322 (2008): 1828–32.
61. W. V. Boynton, D. W. Ming, S. P. Kounaves, S. M. M. Young, R. E. Arvidson, M. H. Hecht, J. Hoffman, P. B. Niles, D. K. Hamara, R. C. Quinn, P. H. Smith, B. Sutter, D. C. Catling, and R. V. Morris, "Evidence for Calcium Carbonate at the Mars Phoenix Landing Site," *Science* 325 (2009): 61–64; M. H. Hecht et al., "Detection of Perchlorate and the Soluble Chemistry of Martian Soil at the Phoenix Lander Site," *Science* 325 (2009): 64–67; David Shiga, "First Liquid Water May Have Been Spotted on Mars," *New Scientist*, February 18, 2009.
62. Michael J. Mumma, Geronimo L. Villanueva, Robert E. Novak, Tilak Hewagama, Boncho P. Bonev, Michael A. DiSanti, Avi M. Mandell, and Michael D. Smith, "Strong Release of Methane on Mars in Northern Summer 2003," *Science* 323 (February 20, 2009): 1041–45.
63. H. C. Tung et al., "Microbial Origin of Excess Methane in Glacial Ice and Implications for Life on Mars," *Proceedings of the National Academy of Sciences* 102, 51 (2005): 18292–96.
64. Mumma et al., "Strong Release of Methane on Mars in Northern Summer 2003," pp. 1041–45.

Part IV

Unveiling the Outer Solar System

Chapter 12

Parachuting onto Another World: The European Space Agency's Huygens Mission to Titan*

Arturo Russo

On Friday, January 14, 2005, a group of nervous and excited scientists gathered in the control room of the European Space Operations Centre (ESOC) in Darmstadt, Germany, to attend an epochal event in the history of planetary exploration and enjoy the crowning achievement of their scientific life. For the first time, a human artifact would be landed on another world in the outer solar system. A spacecraft fitted out with six sophisticated scientific instruments was to be parachuted through the atmosphere of Titan, the largest satellite of Saturn and one of the most intriguing objects in the solar system. The spacecraft had been built by the European Space Agency (ESA) and it was named after the seventeenth-century astronomer and natural philosopher Christiaan Huygens.

The €364 million Huygens probe was ESA's contribution to the Cassini-Huygens mission to the Saturnian system, a joint venture with NASA and the Italian Space Agency (ASI). Launched in October 1997, the Cassini-Huygens spacecraft entered orbit around Saturn in early July 2004. On Christmas day that year, Huygens was released by the Cassini mother spacecraft and reached Titan's outer atmosphere after 20 days and a 4 million kilometer cruise. According to the mission scenario, which the ESOC gathering were eagerly hoping to see realized, a sequence of parachutes would slow it down and its scientific instruments would be exposed to Titan's atmosphere during descent. If the spacecraft survived the impact with the surface, the instruments would hopefully continue to operate, providing additional information for a time that could be anything from a few minutes to half an hour or more. Out in space, Cassini would pick up Huygens's signals, then turn its antenna toward the Earth and relay the recorded scientific data. The scientists and their distinguished guests, including ministers, space agency officials and journalists, were waiting.

Around noon, the news arrived in ESOC that a faint radio signal from the probe had been picked up by the Green Bank radio telescope in West

Virginia, United States. Huygens had then survived the entry phase and was active! Late in the afternoon, the first scientific data, relayed by the Cassini spacecraft, arrived at ESOC. Scientists hurried to analysis, and soon the press got the first stunning images of Titan's surface. A long weekend of intense work—and celebrations—was about to start.

Cassini-Huygens is the largest and most highly instrumented spacecraft ever sent into deep space, and the most ambitious and challenging effort in planetary exploration ever mounted.[1] Conceived in the early 1980s, the mission called for a sophisticated Saturn orbiter to perform more than 60 revolutions around the ringed planet and its satellites over a four-year period, and a probe to be landed on the surface of Titan. The 12 scientific instruments on the orbiter were to conduct in-depth studies of the planet, its rings, atmosphere, magnetic environment, and a large number of its moons. The six instruments on the probe were to provide direct sampling of Titan's atmospheric chemistry and photographs of its surface. NASA provided the orbiter, ESA provided the probe, and ASI provided the high-gain antenna and other hardware systems for the orbiter. The scientific instruments and related investigations were realized by scientific teams in the United States and in ESA's 19 member states. The realization of this mission, from the initial vision in the early 1980s to its completion in 2008, demanded the coordinated effort of thousands of scientists, engineers and specialists in universities, space agencies and industry in Europe and the United States over a quarter of a century.

Both the orbiter and the probe scientific missions have been successfully accomplished. Huygens successfully completed its mission on the very day of its descent through Titan's atmosphere, while the nominal four-year mission of the Cassini Orbiter came to an end on June 30, 2008. However, in April that year, NASA approved a two-year extension, called Cassini Equinox Mission, whose main scientific objectives are more detailed studies of Saturn's moons, in particular Titan and Enceladus; the investigation of seasonal effects on Saturn and Titan; the exploration of new regions of the Saturnian magnetosphere; and the observation of the unique ring geometry of the Saturn equinox in August 2009, when sunlight will pass directly through the plane of the rings. A second extended mission, called Cassini Solstice Mission, was eventually approved, due to go through in September 2017 in order to observe the Saturnian northern summer solstice occurring in May that year. Since Cassini arrived at Saturn just after the planet's northern winter solstice, the extension will allow for the first study of a complete seasonal period.

This chapter deals with the history of Cassini-Huygens from a European perspective. Three aspects will be discussed in particular. First is the changing political and institutional framework that set the stage for the establishment of an important European effort in planetary exploration. During more than two decades, in fact, the European space science community felt that, for technical and financial reasons, Europe could not compete with the important efforts of the United States and the Soviet Union in this field. It was only in the mid-1980s that an ambitious European planetary mission was considered as a realistic possibility, following the successful Giotto mission to comet Halley and the approval of the so-called Horizon 2000 long-term scientific plan. The Cassini mission to the Saturnian system, including a Titan probe, was originally proposed to ESA in 1982 as a collaborative venture with NASA, and the proposal eventually evolved into the actual mission launched 15 years later. Huygens was then the first European mission devoted to planetary science and its approval, in 1988, set the stage for an important ESA effort in this field. In fact, when Huygens performed its epochal descent through Titan's atmosphere, a small European spacecraft was orbiting Mars since a year, and another one had entered lunar orbit two months earlier after a one-year journey driven by innovative solar-electric propulsion technology. Moreover, a Venus orbiter was to be launched in November that year and a highly sophisticated spacecraft was already cruising in space toward a comet rendezvous scheduled in 2014. On a longer term, an ambitious mission to Mercury was included as a pivotal project in ESA's scientific program for the early 2010s. Commenting on this list of achievements and future projects, Roger Bonnet, who had been the ESA director of science in those crucial years (1983–2001), wrote:

> Of course this is a modest list compared with the large number of missions launched by NASA, but it is important to remember that ESA's science budget is nearly one order of magnitude smaller than its NASA equivalent [...] Through these missions, all of them in prominent positions, Europe through ESA has placed itself in the position of a key player in future programmes of planetary exploration [and] ESA has proven to be a reliable partner in international cooperation.[2]

The second aspect of historical interest is the decision-making process, which led to the adoption of the Huygens mission in ESA's Scientific Programme. The founding fathers of European cooperation in space research stipulated that the European space science community at large should remain the only source of ideas and concepts of missions. These are then to be submitted to feasibility studies and discussed by expert groups and advisory committees in a competitive selection procedure concluding with one mission finally approved. The final decision to adopt a scientific mission in ESA's program is thus the outcome of a highly competitive process, involving the various national and/or disciplinary sectors of the space science community; the ESA Science Directorate; the European space industry; the member state space policies; the relations with NASA and/or other potential international partners, and so on. Within this framework, this chapter can be considered a case study in ESA scientific decision-making.[3] The case of the Cassini-Huygens selection is not particularly different from previous ones as regards general methodology. But it presents two related elements of novelty. First, it is the first selection procedure realized within the framework of the newly established Horizon 2000 long-term plan. Second, it is the first planetary mission that entered the selection process on equal conditions with other proposals. We will see how the new institutional framework of the ESA Scientific Programme by mid-1985 was an important element in the eventual success of the scientific constituency supporting the Cassini-Huygens mission.

Finally, the ESA/NASA relationship in the Cassini-Huygens mission will be the third focus of my analysis. This mission, in fact, is an exemplar of ESA/NASA collaboration in space science. Originally conceived by a group of scientists spread across the Atlantic, it was soon evident that only a cooperative effort of both space agencies could make such an ambitious mission to become a concrete reality. However, while scientific cooperation worked smoothly and resulted in the successful achievement of the mission's scientific objectives, it was not so easy to cope with the different political and institutional frameworks in which the two agencies were operating. All the more so because of the mission's very long implementation time, technical sophistication, and high costs. In a sense, the ESA/NASA collaboration in this mission was an important exercise in what the ESA director general in the 1980s, Reimar Lüst, called "equal partnership" in space collaboration:

> It is clear that Europe cannot allow itself to be reduced to a subordinate or subsidiary role in space ventures if it is to maintain its current hard-won position [...] The need for international collaboration on major space undertakings is not disputed, but Europe wishes to enter such undertakings on an "equal partnership" basis, this concept applying at all levels including operational control.[4]

Lüst's emphasis was mainly on the envisaged international effort on the Space Station project, as opposed to the "junior partner" role Europe had had vis-à-vis the United States in the realization of Spacelab. But even in science, following some disappointing experience of the late 1970s, culminating with the ISPM crisis we will deal with below, most European space science policy-makers felt that ESA should establish a core program whose realization did not depend on NASA collaboration, and accept cooperative ventures only in a nonsubordinate condition. In this respect, Cassini-Huygens is peculiar. One can hardly say that it is a case of "equal partnership" from a quantitative point of view, considering the large technical and financial gap between the American and European contributions (even including Italy). But, from a qualitative point of view, a fair equal partnership was indeed established at scientific level; the European political weight was enough important to be a decisive factor in rescuing the mission from the pitfalls of US policy; and the visibility of ESA was granted by the importance of the Huygens mission, both at scientific level and for the general public.

The documentary basis of this chapter includes three kinds of unpublished documents. First are the official documents related to the activity of ESA's committees and working groups, including minutes of meetings, information documents, recommendations, and so on. These documents are sequentially coded and dated according to the various issuing bodies. Second are the unpublished documents related to the activity of the ESA Directorate of Science, such as correspondence, memos, monthly reports, minutes of meetings, and so on. Finally, there are the scientific and technical reports related to mission selection (mission proposals, assessment studies, feasibility studies, etc.). The narrative is divided into three sections. In the first, I will discuss the European Space Agency's institutional framework and the prospects of its scientific program in the 1980s. The second section is devoted to the decision-making process leading to the selection of the Huygens mission in ESA's program. The third covers the development and launch of the mission. Finally, in the last pages I will put the Cassini-Huygens mission in the framework of ESA's present plans for planetary exploration.

The European Space Agency and its Horizon 2000 Scientific Programme

The European Space Agency was officially established in May 1975, taking over the functions of two preexistent organizations: the European Space Research Organisation (ESRO) and the European Launcher Development Organisation (ELDO), both established in 1964.[5] From ESRO, the new agency inherited a strong commitment to space science, supported by a dedicated scientific constituency spread over its ten member states; a solid institutional and managerial framework; and a strong technical infrastructure at the European Space Technology Centre (ESTEC) in Noordwijk, Netherlands, and at the European Space Operations Centre (ESOC) in Darmstadt, Germany. From ELDO, it took over the undertaking to develop a heavy satellite launcher, capable of assuring Europe full autonomy and competitiveness in space. Unlike its sister organization, ELDO did not succeed in its goal and its *Europa* rocket never did reach orbit. The new agency then undertook a completely different design, essentially based on French managerial and industrial effort. This was the *Ariane* vehicle, whose first launch was successfully achieved on Christmas Eve 1979 from the Kourou range in French Guyana. Besides the scientific program inherited from ESRO and the *Ariane* development program, ESA also undertook application satellite programs, in particular for meteorology and telecommunications, and a collaborative effort with NASA for the development of Spacelab, a manned scientific laboratory to be flown on the US Space Shuttle.

With the establishment of ESA, the joint (Western) European effort in space was meant to cover all fields of space activities. There was, however, a significant qualitative difference between the Scientific Programme and the others. The former, in fact, was the only mandatory program for all member states, each of them contributing in proportion to gross domestic product (GDP). All other programs were optional, supported by the participating states to the level of their political and economic interest. This "à la carte system," as it was dubbed at that time, is still effective today. At policy decision-making level, the Scientific Programme is under the responsibility of the

Science Programme Committee (SPC), a delegate body of the ESA Council including representatives of all member states; the other programs are under the responsibility of specific program boards, including representatives of the participating states. At executive level, ESA is managed by a director general (DG), nominated by the council, assisted in his work by a number of directors, each of whom is responsible for one of ESA's programs or for administering part of the agency. Among them, the director of science (D/SCI) is responsible for the definition and implementation of the Scientific Programme.[6]

Within the framework of ESA's program structure, the "special nature of the Scientific Programme" was explicitly recognized from the very beginning as "the common factor through which the whole Organisation was held together."[7] The definition and actual implementation of the Scientific Programme, in fact, involves a political dimension of a different quality from the other programs. The latter are characterized by well-defined objectives, for example, the development of a sound earth observation satellite or a reliable launch vehicle. Once the program and its budget have been approved, it is implemented under the control of the supporting member states. Users' specifications, technical options, and industrial policy considerations are the important elements of the program boards' discussions and decisions. The scope of the Scientific Programme, on the contrary, is essentially decided by the financial envelope approved by the council, within which the European space science community is called to select (and the SPC to approve) specific space missions through a competitive process whose fundamental and essential characteristics have not changed since the ESRO period.

The mission selection procedure, a "planning cycle" in the ESA jargon, is a maieutic process whereby, through successive eliminations, one project is eventually selected and approved for development. The start of a planning cycle is a "call for mission proposal" from the ESA director of science. The proposals submitted by individual scientists or scientific groups are then discussed by the Astronomy Working Group (AWG) and the Solar System Working Group (SSWG), each including some 15 representatives of the various domains in the respective areas of competence. Those missions that are considered interesting from the scientific point of view are then submitted to the Space Science Advisory Committee (SSAC), including a restricted number (typically five) of authoritative scientists plus the chairmen of the working groups who, at this stage, are called to recommend a number of mission proposals for an "assessment study." This is carried out at two levels. On the one hand, a team of consultant scientists, including the proposers, prepares an in-depth scientific definition study of the mission. On the other hand, an internal technical study is conducted in order to assess the feasibility and costs of the mission. The results of these assessment (or mission definition) studies are then presented to the scientific community at an open meeting specifically organized by ESA. A new round of discussions and evaluation within the working groups and the SSAC then narrows down the list of competing missions to about four, which are then subjected to in-depth feasibility study by industry (Phase A study in the ESA jargon) and to a complete mission-design evaluation and costing by ESTEC engineers. The results are again presented to the scientific community. Subsequently, the working groups, for their respective competence, and the SSAC make their final recommendations about which mission(s) should be selected in the ESA program. The director of science then formulates his proposal to the SPC, the legislative body called to take the final decision. Needless to say that it hardly happens that the SPC decision is not in accordance with the D/SCI proposal supported by the SSAC recommendation. The entire cycle lasts a minimum of two years but, as we will see in the case of Huygens, it can be much longer.

During more than 25 years since the beginning of the joint European effort in space in the early 1960s, planetary science and solar system exploration hardly deserved consideration from the ESRO/ESA scientific constituency. Nothing comparable existed in Europe to the important effort of the United States and the Soviet Union. By the early-1980s, American and Russian spacecraft had explored the Moon and the solar system from Mercury to Saturn, ranging from flyby missions to landers and orbiting satellites. Moreover, the American Voyager 2 spacecraft was well on its way to fly by Uranus (January 1986) and Neptune (August 1989). Lunar and planetary missions had not been excluded from the *Blue Book* where ESRO founding fathers had established the institutional and programmatic framework of the new organization, and a Mercury flyby mission had also put forward as a candidate for the 1969 selection, but it had been rapidly discarded as too expensive, "in spite of the desirability of Europe entering the increasingly important field of planetary exploration." It would have absorbed too much of ESRO's limited budget.[8] At that time, in fact, it was more than evident that the program described in the *Blue Book* was far too ambitious and the European space science community had to make a sound selection of the scientific areas in which they wanted to concentrate their efforts, and establish priorities among the various research fields. This was the task of the Launching Programme Advisory Committee (LPAC), the SSAC forerunner in the ESRO institutional framework. Following extensive discussions within the community, the LPAC issued its policy document in February 1970. The conclusions were that four research fields were to be given priority in the incoming decade, namely: (a) fundamental physics, with some priority given to the testing of gravitational theories; (b) plasma physics investigations in the magnetosphere, heliosphere, and polar ionosphere; (c) high-energy (X- and gamma-ray) astrophysics; and (d) special cosmic ray studies such as the determination of isotopic abundances and the measurement of solar particles. Three important research fields, on the contrary, were to be excluded: planetary missions, UV astronomy, and solar physics.[9]

The 1970 policy statement established guidelines for ESRO/ESA's scientific program for a decade. It reflected, on the one hand, the importance of the two physicists' communities, which had been the main protagonists in the first phase of European space research: that interested in the Earth's space environment and that interested in high-energy astrophysics. The former had played a leading role in the definition and early development of ESRO's program, performing many sounding rocket and satellite experiments on the ionosphere and auroral phenomena. Now they looked at the magnetosphere as the new research frontier, with a large variety of experimental possibilities in the study of space plasma and solar-terrestrial relations. The latter were colonizing a new and promising field, where experimental techniques borrowed from cosmic-ray physics could be successfully implemented for studies of high-energy phenomena in astronomical objects. On the other hand, the important losers in the 1970 policy decision were solar physics and UV astronomy, two research fields in which the advent of space technology had stimulated a dramatic breakthrough. The rationale of this decision can be recognized in two main elements. First was the consideration of the important American effort in these fields, based on the OAO (Orbiting Astronomical Observatory) and OSO (Orbiting Solar Observatory) programs, to be followed up by more powerful missions including manned space stations. Second was the weakness of the European astronomers' community in the space research field. Most stellar astronomers and solar physicists were mainly interested in the development of ground-based facilities and they lacked the scientific culture and technical expertise required for developing important space projects. The few groups involved in space research were interested in developing large observatory-like satellites, much at the borderline of ESRO's financial resources and technical capabilities.

Planetary exploration, for its part, never became an important issue in these discussions, and the LPAC policy decision to exclude deep-space missions from the ESRO long-term program was uncontroversial. These missions appeared necessarily expensive, argued the LPAC, and, facing the important effort of the United States and the USSR in this field, "ESRO would experience strong competition with unequal resources."[10] The committee, however, did not exclude the possibility of a low-cost participation of ESRO in a cooperative planetary mission with NASA. This opportunity was offered in 1973, when a proposal for a joint ESRO/NASA Venus orbiter mission was put forward as a candidate for selection, but it lost the competition against two projects that were coherent with the 1970 policy decision: the European contribution to the ESA/NASA International Sun-Earth Explorer program (ISEE-2), and an X-ray space telescope (Exosat).[11]

During the 1970s, most European physicists and geologists interested in lunar and planetary studies ignored ESRO/ESA and managed to get involved in some of the American and Soviet missions. More specifically, they were provided with lunar samples or images from Mars and Venus to conduct independent scientific investigation. A few scientific groups were thus established, in particular at the University of Rome (Marcello Fulchignoni), the University of Paris-Orsay (Philippe Masson), the German Aerospace Research Centre in Münich (Gerhard Neukum), and University College of London (John Guest). These groups had hardly any contact with ESRO/ESA, whose "ban against planetology" was considered with some distrust.[12]

For the new generation of European planetary scientists, in fact, the Giotto mission to comet Halley in the early 1980s represented the first real opportunity to meet ESA.

The circumstances that led to the selection of Giotto in 1980 have been discussed elsewhere, and only two important elements will be recalled here.[13] First, this decision came after a new long-term planning exercise undertook in 1978, whose first aim was to make a case for a substantial increase in the level of resources for the scientific program. It came to nothing as regards this plea but, contrary to the 1970 policy statement, it concluded that "Europe [should] follow a policy in which all the fields of European scientific excellence in space are developed and supported." In other words, while being aware that not all disciplines could be supported if additional funds were not available, the SSAC agreed that the choice should not be made a priori and once and for all, and the competitive selection procedure should be open to any proposal, provided that it was scientifically sound, technically feasible, and financially affordable.[14]

Second, the selection of Giotto arrived at the conclusion of a dramatic competition at all decision-making levels. The proposal for a cometary mission, in fact, was strongly opposed first by the supporters of a lunar mission proposal, then by the advocates of a satellite for magnetospheric studies, and finally by the astronomical community interested in developing a large satellite devoted to accurate astrometry measurements. The latter was strongly supported by France, where the project had been originally conceived and preliminary studies performed, while the cometary mission was strongly supported by Germany for just the same reasons. This, of course, added no little political flavor to the selection procedure. The eventual outcome in favor of Giotto was largely the result of the firm commitment of the ESA director of science, Ernst Trendelenburg, to undertake the cometary mission. This commitment was partly inspired by public relations considerations. Visiting a famous comet and sending close-up images of its nucleus to TV screens all over the world was certainly more appealing than accurately measuring star positions, and would add glamor to the European space effort, which lacked space walks and footsteps on the Moon. If, after all, the ESA program depended on taxpayers' money, why not offer them a colorful vision of the wonders of nature in addition to esoteric papers in scientific journals?

In January 1983, the ESA Council unanimously approved the appointment of the French physicist Roger Bonnet to the post of director of science, with effect from May 1. Bonnet had been trained in the cradle of French space science, that is, Jacques Blamont's Service d'Aéronomie Spatiale at Verrières-le-Buisson. He obtained his PhD in 1968, amid the student revolt that swept through French universities and scientific institutions, and one year later he became the first director of the newly created Laboratoire de Physique Stellaire et Planetaire, also located in Verrières-le-Buisson.[15] Bonnet knew well the functioning of ESA's scientific program, both from his direct involvement in some of ESRO/ESA projects and in his capacity of SSAC chair from 1978 to 1980. In fact, he had been responsible for the important report on the development of European space science in the 1980s discussed earlier.

Taking up his duties, Bonnet undertook to work out a new approach to scientific planning within ESA, partially amending the established procedure of introducing a new mission every second or third year. The basic idea was to establish for once a long-term program including a firm commitment in a few predefinite scientific areas. This was essential, argued Bonnet, in order to undertake the long-lead technological developments required by those ambitious missions that had been proposed by the scientific community but not accepted because their feasibility could not be proved during a single planning cycle. Moreover, the definition of a balanced long-term program would present three other advantages. First, the various sectors of the European space science community would know ahead in which direction they had to invest their efforts and in which time frame their projects would eventually be realized. Second, a clear definition of medium- and long-term objectives would help in matching the available resources to financial requirements. In particular, a fixed envelope would be established for each planned mission, thus forcing the scientific community to limit its ultimate ambitions, and the ESA management to ensure more efficient control over program development. Finally, by establishing its own long-term strategy, ESA would secure a stronger negotiating position in the international scenario.

The new plan for European space science, called "Horizon 2000," was worked out in 1984 and eventually approved in January 1985 by the Ministerial Conference of ESA member states in Rome.[16] (See Figure 12.1) The basic philosophy of Horizon 2000 was to establish two classes of projects. The first included four predefined "cornerstones," that is, ambitious and technologically challenging missions, costing about two annual budgets of the scientific program and to be realized according to a phased schedule over a 20-year period of time. The second class included ten medium-sized missions (M-missions), costing about one annual budget, to be selected through the usual selection procedure. The cornerstones were devoted respectively to Solar-Terrestrial Physics, X-ray Astronomy, Planetary Science, and Infrared Astronomy. The M-mission program included five projects already under development and another five to be selected in the future. Approving Horizon 2000, ESA member-state governments undertook to increase the level of resources of the Scientific Programme by 5 percent annually in real terms (i.e., after adjustment for inflation) over the period 1985–1994. One can hardly underestimate the importance of this decision. This was the first time in 15 years that the ESA scientific budget had been increased, indeed a granted 5 percent annual increase above inflation over a ten-year period. As a mandatory activity, in fact, any decision to increase the level of resources of the Scientific Programme required unanimous approval in the ESA Council, a result that had not been possible since 1971. Moreover, the cornerstone philosophy established a solid stability framework for long lead planning: scientists, ESA managers, and industry knew ahead of time in which direction they ought to invest their efforts and pursue technological developments.

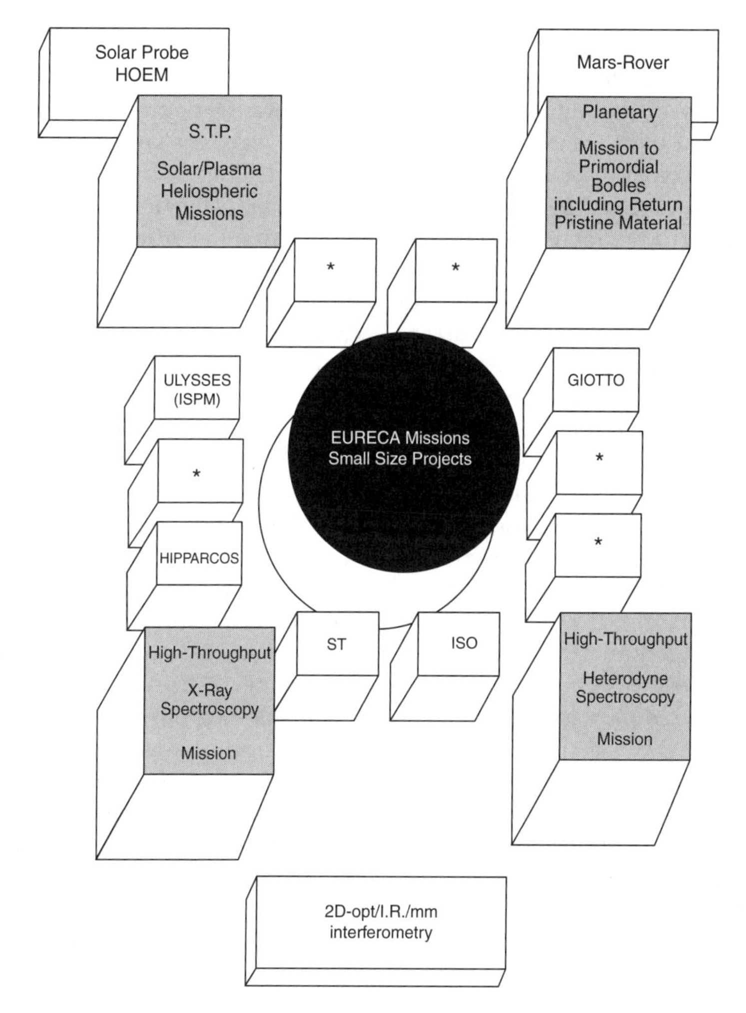

Figure 12.1 The European Space Agency's Horizon 2000 plan laid out options for the twenty-first century. (Credit: ESA.)

Finally, the establishment of a European long-term plan extending over two decades changed the shape of cooperation between ESA and NASA, as it was agreed that the cornerstones had to remain under European leadership and be consistent with ESA's own technical and financial means. Cooperation could eventually bring new, added capabilities to these purely European missions or be limited to medium-sized projects.

The decision to devote a cornerstone to planetary science represented both a novelty and a confirmation for ESA. The novelty was in the recognition that planetary science and solar system exploration should be a pivotal scientific field in the European space framework for the incoming new century. The confirmation was in the selection of primordial bodies (comets and asteroids) as the "area where Europe could take the lead, following the Giotto mission." The latter, scheduled for launch a few months after the approval of Horizon 2000, was not only the first European deep space mission, but also the very first mission aiming at a close investigation of cometary phenomena.[17] In fact, the original objective of the planetary cornerstone was to return a comet nucleus sample, the next logical, although ambitious step in cometary exploration after Giotto's flyby mission. This objective, however, could not be maintained and the project eventually approved in 1993 consisted in a mission to explore the nucleus of a comet by an orbiter and a small lander. Called Rosetta, after the famous Egyptian stone, which provided key information for the decipherment of hieroglyphic writing, it was launched in March 2004 and is now cruising toward its rendezvous with comet 67P/Churyumov-Gerasimenko scheduled for 2014.

Within the framework of Horizon 2000, planetary science was no longer under a ban, and sound proposals could be submitted and supported for selection in the flexible part of Horizon 2000, that is, the M-mission program. In fact, as astronomy and plasma physics were well represented in this part of the plan by already approved missions, Bonnet felt a moral commitment to foster planetary missions within the scheduled procedures for the selection of the new M-missions.[18] An important step in this direction was the recruitment of Marcello Coradini as the new coordinator of ESA solar system missions and executive secretary of the SSWG. Coradini had been trained in Marcello Fulchignoni's planetology group at the University of Rome and had started his scientific career in the early 1970s analyzing Mars data from the US Mariner 9 mission. Ten years later, Fulchignoni's group had evolved into the important planetology section within the Istituto di Astrofisica Spaziale of Italy's National Research Council (CNR), and Coradini had been loosely involved in the realization of the Giotto camera, in which Bonnet was also involved. Taking up his duties in ESA in early 1987, the mandate Coradini received by the D/SCI was "to build up a programme of solar system exploration."[19] The first battleground would be the selection of M1, the first new medium-sized mission, whose planning cycle was already under development.

The Cassini Project and the Birth of Huygens

The idea of a cooperative ESA/NASA mission to the Saturnian system can be traced back to the early 1980s, on the wave of the successful Voyager missions to the two major planets of the Solar System, Jupiter and Saturn. Both launched in 1977, Voyager 1 encountered Jupiter in March 1979 and Saturn in November 1980, while Voyager 2 encountered Jupiter in July 1979 and Saturn in August 1981. Taking advantage of a favorable planetary alignment, Voyager 2 was eventually extended to carry out the exploration of the Uranus and Neptune systems, encountered in January 1986 and August 1989, respectively.[20]

One of the most important discoveries of Voyager 1 was the intriguing composition of the atmosphere of Titan, Saturn's largest moon and the second largest in the solar system after Jupiter's Ganimede. The existence of an atmosphere on Titan was first suggested in 1908 by the Spanish astronomer José Comas Solà, who observed disc edge darkening features and concluded that they were due to an atmosphere.[21] This was confirmed in 1943–1944 by Gerard Kuiper of the University of Chicago, who discovered that a relatively thick atmosphere was in fact present and detected methane in it by spectroscopic observations.[22] Until the mid-1970s, it was believed that methane was the major constituent of Titan's atmosphere, but a theoretical model proposed later in the decade by Donald Hunten of the University of Arizona suggested that other constituents could also be present, in particular molecular nitrogen.[23] The problem of the composition of Titan's atmosphere thus provided good reasons to plan a close approach to the satellite during the Voyager 1 Saturn flyby, in order to carry out a radio occultation experiment to probe its atmosphere.[24] The results were exciting: "Titan [. . .] was revealed as an entirely new world, a unique object."[25] Not only was it confirmed that molecular nitrogen was the main constituent of the atmosphere, with a small percent of methane, but the infrared spectrometer on Voyager showed that many organic molecules were also present. The surface of Titan was completely obscured from the Voyager camera by a thick orange/brown smog made of a mixture of various hydrocarbon and nitrogen compounds, including ethane, ethylene, propane, and hydrogen cyanide. In fact, the dissociation of methane and nitrogen molecules, driven by solar UV radiation, cosmic rays, and precipitating electrons from Saturn's magnetosphere, gives rise to a complex organic chemistry in Titan's atmosphere and on its surface, by which the fragments of the parent molecules recombine to make a large variety of carbon compounds.

In the eyes of the Voyager scientists, this planet-like satellite, with its thick atmosphere mostly made of nitrogen and methane, resembled what our Earth should have looked like some four billion years ago, before organic life started to colonize its surface and produce oxygen by green-plant photosynthesis.[26] The fundamental difference between the early Earth and Titan is the low temperature on the latter's surface (-179°C), which makes it impossible for the presence of liquid water and then the emergence of life. On the contrary, the intense organic chemistry at work in the atmosphere of primitive Earth did have a chance to evolve into prebiotic chemistry and eventually into biology. Subsequent erosion, plate tectonics, and the evolution of life itself have obliterated all records of those conditions and processes on our Blue Planet, but Titan could provide its human inhabitants of today with an opportunity to travel back in time, if they were able to travel in space up to there: "Titan offers us a kind of controlled experiment to study pre-biotic chemistry on a planetary scale. Titan is a world where organic chemistry has been proceeding for 4.5 billion years; at low rates, obviously, because of the low temperature, but for very long times."[27]

Among the scientists who were particularly excited about the striking results coming from the Voyager infrared spectrometer were Tobias (Toby) Owen and Daniel Gautier. The former, then at the State University of New

York at Stony Brook and currently at the university of Hawaii, had been studying planetary atmospheres and comets since the mid-1960s, using ground-based telescopes and measuring instruments onboard Apollo and Viking spacecraft. The latter was an expert in planetary atmospheres at the Meudon Observatory in Paris. For both of them there could be no doubt that one had to plan a dedicated mission to Titan. At that time, NASA was planning future scientific missions and Owen was in charge of the Outer Planets Group of the Solar System Exploration Committee. Following the Agency's 1977 decision to develop the Galileo mission to Jupiter, the exploration of the Saturnian system appeared as the next logical step. Owen and his group discussed a mission concept based on a Saturn orbiter carrying two atmospheric entry probes, one to Saturn and another to Titan. However, the financial difficulties resulting from the escalating costs of the space shuttle program made it impossible to realize such an ambitious undertaking, and a less-expensive strategy was then elaborated by the NASA Ames Research Center, consisting in four successive missions to Titan, Saturn, Uranus, and Neptune, respectively. In each mission, a probe would be sent onto the planet or satellite by a relatively simple flyby carrier. Gautier, for his part, unsuccessfully tried to convince the French Centre National d'Etudes Spatiales (CNES) to undertake building a Titan probe, as the project was considered far too expensive for France alone. At that time, the idea of a possible US/European cooperative mission was also discussed between the two scientists.[28]

A promising opportunity finally arose in Europe in July 1982, when ESA's Directorate of Scientific Programs issued a call for mission proposals to start the selection process for the scientific mission(s) to be launched at the beginning in the following decade. Gautier decided to answer the call, working out a proposal for a mission to Titan. While developing this idea, he received a telephone call from Wing-Huen Ip, a planetary scientist of Chinese origin at the Max-Planck-Institut für Aeronomie in Lindau who had worked on Pioneer's and Voyager's data on Saturn.[29] Ip proposed that he and Gautier prepare a joint a proposal for a Saturn/Titan mission, and the latter was happy to accept. In the following months they built up a group of interested scientists and prepared a mission proposal, which was submitted to ESA in November under the name "Proiect Cassini."[30] It foresaw the realization of a Saturn orbiter carrying a probe to be released into the atmosphere of Titan. The project was to be realized by an ESA/NASA collaboration, with Europe providing the spacecraft to orbit Saturn while NASA would contribute the Titan probe and the tracking by its Deep Space Network. The name Cassini was proposed by Ip after the Italian astronomer Giovanni Domenico Cassini (1625–1712), called in 1669 by king of France Louis XIV to supervise the construction of the new Paris Observatory, of which he became the first director. In Paris, Cassini undertook detailed studies of the Saturnian system, discovering four new satellites: Iapetus (1671), Rhea (1672), Tethys (1684), and Dione (1684). He also discovered that Saturn's rings are separated into two parts by a gap, now called "Cassini Division," and (correctly) suggested that they were composed of myriads of small particles.[31]

The "Project Cassini" proposal was presented by a group of 29 European scientists coordinated by Gautier and Ip. Eighteen scientific institutions from seven ESA member states were represented in the group, whose membership, according to their own presentation, included representatives of four disciplines: atmospheric science (9), planetology (9), interplanetary and magnetospheric physics (9), and exobiology (2). The underlying idea, in fact, was to study the whole of the Saturnian system, including specific objectives for each of the types of bodies in the system—the planet and its rings, Titan, the icy satellites, and the magnetosphere—as well as the interactions that occur among them. This concept of "system science" would have been an important element in fostering a large support behind the Cassini project within the European planetary science community, and promoting the mission through ESA's highly competitive selection process.

Cassini was one among 20 different mission proposals received by ESA's Scientific Directorate in November 1982, 8 in the field of solar system science and 12 in the field of astronomy. All proposals were discussed by the Astronomy and Solar System Working Groups and by the SSAC, and eventually 5 of them were selected for an assessment study phase in the first half of 1983. The Cassini project was not among them, pending a clarification about the possibility of a collaborative venture with NASA.[32] At that time, Gautier, Ip, and Owen had established good relations among themselves, and a proper institutional framework was being established in support of the Cassini concept. In June 1982, the Space Science Committee of the European Science Foundation and the Space Science Board of the US National Academy of Sciences had set up a joint working group to study possible cooperation between Europe and the United States in the field of planetary science. One of the cooperative projects discussed (and eventually recommended) by the working group was a Saturn orbiter and Titan probe mission.[33] In 1983, following the suggestions of Owen's Outer Planets Group, the Solar System Exploration Committee of NASA's Advisory Council recommended that NASA should include a mission to the Saturnian system consisting of an orbiter based on the Galileo spacecraft and a Titan probe supplied by an international partner.

While a solid scientific constituency supporting the Saturn/Titan mission was building up on both sides of the Atlantic, the relationship between ESA and NASA could not be worse! In fact, all ambitious plans for scientific cooperation between the two agencies had fallen apart. The project for a joint cometary project aborted in January 1980, following the NASA decision not to support the Solar Electric Propulsion system that would have driven an American spacecraft to rendezvous comet Tempel-2 in 1988, carrying a European probe to be released in 1985 for a rendezvous with Halley. By the end of 1981, two other cooperative projects were also abandoned: a Spacelab mission devoted to solar physics and a new version of a lunar mission already studied within ESA.[34] Much more dramatic was the American decision to withdraw from the International Solar Polar Mission (ISPM), an ambitious cooperative mission under development since 1977. This consisted in the simultaneous launch of two spacecraft, built in Europe and the United States, respectively, into a trajectory that would take them out of the ecliptic plane, one over the north pole of the Sun, the other over the south pole. The dual spacecraft mission was originally to be launched in February 1983 on the Space Shuttle, but in 1980 NASA announced a two-year delay because of technical and financial difficulties with the development of the Shuttle itself. One year later, the American space agency informed its European partner that they would not continue the development of their spacecraft in the ISPM. NASA's unilateral decision jeopardized the concept of a two-spacecraft mission and severely impaired the fulfillment of the mission's scientific objectives. Not only did it become impossible to perform stereoscopic and imaging observations, only possible from the NASA spacecraft, but also one half of the instruments being developed for flying on the dual mission would not be used, thus eliminating some 80 US and European investigators right away from the project. Strong political and diplomatic actions were undertaken by the ESA executive and by member-state representatives in order to reverse NASA decision. These, however, came to nothing. In the event, it was decided that ESA should proceed with a single spacecraft mission, renamed Ulysses.[35]

The case of the ISPM mission is particularly revealing of the difficulties in scientific cooperation between ESA and NASA. Both space agencies needed such cooperation, as a consequence of the budgetary limitations of the 1970s. In 1974, the US Congress stated that "substantial participation of other nations" was a sine qua non condition for approving NASA's Space Telescope project.[36] Similarly, ESA's scientific advisory bodies could hardly design ambitious scientific missions that could dispense with NASA participation. However, in spite of the good scientific and technical relations established in that period, many differences existed between the two agencies as regards their institutional and political framework. The difference in the budget procedures was the most striking one. Decision-making could be very long for ESA, but once a project had been approved its financial allocations were also approved in terms of a certain cost-to-completion. In a way, provided no cost escalation occurred, the project became legally binding for member states and there was no threat of cancellation. NASA, on the contrary, was a national agency whose overall program and budget had to be negotiated annually with the federal government and the Congress. Funds could always be shifted from one program to another on the basis of political considerations, congressional lobbying, or national security priorities. A typical memorandum of understanding (MOU), like that signed for the implementation of the ISPM mission, is particularly revealing. It included the (obvious) statement that its applicability was always subject to the availability of funds for both parties, according to their "respective funding procedures." This condition applied with near certainty to ESA after the mission had been approved in November 1977. It was different on the US side, however. Here the final go-ahead was given by Congress in early 1978, with the inclusion of ISPM in the FY 1979 budget. Two years later, as a consequence of President Jimmy Carter's budget cuts, NASA's research budget was dramatically reduced in order to protect the escalating space shuttle program, and the mission had to be delayed by two years. The election of Ronald Reagan as Carter's successor brought about fundamental changes in the budget process, which led to a further decrease in the space science budget and, ultimately, to the unilateral cancellation of the American spacecraft. Here is how Bonnet and his right arm in ESA, Vittorio Manno, recalled the event:

> The outrage and incredulity in Europe were great. Outrage at the way the cancellation had been carried out, and incredulity that an international agreement would be cancelled at all. This reflected ESA's stunned realisation of the fundamental difference in attitude between the two organizations about the sanctity of a Memorandum of Understanding. In Europe [. . .] the MOU was considered as legally binding on its Member States, while it became painfully clear that this was not the case for the US administration. That fundamental difference was to cast its shadow on all present and future international agreements between the two organizations.[37]

In the aftermath of the ISPM affair, the prospects of launching a new important cooperative venture with NASA was hardly regarded with enthusiasm at ESA Headquarters, where it was clear that "Europe would no longer accept being considered a subordinate participant."[38] Therefore, it was not without some reservation that in October 1983 the SSAC recommended that joint ESA/NASA assessment studies should be undertaken in 1984–1985 on the Cassini mission and two other cooperative projects that had been discussed in the previous months, that is, an ultraviolet astronomy mission called Lyman (after the American physicist Theodore Lyman, the discoverer of the ultraviolet lines in the hydrogen spectrum), and a radio astronomy mission called Quasat. The latter was based on a spacecraft carrying a large radio antenna to be used in conjunction with a network of ground-based very-long-baseline-interferometry (VLBI) networks in Europe, the United States, the USSR, and Australia.[39] As was expected, Gautier, Ip, and Owen led the joint ESA/NASA study team on Cassini.

The assessment study of the Cassini project was conducted between April 1984 and June 1985 by a team of 13 scientists, 9 of whom were from the United States (table 12.1). Three members of the European contingent were

Table 12.1 The science team for the assessment study of the Cassini mission (1984–1985)

M. Allison	Goddard Institute for Space Studies, New York
S. Bauer	Karl Franzens Universitat, Graz
J. Cuzzi	NASA Ames Research Center
M. Fulchignoni	Istituto di Astrofisica Spaziale, Rome
D. Gautier	Observatoire de Paris
D. Hunten	University of Arizona, Tucson
W. Ip	MPI fur Aeronomie, Katlenburg-Lindau
T. Johnson	Jet Propulsion Laboratory, Pasadena
H. Masursky	US Geological Survey, Flagstaff
T. Owen	State University of New York, Stony Brook
R. Samuelson	NASA Goddard Space Flight Center
F. Scarf	TRW, Redondo Beach
E. Sittler	NASA Goddard Space Flight Center

among the 29 authors of the original 1982 proposal: Gautier, Ip, and Siegfried Bauer, a professor of meteorology and geophysics at the University of Graz. The fourth was Marcello Fulchignoni, the pioneer of planetology in Italy.[40] Reversing the idea of Gautier and Ip's original proposal, the Titan probe was soon identified as ESA's potential contribution to the international Cassini mission. It was within the technical capabilities of the European space industry, whose experience in planetary missions was limited to that acquired in the Giotto mission, and the estimated costs were within the budget allocated to the first M-mission of the Horizon 2000 program. The objectives of the Cassini mission were summarized thus in the study report:

> This advanced, joint ESA/NASA mission basically consists of multiple (30) orbital tours through the Saturn-system by an Orbiter craft, which will carry a Titan Probe to be ejected upon arrival for entering and traversing the cloudy atmosphere of Titan down to the surface. Such a scheme permits a thorough investigation over several (4) years of these two bodies in particular and to renew—with much increased resolution—the assault on the complex Saturnian system of satellites and rings, as well as Saturn's magnetosphere, obtaining answers to many of the fundamental questions posed by the earlier explorations. Cassini should, namely, enable us to study much more thoroughly the interior, atmospheric structure and chemistry, dynamic meteorology, cloud patterns and aerosols, ionosphere and exosphere of the two primary targets (Saturn and Titan); plus the general characteristics of all the satellites and their diversified surfaces, including internal/superficial activities, their histories and relationships with the outside environment; also, the structure, composition, and physical processes of the complex ring system, focusing on the dynamics of the constituent particles and embedded moonlets (or mini-satellites), their evolution and interactions with the magnetosphere and the whole Saturnian system (and even solar system); as well as the overall characteristics of the huge immersing magnetosphere with its plasma and many other physical processes and interactions with solar wind, rings and individual satellites.[41]

The Saturn orbiter, provided by NASA, would be based on the Mariner Mark II (MMII) spacecraft, a family of spacecraft designed by the Jet Propulsion Laboratory (JPL) for planetary exploration. Its scientific payload would include an imaging system, a magnetometer, a Titan radar mapper, different kinds of spectrometers, and so on. The Titan probe, built in Europe under ESA responsibility, would consist of two major components, an umbrella-shaped deceleration system, and a descent module (figure 12.2). The former would provide aerodynamic braking and thermal protection during atmospheric entry, the latter would descend through Titan's atmosphere by a parachute. Scientific data would be collected during the atmospheric descent phase and for a few minutes after its impact on the surface. During the probe mission, the orbiter would act as a radio-relay station to acquire its data for retransmission to Earth. The combined orbiter/probe spacecraft, associated with the upper stage of a Centaur rocket, was to be launched in 1994 into Earth orbit by the Space Shuttle. Subsequently, the Centaur upper stage would inject the spacecraft onto its interplanetary orbit.

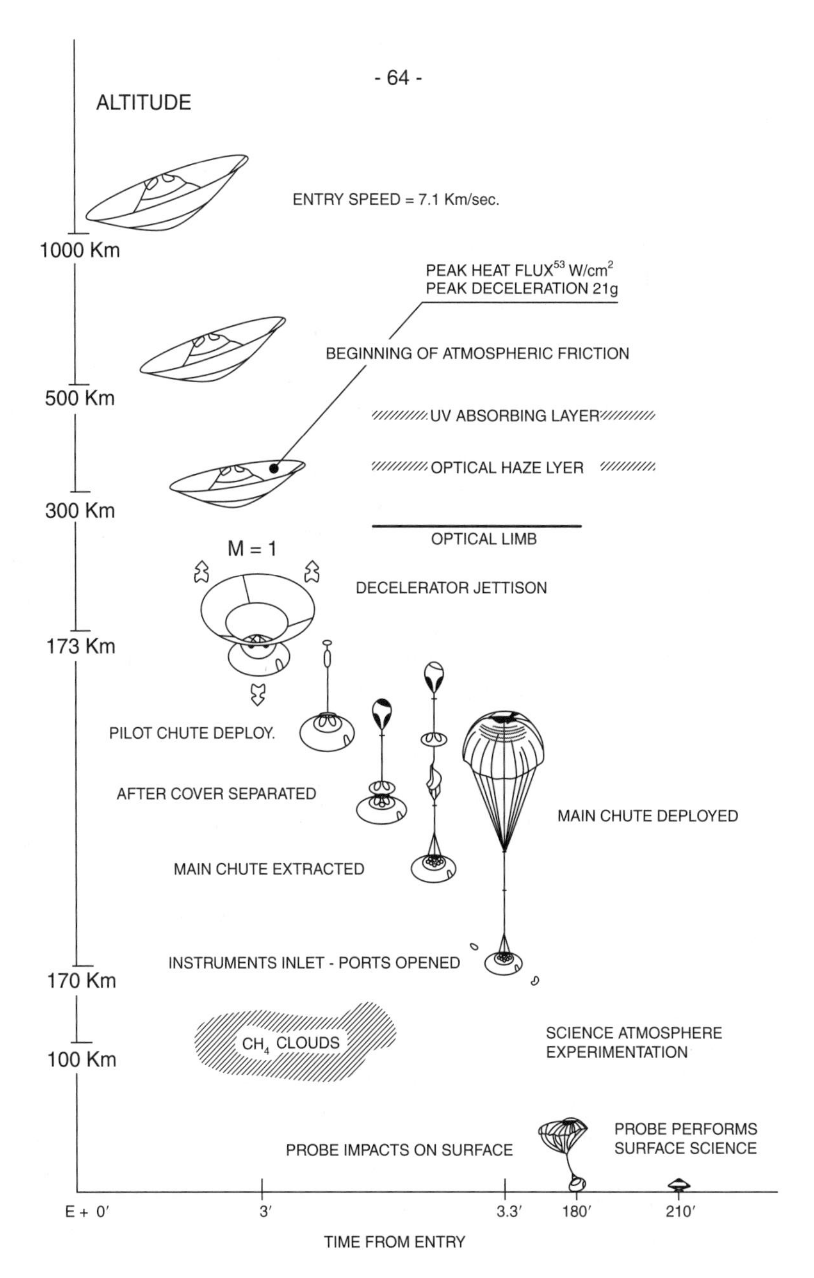

Figure 12.2 The Titan probe concept as executed with the Huygens lander. (Credit: ESA.)

The assessment study report was discussed by the SSWG on January 9, 1986, and the working group unanimously recommended that an industrial Phase-A study should be carried out for the European Titan probe of the Cassini mission to Saturn.[42] The following day, the SSAC was called to recommend which Phase-A studies should be supported by ESA, in view of the final selection of the first M-mission in Horizon 2000, scheduled for 1988. Besides Cassini, two other candidate missions were on the table, namely, the astronomy missions Lyman and Quasat recommended by the Astronomy Working Group (AWG). While all three missions were considered to be of outstanding scientific value and qualified for more detailed feasibility study, two questions had to be taken into account. First, the availability of resources would allow only two parallel Phase-A studies to be undertaken in 1986. Second, since all three missions were in collaboration with NASA, a decision should also take into account the decision process on the other side of the Atlantic. Now, Lyman and Quasat, if approved by NASA, could be included in the US Explorer Program by a decision expected in the next 6–8 months. Cassini, on the contrary, needed Congressional approval and a decision was not expected before 1990.

The discussion in the SSAC clearly showed the different views and feelings of those members coming from the astronomy community and those with a scientific interest in solar system exploration.[43] The AWG chairman, J. Leonard Culhane, stressed that it was impossible to establish a priority between the two astronomy missions and the need for immediate Phase-A studies for both of them was "urgent, compelling and timely." A delay might result in the loss of one or both missions, he argued, "because the time factor was extremely important [vis-à-vis NASA decisions]." And the SSAC chairman, the Dutch astrophysicist Johann Bleeker, insisted that it was very important to embark on Phase-A studies on both astronomy missions in order to influence the NASA planning of the next Explorer program. The time factor, he argued, was not so important for Cassini, which could be a candidate in the next round of Phase-A selection in early 1987. "This might be a psychological blow to the planetary community—acknowledged Bleeker—but the 1988 selection could still include Cassini and nothing would be lost." For the advocates of the Saturn/Titan mission, however, psychology could not be underestimated. There was "an overwhelming number of planetary physicists behind the Cassini mission," reminded Hugo Fechtig, an expert in interplanetary dust involved in the Giotto mission, and the SSWG chairman Martin Huber stressed that the planetary community had already been disappointed by the recent decision to abandon the Kepler mission to Mars in order to start immediately the SOHO/Cluster cornerstone.[44] The psychological aspect could not be neglected, he concluded: "If the two astronomy missions were selected for Phase-A studies, then Cassini must also been included. Clearly a problem of credibility existed." Credibility was also important for Roger Bonnet, who argued that the whole schedule problem was very critical:

> In order not to worsen the case, and since there was some momentum on this mission [it is important] not to destroy this by a delay in the Phase-A studies. Any delay on the start of a Phase-A on Cassini should be very clearly explained to NASA and the US scientific community involved, so that it would not be interpreted as a lack of interest on ESA's side.

In the event, by a majority vote, the SSAC recommended that Lyman and Quasat should be subjected to Phase-A studies starting immediately, while a Phase-A study on Cassini should be deferred until 1987, pending a clarification about NASA planning. The committee, however, underlined that in case the outlook for a timely mission emerging from discussions with NASA became unsatisfactory, they "would particularly welcome suitable proposals in the area of planetary sciences from among the new mission proposals [being submitted in response to ESA's call issued in July 1985]."[45] This sentence deserves a comment. At that time, the Giotto spacecraft was successfully approaching the Halley comet, with the rendezvous scheduled in two months' time. It was evident that Europe could play an important role in the field of solar system exploration, and the widespread support to the Cassini mission demonstrated that a competent and motivated planetary science community had established and claimed proper recognition in the ESA program. After the approval of Phase-A studies of two astronomy missions, it was clear that at least one of them would certainly be a strong candidate for the 1988 selection. The SSAC felt that should Cassini be abandoned because of NASA planning, another candidate in the planetary field should also be on the decision-makers' table.

The SSAC recommendation was finally endorsed by the SPC, but the decision to start a Phase-A study of the Cassini project in 1987 was far from being unanimous. Only six delegations voted in favor, four voted against (Belgium, Denmark, Italy, and the United Kingdom), and one abstained (Spain). While expressing their appreciation for the scientific value of the project, they considered that one could not rule out the possibility that by 1987 there might be other projects of equal interest, which Cassini should compete with.[46]

By autumn 1986, the situation appeared completely different. The Challenger accident in January that year, in fact, had far-reaching consequences on NASA's planning and the envisaged cooperation on Lyman and Quasat was no longer possible in the short term. The very feasibility of these missions was jeopardized, unless they could be redefined in a purely European frame or other cooperative ventures could be worked out (possibilities with Canada and/or Australia were under consideration). On the contrary, the support for the Mark Mariner II program of planetary exploration had been confirmed and Cassini was planned to be the second mission in this program, following a Comet Rendezvous Asteroid Flyby (CRAF) mission. In this situation, the SSAC recommended (and the SPC approved) that the assessment study on Cassini be updated in order to take into account the new mission scenario (in particular a 1995 launch by a Titan vehicle instead of the Space Shuttle), and a Phase-A study be initiated without delay.[47] The SSAC also confirmed its recommendation for Phase-A studies of Lyman and Quasat, albeit with reduced scientific objectives and performance requirements, in order to make their implementation possible in a purely European framework.[48]

Besides these three projects, coming from the mission proposals submitted in 1982, two other projects were recommended for Phase-A studies among those submitted in response to the call for mission proposals of July 1985.[49] The first was a gamma-ray astronomy mission called GRASP (Gamma-Ray Astronomy with Spectroscopy and Positioning), to be realized in a purely European context and launched by an Ariane vehicle.[50] The second project was Vesta, a cooperative endeavor of ESA, the French CNES, and the Soviet space agency Intercosmos (IKI). The project, named after one of the largest asteroids, foresaw the launch of two spacecraft aimed at visiting up to eight small bodies, possibly including one or two comets, over a seven-year period. Each of the two Vesta spacecraft consisted of a flyby module for flying by several asteroids at distances of the order of 500–2,000 kilometers, and an approach module that would be jettisoned in the vicinity of a selected asteroid and release two penetrators, which would anchor themselves to the target. All five projects would be candidate for the 1988 mission selection.

The industrial Phase-A study of the Titan probe in Cassini was conducted from November 1987 to September 1988 by an industrial consortium led by Marconi Space System. It was coordinated from the ESA side by Jean-Pierre Lebreton (study scientist) and George Scoon (study manager). From the American side, the main actors were from JPL: Wesley Huntress, a planetary scientist who would become NASA's associate administrator for space science in 1993, John Beckman (Cassini study manager), and Ronald Draper, the Mariner Mark II project manager. A joint science working group including European and American scientists supported the study. The composition of this group was the same as for the assessment study, with the addition of four new members (table 12.2).[51]

All Phase-A study reports (Cassini, Lyman, Quasat, GRASP, and Vesta) were discussed by the space science community during an important meeting attended by over three hundred scientists, on October 25, 1988, in Bruges.[52] In the following two days, the working groups and the SSAC were called to conclude the decision process by recommending which of these missions should be selected for inclusion in the ESA program. For the SSWG the choice was between Cassini and Vesta, a very difficult choice indeed, as both missions were dedicated to planetary exploration and considered "excellent and scientifically highly interesting."[53] The decision, of course, was a matter of politics as well as of science. From the scientific point of view, a close-up study of a number of asteroids and comets was as interesting and exciting as parachuting a (European) probe onto a planetary body in the outer solar system. The Vesta mission promised to pursue and extend the small-body exploration program that ESA had begun with the Giotto flyby of comet Halley and was presented as the forerunner of Rosetta, planned for the turn

Table 12.2 New members of the Cassini Science Team for the Phase A study (1987–1988)

M. Blanc	Centre de Recherche en Physique de l'Environnement, St. Maur
S. Calcutt	Dept. Atmospheric Physics, Oxford University
P. Nicholson	Cornell University, Ithaca
B. Swenson	NASA Ames Research Centre, Moffet Field

of the century. Cassini, for its part, would lead Europe to the very frontiers of solar system exploration and the European industry would acquire unique knowledge in the domain of atmospheric entry probes. The two missions, however, were very different as regards their political support, scientific constituency, and international framework.

Cassini had been conceived from the very beginning as an ESA/NASA collaborative project, in which ESA visibility was secured by the fact that the Titan probe would be built in Europe and its mission operated by ESOC. The scientific data from the probe would also be transmitted to ESOC, where the principal investigators (PIs) would be present during its historic descent onto Saturn's most interesting moon. A large and variegated scientific constituency had gathered in Europe behind Cassini, and looked at this ambitious mission as the well-deserved red-carpet entry into the field of planetary exploration, after the many disappointments of the past. They had two influential advocates in the ESA scientific advisory structure: that is, the SSWG chair Siegfried Bauer, whom we have met as one of the very first proponents of the Cassini concept, and the SSAC chair Hans Balsiger, a space physicist from the University of Bern who had been involved in ESA's GEOS and Giotto missions. Moreover, the American planetary science community was also strongly interested in the European approval of Cassini, in support of their eventual lobbying activity to have the mission approved by the Congress. Last but not least, Europe's involvement in the Cassini mission through the realization of the Titan probe would provide the European industry with unique and invaluable experience in the technology of planetary reentry vehicles.

In support of Vesta there was the powerful lobby of CNES. The mission, in fact, had been conceived in 1984 as a French/Soviet collaboration, the CNES being responsible for building the two spacecraft and IKI for the launch. The contribution of ESA had been solicited in response to the 1985 call for proposals, when the mission was at an advanced stage of definition, and many people, including Bonnet, considered that the agency would play only a subcontractor role to CNES. The SSWG itself had not been unanimous in recommending a Phase-A study on Vesta, as most members would have preferred the alternative mission proposal CAESAR (Comet Atmosphere Encounter and Sample Return), supported by a wide scientific community interested in further cometary studies after Giotto. The cost of CAESAR, however, was well beyond the financial limits set for an ESA medium-sized mission and no partner for a cooperative project could be identified. A majority vote for a Phase-A study on Vesta had finally been reached, and endorsed by the SSAC, but only with the proviso that a larger involvement and more visibility be granted to ESA, and a comet flyby be included in the baseline mission scenario.[54]

The discussion within the SSWG was lively and impassioned, the supporters of each mission advocating their respective merits and advantages. Among the SSWG members there were, on the one hand, W. Ip and M. Blanc, two obvious supporters of Cassini, and, on the other hand, P. Masson and J. A. M. McDonnel, members of the science team that had supported the

Phase-A study of Vesta. A list of "mission criteria" was drawn, according to which the working group should review and compare the two candidate missions. After this, the chair made a "tour de table" asking each member to express informally his/her preference. Following a further exchange of views and comments, a "second tour de table" was held, from which a clear majority consensus emerged in favor of Cassini. A formal vote was finally called, by which the SSWG approved a resolution recommending Cassini as the candidate project in the field of solar system science for the selection of ESA's next scientific project.[55]

It was up to the SSAC to make the final choice between Cassini and GRASP, the candidate project recommended by the Astronomy Working Group against Lyman and Quasat. The gamma ray mission, in fact, followed the long-standing ESA tradition in the field of high-energy astrophysics (COS-B and Exosat) and anticipated the ambitious XMM cornerstone mission in X-ray astronomy. Moreover, GRASP was a fully European mission, indeed the only purely European mission among the five candidate projects. As regards the two other projects, Quasat could hardly be realized within the established cost envelope and the envisaged ESA/NASA/Canada collaboration on Lyman was still to be clarified.[56]

The discussion within the SSAC covered all aspects of the important choice to be made.[57] On the one hand, the supporters of GRASP claimed a well-established European tradition of scientific excellence. On the other hand, the Cassini advocates stressed the importance of opening a new, fascinating territory to European space science. Moreover, by paying the ticket for the Titan probe, ESA would provide European space scientists with access to the NASA-built Saturn orbiter. Among the supporters of the gamma ray mission we can certainly number the AWG chair Len Culhane, an X-ray astrophysicist from University College London, and Claudio Chiuderi, an astronomer from the University of Florence. The case of Cassini was made out by the SSAC chair Hans Balsiger, the SSWG chair Sigfried Bauer, and David Southwood, a space physicist from Imperial College London then involved in the Galileo spacecraft magnetometer team. They succeeded in convincing the other astronomers in the SSAC membership, namely, Jean-Loup Puget, an IR astronomer from the Institut d'Astrophysique Spatiale in Orsay, and Peter G. Mezger, a radio-astronomer from the Max-Planck-Institut für Radioastronomie in Bonn. Once a "tour de table" consultation showed that a majority consensus existed in favor of the Cassini project, the SSAC agreed to recommend unanimously that the Saturn/Titan mission be selected as ESA's next scientific project.[58]

Following the SSAC recommendation, the ESA director general requested the SPC to select Cassini as ESA's next scientific project. Well aware of the controversial character of this decision, the DG acknowledged "the great disappointment of the communities supporting the non-selected projects." However, he continued, "This is the time honoured competitive procedure for scientific selection and has been applied systematically and successfully in the past. The scientific community knows and has accepted this procedure throughout the years."[59]

The member-state delegations in the SPC finally endorsed the SSAC's recommendation, but some of them wanted to express their strong interest in other candidate projects and insisted that all of them be submitted to a new competitive selection procedure, should the Cassini project fail to be approved by the US Congress. The French delegation, in particular, regretted that ESA was selecting a major project "once again depending on NASA, given the difficulties that were known to arise in such cooperation." Vesta, on the contrary, could have provided the European space science community with an opportunity to establish a cooperative venture with the Soviet Union, "thus restoring a measure of balance in international cooperation."[60]

At this SPC meeting it was also decided that the Titan probe should have its own name, in order to underline that it was a separate, European-built component of the mission. Reminding delegations that the Saturn moon Titan had been discovered in 1655 by Christiaan Huygens, Roger Bonnet proposed the name "Huygens" for the Titan probe that Europe would contribute to the Cassini mission. The Netherlands delegation very much appreciated this homage to the Dutch astronomer and natural philosopher who had been one of the founding fathers of modern science. The chairman of the SSAC (H. Balsiger), for his part, said he was happy to accept the proposal on behalf of the European space science community.[61] Huygens thus entered the ESA Scientific Programme as Europe's first planetary mission and the first medium-sized mission (M1) in the Horizon 2000 long-term program.

Ups and Downs of Equal Partnership

During the Phase-A activities in Europe, important developments had occurred in the United States. In early 1988, the Cassini scientific constituency succeeded in convincing NASA policymakers to combine the Saturn/Titan mission and the CRAF mission into a single program, to be submitted to Congress for fiscal year 1990 as a new start. In fact, maintaining the original schedule of two different programs, CRAF first and then Cassini, would have implied a delay of four years. On the contrary, by using standardized equipment for both missions within a single program, a dramatic cost saving could be achieved. NASA could get two missions for the price of one and a half, argued Owen and his colleagues, and all sectors of the planetary science community would be satisfied in one stroke. The argument went through the chief of solar system programs, Geoffrey Briggs, and the associate administrator for space science, Lennard Fisk, up the NASA administrator James Fletcher, who agreed to submit the CRAF/Cassini program to the White House and the Congress for inclusion in the NASA budget.[62] These developments were regarded with understandable satisfaction in Europe: "The CRAF/Cassini programme is sailing safely through the perilous waters of the several [Congress] Committees," wrote Coradini to the SSWG members. He further explained that

> [t]he House Appropriations Committee has included, as from NASA request, 30 M$ for the initiation of the CRAF/Cassini Program, setting a financial upper limit for this program of 1.6 B$. Should problems arise, leading to

> [...] a cancellation of one of the missions, CRAF will be the one to be terminated. These are very positive signs which make us feel confident of a final success.[63]

In the event, the Congress approved the dual CRAF/Cassini program in October 1989, one year after ESA's decision on Cassini-Huygens. At this time, the mission was set for launch in April 1996, with arrival at Saturn in December 2002. During its cruise, the spacecraft would have the opportunity to fly by an asteroid one year after launch and obtain important scientific information about its properties. Moreover, Cassini would fly by Jupiter in February 2002, and its instrumentation would allow new studies of the planet's magnetosphere and atmosphere two years after the end of the Galileo mission.

According to Roger Bonnet, the Cassini-Huygens mission can be considered as "a paradigm of international cooperation."[64] The selection of the scientific instruments on board the orbiter and the Titan probe was an important aspect of this paradigm. In fact, it was decided that payload selection would be subjected to coordinated planning and that each payload should include instruments provided by scientific groups from both sides of the Atlantic. As usual in space missions, each instrument would be realized under the responsibility of a PI, supported by a team of coinvestigators (CoI). Moreover, a number of interdisciplinary investigations was envisaged, both in the orbiter and the probe, to address the wider Saturnian system using a combination of instruments under the responsibility of an interdisciplinary scientist (IDS). Finally, the orbiter payload would include five facility instruments, each operated by a scientific team coordinated by a team leader. Both ESA and NASA would set up a committee for scientific evaluation of proposals received for the probe and the orbiter, respectively, but it was agreed that each party should appoint one-third of the members of the other's evaluation committee.[65] The selection procedures of ESA and NASA were not very different from each other, but NASA was financially responsible for the orbiter payload while, following the usual ESA rules, the Huygens instruments were to be paid by the scientific groups providing them.

On October 10, 1989, with the mission finally approved, NASA and ESA issued parallel Announcement of Opportunity (AO) for instrument proposals and interdisciplinary investigations on the Saturn orbiter and the Titan probe, respectively.[66] The American space agency received over two hundred proposals, which were reviewed by a 60-member peer review panel, comprising two-third American and one-third European participation. ESA representatives were invited to participate as observers in all meetings and a joint NASA/ESA meeting was held in July 1990 to discuss candidate instruments. A payload recommendation was then formulated in November by Fisk and approved by the NASA administrator. In the event, the Cassini orbiter included seven PI instruments, five facility instruments, and six interdisciplinary investigations. Europe was represented by two PIs and one IDS. The European contingent also included eighteen facility instrument team members and fifty-seven CoIs.[67]

A similar procedure was adopted in Europe. In response of the ESA Announcement of Opportunity for Huygens, twenty proposals were received, including six with European PIs and fourteen with US PIs. A peer review committee (PRC) was set up by the SSWG, composed of nine members, six Europeans and three Americans, and chaired by the SSWG chairman Philippe Masson, a planetary geologist from the University of Paris Sud (Orsay). NASA representatives attended the PRC meetings and endorsed its conclusions.[68] The PRC concluded its work in June 1990, recommending six instruments on the Huygens payload, four with a European PI and two with an American PI (table 12.3). The Titan science mission also included three interdisciplinary investigations, coordinated by two European and one American IDS, respectively (table 12.4).[69] The PRC recommendation was endorsed by the SSWG and the SSAC, and eventually approved by the SPC in September 1990.[70]

Table 12.3 The Huygens scientific payload approved in 1990

Acronym	instrument	Science objectives	Principal investigator
HASI	Huygens atmospheric structure instrument	Atmospheric temperature and pressure profile, winds and turbulence; electric properties of the atmosphere	Marcello Fulchignoni Istituto Astrofisica Spaziale, Rome (I)[a]
GCMS	Gas chromatograph and mass spectrometer	Chemical composition of gases and aerosols in the atmosphere	Hasso Niemann NASA/GSFC, Greenbelt, MD (USA)
ACP	Aerosol collector and pyrolizer	Aerosol sampling	Guy Israel Service d'Aéronomie, Verrieres-le-Buisson (F)
DISR	Descent imager and spectral radiometer	Atmospheric composition and cloud structure; aerosol properties; surface imaging	Marty Tomasko University of Arizona (USA)
SSP	Surface science package	Titan surface state and composition at landing site	John Zarnecki University of Kent (UK)[b]
DWE	Doppler wind instrument	Wind profile	Michael Bird University of Bonn (D)

[a] Now at the Observatoire de Paris-Meudon and Université de Paris 6 (F).
[b] Now at the Open University, Milton Keynes (UK).

Table 12.4 The Huygens interdisciplinary investigations approved in 1990

Investigation	Interdisciplinary scientist (IDS)
Aeronomy of Titan's atmosphere	Daniel Gautier, Observatoire de Paris—Meudon (F)
Study of Titan's atmosphere-surface interactions	Jonathan Lunine, University of Arizona (USA)
Study of Titan's chemistry and exobiology	François Raulin, Université de Paris—Val de Marne (F)

With the definition of the scientific payloads and the selection of the groups responsible for building the instruments, the Cassini-Huygens mission was well on its way toward launch. In early 1991, the launch date was moved forward to November 1995 and a new trajectory planned, including a Venus slingshot and a close flyby of another asteroid en route to Jupiter. On the NASA side, the project was being implemented at JPL under the responsibility of project manager Dick Spehalski and project scientist Dennis Matson. In Europe, the Huygens development team was based at ESTEC under the leadership of Hamid Hassan, while Jean-Pierre Lebreton acted as project scientist. For the implementation of the scientific aspects of the mission, a Cassini Project Science Group (PSG) was set up, composed of all PI and IDS, the team leaders of the orbiter facility instruments, and the ESA and NASA project scientists. A Huygens Science Working Team (HSWT) was also established by ESA, including the probe PI and IDS, with the task of providing scientific advice to the Huygens project manager.

The way, however, proved to be fraught with difficulties arising less from technical problems than from political uncertainty in the United States. In October 1991, in fact, the US Congress dwarfed the 13 percent increase for the NASA 1992 budget requested by the White House, approving only a 3 percent increase, hardly enough to compensate for inflation. In this framework, the budget for the CRAF/Cassini program was set at $211 million, dramatically less than the requested $328 million.[71] As a consequence, in order to ease its annual funding, NASA decided that the launch of Cassini should to be delayed until October 1997, an option that the ESA director general, Jean-Marie Luton, considered as "unacceptable!" "Delaying the mission until 1997 would not only cost ESA an additional $30 million, but would increase the danger of even greater delays since the next launch opportunity after 1997 occurs only at the end of the century in 1999, that is to say 4 years after the presently agreed launch date."[72] Luton also regretted that NASA had planned the delay without any consultation with its European partner. "This is not in the spirit of the MOU whose implementation, until now, had offered the basis for a fairly smooth interface between the two sides of the project."

The DG's initiative reflected the deep concern of the European space science community. The SSAC chairman, David Southwood, interpreted their feelings in a letter to his American counterpart, the chairman of NASA's Space Science and Applications Advisory Committee, Berrien Moore:

> A slip to a 1997 launch for Cassini creates an intolerable stress on our programme by increasing our cost for the Huygens probe up to launch (the critical figure for us) by about 15 % [...]. Although I appreciate the perennial problems generated by the year-by-year approach to budgeting in the US system, please understand in turn our predicament where we work to fixed costs to completion assigned when the mission is approved. We have dragooned, cajoled and otherwise persuaded our member states into not only agreeing the central funding for the Huygens probe but committing funding for Huygens and Cassini instrumentation by emphasising the importance of not delaying

> the NASA timetable and they have responded positively. It is appalling within a year of the selection to be facing a serious budgetary crisis that is entirely not of our making. The situation [is] very serious for the European programme right now but I cannot emphasise enough the change [it] can easily induce in the climate of cooperation.[73]

The specter of the Ulysses affair seemed to be back ten years later. "It [is] a case of 'once bitten, twice shy,'" commented the Swiss delegation to the SPC, urging the ESA executive "to take a firm stand in defending the interests of the [European] scientific community and in insisting that commitments be honoured."[74] The credibility of NASA scheduling and its very reliability as a partner were explicitly challenged: "It should be impressed upon NASA that international cooperation was a valid proposition only as long as the partners honoured their commitments," said the Netherlands delegation.[75] Commenting on the situation during a dramatic SSAC meeting, where the possibility of cancelling the Huygens project was also taken into consideration, the Dutch astronomer H. Habing (supported by the Italian C. Barbieri) considered that "cooperation with the US had always been fundamental but there was a breaking-point at which it was no longer possible to continue, and ESA was very close to this point."[76]

Despite all efforts, it was hardly possible for ESA and the European scientists to influence the US Congress debates on such an important item in the federal budget. In autumn 1989, the Congress had approved the CRAF/Cassini program within a budget limit of $1.6 billion, but the cost of Cassini alone was now estimated at about $1.7 billion and the actual implementation of the whole program was definitely jeopardized. In fact, on January 29, 1992, the US president George Bush definitely cancelled the CRAF mission in its 1993 budget request for NASA. The latter, for its part, started planning a dramatic reconfiguration of the Cassini project—a "descoping exercise," in the space jargon—in order to reduce its overall costs and obtain Congress approval.[77] By May that year, the head of the Scientific Projects Department in ESTEC, David Dale, told concerned SSAC members that "NASA was confident that the 1.4 B$ necessary to undertake the mission would be approved in the FY 93 budget."[78]

The JPL engineers made important changes in the orbiter design. In the originally envisioned Cassini spacecraft, the remote-sensing instruments and the particles-and-fields instruments were to be mounted on a scan platform and a turntable, respectively, in order to keep them pointing at a target and continuously taking data without the vehicle needing to maneuver. Moreover, a separate steerable antenna would provide the communications link with the Huygens probe, thus allowing the orbiter remote sensing instruments to observe the entry of Huygens into Titan's atmosphere. These features did not survive in the "descoped" design: the instruments would be bolted to the craft and the orbiter-probe communications link would be realized by the same antenna used for communications link to Earth. The obvious consequence was that the entire spacecraft had to be turned toward a target to make observations and toward Earth to transmit data. Moreover, since the antenna would not be facing the Earth at all times, a buffer system was required to hold the data until the observations were complete and the antenna could be turned back toward the Earth for downloading. In other words, the expensive engineering complexity of the original design was traded for later operational complexity. At the same time, in order to reduce the operation costs during the first few years in flight, all plans for acquisition of scientific data en route to Saturn (asteroid flyby, Jupiter flyby, and cruise science) were given up. The results of JPL's descoping exercise were not peacefully accepted by the scientific community, much disappointed over the removal of the scan platform from the orbiter spacecraft. But, as Toby Owen put it when Cassini was successfully going into Saturn orbit, "[T]he alternative was far worse: no mission!"[79]

A new major political crisis burst in 1994, when the US Congress started discussing NASA's 1995 budget.[80] In May that year, the Senate's Appropriations Committee allotted to the panel that supervised the NASA budget $700 milion less than the amount requested. The consequence, warned the panel chair Barbara Mikulski, was that, in order to save the space station program, either Cassini or the X-ray astronomy mission AXAF had to be cancelled.[81] The prospects for Cassini were not encouraging as the mission did not enjoy the support from NASA's administrator Daniel Goldin. From the very beginning of his tenure, in April 1992, Goldin had enunciated publicly his approach to the US space program, a new policy he synthesized by the well-known slogan "faster, cheaper, better." With the overall space budget on the decline after the Cold War was supposed to be over, he felt that NASA should undertake smaller and cheaper missions, which could be realized in a short period (typically 8 years, i.e., two presidential terms), with quicker scientific return in case of success and less financial risk in case of failure. Cassini-Huygens was the perfect antithesis of this policy: a mammoth mission conceived ten years earlier which would not provide scientific results until more than ten years later. Its cost was to be measured in billions of dollars and a failure at launch or beyond would have been devastating. Its management suffered from the complexity of a large international cooperation and an 18-instrument payload coming from countless scientific institutions in two continents.[82]

Goldin did not hide his distrust toward the Cassini mission and, apparently, he was prepared to cancel it, or at least to submit it to a drastic revision in order to reduce its cost.[83] Reporting to Bonnet on a meeting of NASA's Space Science and Applications Advisory Committee, the SSAC chair Lodewijk Woltjer wrote with worry:

> It is not at all obvious what will happen if a choice had to be made between AXAF and Cassini. International cooperation remains a secondary consideration, compared to the relative weight of the various pressure groups. Mr. Goldin stresses his "faster, cheaper, better" policy, which was echoed by everyone. All missions seem to become cheaper, whether on the basis of technological developments or of wishing it to be so, is less clear. [...] The overall

> impression is one of total uncertainty. According to Mr. Goldin, everyone should support the whole programme (including Space Station) or everything will go down. In addition, he would immediately cancel any programme that runs above budget. On a question if this applied also to the Space Station, the answer was rather long. [. . .] My impression is that it would be very imprudent for ESA to depend too strongly on NASA for its long term future projects.[84]

The message was clear to Bonnet and his staff, as well as to Gautier and the other European scientists who were working hard at that time to build the instruments.[85] In this situation, a strong action was then undertaken to save the mission. At the scientific level, the European planetary scientists joined the effort of their American colleagues to advocate the mission in any scientific and political forum where NASA policy was discussed. They found an influential supporter in Carl Sagan, a well-known astronomer and author of best-selling popular scientific books.[86] The important journal *Space News* published an editorial in which the cancellation of Cassini or AXAF was considered as "an unacceptable option for reducing NASA's budget."[87] Not less important were the initiatives at institutional level. The Italian Space Agency (ASI), which was building the antenna and the radio frequency subsystem for the Cassini spacecraft in the framework of a NASA/ASI bilateral agreement, reportedly told the Americans that Italy's important contribution to the International Space Station was at stake.[88] An unprecedented initiative was undertaken by the ESA director general, Jean-Marie Luton, who sent a formal letter to the US vice president, Al Gore. Here he stressed the strong commitment of ESA in the Cassini mission (around $300 million, of which two-thirds was already spent, not including the approximately $100 million contribution of Italy) and concluded: "Europe therefore views any prospect of a unilateral withdrawal from cooperation on the part of the United States as totally unacceptable. Such an action would call into question the reliability of the US as a partner in any future major scientific and technological cooperation."[89] Finally, a strong diplomatic action was taken by ESA member-state governments, whose ambassadors in Washington appealed to the State Department to warn about the devastating consequences of the eventual cancellation of the Cassini mission.[90]

In any event, President Bill Clinton came to NASA's aid by redirecting other spending in order to shore up the agency's science budget. Mikulski's Panel therefore could grant NASA enough money to support both the space station and the major scientific missions.[91] Having thus survived the attack in the Congress, Cassini could not spare itself of Goldin's persisting distrust. In 1995, the NASA administrator demanded that the whole Cassini program, including the Huygens probe and the antenna being developed in Italy, be submitted to an independent review by a team of external experts. This was felt in Europe as an unacceptable violation of the cooperative spirit, all the more so as the US laws prevented the Europeans from having access to any American technical know-how related to defense. "After some tense meeting, we eventually accepted the principle of that review," recalled Bonnet a few years later. The review team was led by Herbert Kottler, head of the Aerospace Division of MIT's Lincoln Laboratory, and Bonnet recognized that their work eventually turned into a positive experience, both because the ESA contribution came successfully out of it, and because it created confidence and strengthened the cooperative spirit.[92]

A Long Journey

Following the 1994 political crisis and the 1995 review, the Cassini development program went on smoothly (figure 12.3). During 1996, the 319-kilogram Huygens probe was integrated and tested at the Dornier facilities in Ottobrunn near Munich. In April 1997, following the satisfactory conclusion of the Flight Acceptance Review, it was shipped to Kennedy Space Center to be fitted to the Cassini main spacecraft. By this time, the large scientific community involved in the mission (more than 250 scientists from 17 countries) had defined in detail the scientific objectives of the mission and determined, together with engineers and mission planners, the orbital characteristics that offered the best opportunities for the various instruments. The launch was originally scheduled for October 6, 1997, but it had to be delayed because an accident in the cooling system when the spacecraft was already mounted on the Titan IV rocket damaged the insulating foam of the Huygens probe.[93] It was eventually scheduled for October 15, 1997.

In the few months before launch, the mission came under attack by the antinuclear movement. The target of the "Stop Cassini" campaign launched by the Florida Coalition for Peace and Justice (FCPJ) was the spacecraft's electrical power source, based on three radioisotope thermoelectric generators (RTGs) powered by 32.7 kilograms of plutonium dioxide. Moreover, the orbiter and the Huygens probe also carried 129 radioisotope heater units (RHUs) with a total of 0.35 kilograms of plutonium dioxide for temperature regulation. Activists initiated a mass letter-writing campaign to politicians, and some leaders of the protest threatened to stop the launch by invading the launch pad at Cape Canaveral. A few Congressmen joined the protest, sending letters to NASA and the White House to halt the launch. Almost daily events (public debates, press conferences, rallies, etc.) were organized by the FCPJ in September and early October, including rallies at the United Nations in New York (September 20) and the White House in Washington (September 28), and culminating in an international demonstration announced at Cape Canaveral on October 4, where "affinity groups will attempt to enter [the] Air Force Station and sit on the Cassini launch pad in order to stop the launch." That same day, a "Cancel Cassini" demonstration was also called in Darmstadt, including "a march to the [ESOC] offices of the European Space Agency."[94]

The "Stop Cassini" protests were important enough to convince Bonnet to write a nervous letter to all European scientists involved in the mission, inviting them to respond "to the legitimate concerns of the public or even the science community, or to smooth out any emotional statement on issues which have thoroughly and seriously analysed by the US and European mission

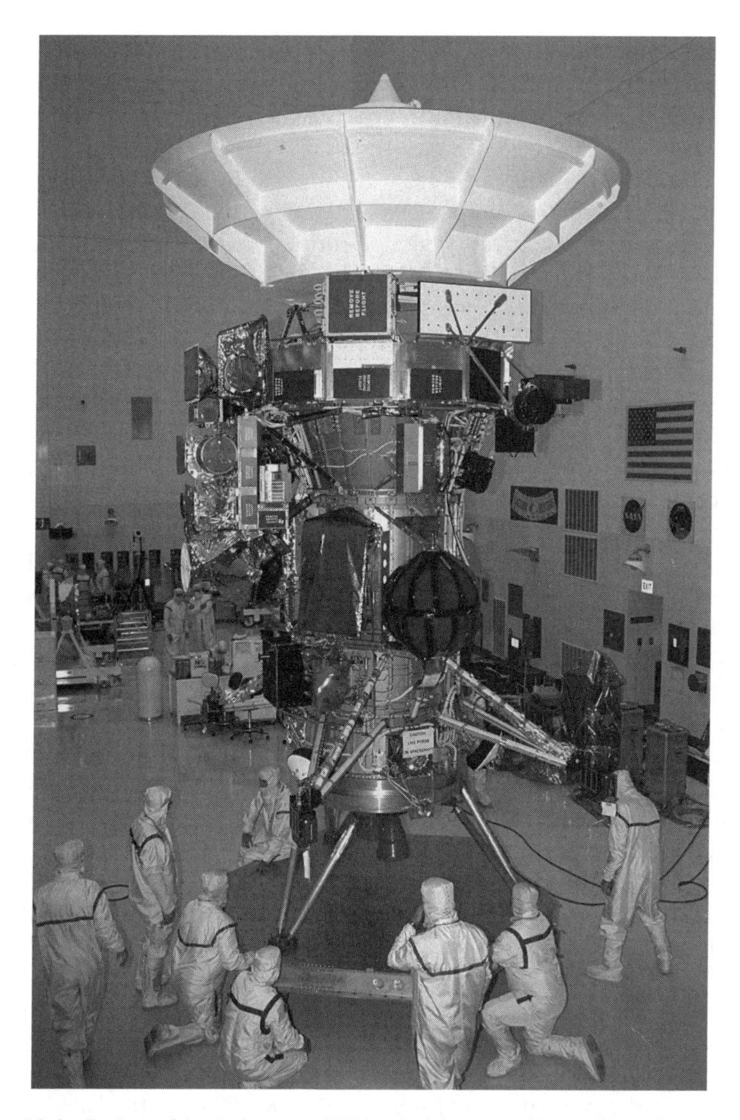

Figure 12.3 Jet Propulsion Laboratory (JPL) technicians reposition and install the Huygens probe on the Cassini orbiter in the Payload Hazardous Servicing Facility (PHSF). (Credit: NASA/JPL image no. KSC-97EC-1098.)

planners." In order to help them in this effort, Bonnet attached "fact sheets" about the RTG technology and the need of resorting to such devices for deep-space missions, when solar power is not feasible.[95] One of the authors of the original Cassini proposal, Wing-Huen Ip, went as far as to write a letter to the US president Bill Clinton to express his and his colleagues' concern about the possible disrupting effects of the protests.[96] In the event, following NASA's formal request for "nuclear launch safety approval," the director of the White House Office of Science and Technology Policy authorized the

NASA administrator to proceed with the launch of the Cassini spacecraft, "having reviewed the final Safety Analysis Report of the Department of Energy and the Safety Evaluation Report of the Interagency Nuclear Safety Review Panel."[97]

The Cassini-Huygens mission was successfully launched from Cape Canaveral Air Force base on October 15, 1997, at 4:43 hrs local time. About five hundred representatives of the scientific, engineering, and industrial teams in Europe, which had created the Huygens probe attended the launch.[98] Eight days later, at 10:09 Central European Time, ESOC established connection with the spacecraft and a happy bunch of scientists and engineers were reassured that Huygens was "alive and well in all respect." All engineering systems and subsystems were performing nominally and all experiments were functionally capable of accomplishing their task: "Six experiments, six green lights," claimed Huygens project scientist J.-P. Lebreton. And the project manager H. Hassan finally announced: "We will now let Huygens go back to sleep, except for the planned six monthly checkouts. The Probe will remain in that condition for the seven-year journey to Saturn. But we now have every reason to expect a successful outcome of this unprecedented mission."[99]

Huygens's journey was not as peaceful as expected.[100] In February 2000, when Cassini was somewhere halfway between Venus and Jupiter, a test of the radio relay link between the probe and the Cassini spacecraft showed that the receiver presented unexpected behavior when a simulated Doppler effect was applied to the carrier. Further laboratory analysis confirmed that the Doppler frequency shift caused the data signal from Huygens to fall outside the narrow bandwidth of Cassini's receiver.[101] The news could not be worse: it knelled the loss of the Huygens mission! In fact, the scientific data gathered by the probe were to be relayed to Earth through a special dedicated radio communication system on Cassini. The relay link would start just after the release of the probe and end when Cassini passed Titan and lost sight of Huygens (about three hours later). During that time, the relative speed between the probe and the Cassini spacecraft would be very high, what implied a significant Doppler shift in the signal received by the latter. In other words, even though Huygens duly sent its scientific data during its descent on Titan, they could not be recorded by Cassini's receiver.

The Huygens communication system had been provided by the Italian firm Alenia Spazio and it had been tested both in Alenia's laboratories and at Alcatel, the French prime contractor for the Huygens probe. All tests had been successful, but apparently the system had never been tested in the actual mission scenario, when differences in relative speeds and trajectories would be important. And, what was more awkward, the shortcoming with respect to the Doppler shift did not surface during any of the ESA, NASA/JPL, or independent reviews performed during the project life cycle.

Once the problem clarified in all its gravity, a Huygens Recovery Task Force (HRTF) was set up, in order to study possible recovery options and rescue the mission. The HRTF included experts from NASA and ESA, and was cochaired by Kai Clausen of ESTEC and Leslie J. Deutsch of JPL. During six months of hard work, the team performed extensive ground and flight testing, modeling, and simulation to understand the failure mechanism in the radio relay.[102] They elaborated models to predict the corruption of science data under several scenarios and discussed these predictions with the science teams to determine the overall performance in each scenario. Finally, the HRTF recommended a solution, which, they claimed, would return 100 percent of Huygens science data. This involved an important change in the Cassini trajectory in order to reduce the Doppler shift to the point that the spacecraft's receiver could track the Huygens signal during the descent. According to the new scenario, Cassini would fly by Titan at sixty-five thousand kilometers rather than the planned twelve hundred kilometers, and Huygens would enter the Titan's atmosphere on January 14, 2005, seven weeks later than originally planned.[103] In June 2001, the new ESA director of science, David Southwood, and the NASA associate administrator for Space Science, Edward Weiler, endorsed the HRTF recommendation and directed the Cassini and Huygens teams to proceed with the implementation of the new mission scenario.[104] By November that year, following five days of extensive tests, scientists and engineers on both sides of the Atlantic were confirmed that Huygens would be able to fulfill its mission.

The fulfillment of the Cassini-Huygens mission is not yet a matter for history, but current scientific news. The spacecraft entered orbit around Saturn on July 1, 2004, after a seven-year cruise through the solar system, including four gravity-assisted swing-by maneuvers by Venus (April 26, 1998, and June 24, 1999), Earth (August 18, 1999), and Jupiter (December 30, 2000). On Christmas day, 2004, Huygens was successfully released by the Cassini mother spacecraft and reached Titan's outer atmosphere on January 14, 2005, starting its descent through hazy cloud layers at 9:06 UTC from an altitude of about 1,270 kilometers. Six minute later, following the deployment of the main parachute and the separation of the heat shield, the transmitter was turned on and instruments started taking data at an altitude of about 150 kilometers. The first evidence that Huygens was alive arrived at about 10:25 UTC, when the Green Bank radio telescope in West Virginia, United States, picked up the faint radio signal from the probe. The first scientific data, relayed by the Cassini orbiter via NASA's Deep Space Network, arrived at ESOC at 16:19 UTC.[105] The excitement of the moment was captured by the scientific magazine *Nature*:

> Scientists waiting anxiously for the data to arrive [...] hugged each other when the first signals arrived during the morning, showing that the mission, 20 years in the planning and execution, was functioning. Many had worked most of the previous night preparing for the data's arrival. Their delight grew as they began to look at what the data—relayed in compressed from the mother ship Cassini in the late afternoon—seemed to be telling them [...] It was 15 hours from the start of the data bump to the press conference that brought the world the first images of Titan. And they were perhaps the most intense

> times in the professional lives of the six Huygens investigators and their teams [...] Celebration and concentrated work ran in parallel. High-ranking guests, including ministers and space-agency officials, banqueted in ESOC's main building as scientists devoured data in the building next door.[106]

Following more than 22 years of studies, planning, development, and interplanetary travel, Huygens's primary mission was finally fulfilled during the 147-minute parachute descent, with its six instruments studying the physical and chemical properties of Titan's atmosphere all the way from an altitude of about 150 kilometers to the ground level (figure 12.4). After landing, the surface phase of the mission lasted about 72 minutes, considerably longer than had been anticipated.[107] No scientific measurements were performed before Titan arrival and Huygens was dormant for most of the

Figure 12.4 Image of the surface of Titan taken by the European Space Agency's Huygens probe when it touched down. First images released by the ESA depict sinuous drainage channels leading to an apparent shoreline. What's draining? Possibly liquid methane. The orange landscape around the Huygens landing site is littered with little rocks, rounded and smooth like river-rocks on Earth. One of the images seems to show tendrils of ground fog made not of water but perhaps ethane or methane. (Credit: NASA/JPL/ESA image no. PIA07232.)

3.5-billion-kilometer cruise with its mother spacecraft. Cassini listened to Huygens for about 3 h 40 min, then it was commanded to turn its antenna toward Earth for transmission of the stored probe data. More than 474 megabits of data were thus received, including some 350 pictures collected during the descent and on the ground, which revealed "a landscape apparently modelled by erosion, with drainage channels, shore-like features, and even pebble-shaped objects on the surface."[108] A flow of scientific papers started to be published both in Europe and the United States, reporting the results obtained by the various instruments carried onboard the Cassini mother spacecraft and the Huygens probe.[109]

Epilogue

Following the pioneering Giotto mission, the Cassini-Huygens mission established a solid base for the ESA involvement in planetary science. This involvement was eventually characterized by two different kinds of undertaking. First are the two large cornerstone projects Rosetta and BepiColombo. The former, as we have seen, was included in the Horizon 2000 program, following the successful Giotto mission. It is basically a pure ESA mission, with only three Americans among the 26 PIs who have contributed to the scientific payloads of the orbiter and the lander. The latter, named after the Italian scientist Giuseppe (Bepi) Colombo, is a mission to Mercury consisting of two orbiters, one for the study of the planet's surface and internal composition and another for magnetospheric studies. It was included as a new planetary cornerstone in the so-called Horizon 2000 Plus program, approved in 1994 as an extension of Horizon 2000 up to 2016.[110] The mission is scheduled for launch in 2015, with arrival on Mercury in 2022. BepiColombo is a collaborative venture (under ESA leadership) between ESA and the Japanese Aerospace Exploration Agency (ISAS/JAXA). The latter will be responsible for the realization of the Mercury Magnetospheric Orbiter, with four of its five instruments being provided by Japanese PIs (one by a European PI). ESA will be responsible for the realization of the Mercury Planetary Orbiter, whose 11 instruments will all but one be realized in ESA member states (one in Russia).

Both Rosetta and BepiColombo are large and ambitious missions, highly expensive and technologically challenging.[111] They give concrete form to the European commitment in an important autonomous effort in planetary exploration, as established in the Horizon 2000 philosophy. The selection of Mercury as the next major target was not uncontroversial. During the discussions for the definition of the Horizon 2000 Plus plan, in fact, the planetary cornerstone was a matter of hot debate. It was clear that European missions to the outer solar system could not be considered because the required RTG technology had not been developed in Europe, and could hardly be developed in the future because of political constraints. As regards the Moon, ESA was studying at that time the feasibility of a "Moon Programme" on an optional basis, therefore independent from the mandatory Scientific Programme.[112]

This left the terrestrial planets, more specifically which of them to choose for a major ESA effort in the new millennium. The choice, in fact, was between Mars and Mercury, as Venus had just been visited by NASA's Magellan mission (1990–1994). Opinions were very much divided and, eventually, this was the only issue about which the survey committee preparing the Horizon 2000 Plus plan was called to a vote.[113] Apparently there were two main reasons for Mercury to prevail. First, the scientific community interested in Mars considered that an ambitious cornerstone mission would take too long in order to be realized (certainly after Rosetta!) and then preferred to stake on a medium-sized mission then under study within the ESA planning cycle (Intermarsnet). Second, both planetary scientists and magnetospheric physicists were interested in Mercury, thus making two communities supporting a mission to this planet.[114]

The second kind of planetary projects undertook by ESA at the beginning of the new century is quite different from the massive and costly cornerstone missions, as well as from the "ordinary" medium-sized missions. They are two "express" missions to Mars and Venus, respectively, and a small lunar orbiter. They came out from an important revision of the managerial and industrial procedures of scientific projects carried out in ESA in the depressed financial context of the 1990s.[115] In fact, the October 1995 ministerial meeting of ESA member states in Tolouse froze the scientific budget for the next three years, with no increase for inflation. Moreover, in June 1996 the Cluster mission, aimed at studying the Earth's magnetosphere using four identical spacecraft flying in a tetrahedral formation, was lost in the launch failure of the new Ariane 5 vehicle. Bonnet and his staff felt they had a moral commitment to work out a recovery mission for Cluster, but this required new funding and a dramatic revision of the implementation plan of Horizon 2000 Plus. The outcome of many discussions in January–April 1997 was a new approach for implementing scientific projects. This included three main elements: first, the introduction of new management methods in order to significantly reduce the development and industrial costs; second, the concept of "Flexi" (F-) missions, that is, faster and cheaper missions to be decided in a shorter decision-making period and based on the reuse of existing hardware, spacecraft subsystems, and contractual arrangements with industry; third, the realization of small or micro satellites aimed at demonstrating key technologies related to a cornerstone project.

In this new framework, Mars Express came out as the first F-mission. It was selected in 1997 without a formal competitive process, taking advantage, on the one hand, of spare instruments developed for the ill-fated Russian Mars-96 mission and still available on the shelf in several European laboratories; on the other hand, of spacecraft elements and subsystems being developed for the Rosetta mission.[116] Its cost was fixed at 150 million Euro (1996 rates), that is, about half the cost of a typical M-mission, and its development from approval to launch also required about half the time. Launched on June 2, 2003, Mars Express arrived at the Red Planet on Christmas day that year. The mission was planned to last one Martian year (687 Earth days), but it has eventually been extended until end 2014. Venus Express followed the path laid out by its predecessor. It was selected in November 2001 following a call of ideas to reuse the Mars Express platform, and eventually approved for implementation one year later. Then it took three years to launch of mission (November 9, 2005), thanks to reuse of the same design as Mars Express and the same industrial teams that worked on that mission. Moreover, five of its seven scientific instruments had been developed for Mars Express (three) and Rosetta (two). Venus Express entered Venusian orbit on April 11, 2006. The end of the mission is scheduled to end in 2014.

The lunar-orbiting mission was selected as the first Small Mission for Advanced Research in Technology (SMART-1), with the objective of flight testing the Solar Electric Propulsion (SEP) technology for future deep space missions, in particular BepiColombo. Approved in 1999, its cost was set at 84 million Euro (1999 rates), a very low figure, which meant that a low-cost launch and a new procurement and management approach had to be adopted. In fact, the SMART-1 spacecraft was launched on September 27, 2003 as an auxiliary payload on a commercial Ariane-5 launch. It arrived in lunar orbit in November 2004 and its battery of miniaturized instruments carried out lunar science investigations until September 3, 2006, when the mission ended through lunar impact.

The hoped-for future of space science in Europe is written in a document titled *Cosmic Vision*, the result of a new (2004–2005) exercise in long-term planning undertook by the European space science community to address ESA's Scientific Programme in the 2015–2025 decade.[117] In the field of planetary science, the vision includes three main objectives within the framework of a general effort to understand "How does the Solar System work [and] what are the conditions for planet formation and the emergence of life." First is the Jupiter exploration program (JEP), including a series of multiple spacecraft entering the system and a specific study of Jupiter's icy moon Europa by an orbiter and, possibly, a lander. The second objective, as to be expected, regards the study of asteroids and other small bodies by a sample return mission to a near-Earth object. Finally, the third objective is Mars, whose exploration should be pursued by technologically advanced landers and sample return missions. As a matter of fact, Mars exploration is the pivotal element of ESA's "Aurora" space exploration program, an optional program approved by the ESA Council meeting at ministerial level in Berlin in December 2005, after four years of preparatory studies. The scope of the Aurora program, of which Italy is the main contributor, is to develop a major European undertaking in robotic and human exploration of the solar system, the first step being the so-called Exo-Mars program in collaboration with NASA. While being an optional program, the part of Aurora dedicated to robotic exploration has been joined to the mandatory scientific program under the new Directorate of Science and Robotic Exploration.[118]

Concluding the review of ESA's current (and future) programs in planetary science, one can appreciate, I hope, the central role of the Huygens mission in the history of European space science. Let me just recall two main points. First is the turning point Huygens represented in the ESA eventual commitment in planetary research and solar system exploration. The European Space Agency was a latecomer, but it entered this field through the main door, first paying a visit to a famous comet and then parachuting a sophisticated scientific robot onto a mysterious body in the outer solar system. This helped to create and strengthen a solid scientific community, set up a reliable managerial structure, and consolidate an industrial competence in key space technologies (e.g., those regarding the entry into a planet's atmosphere). Pride and self-reliance are not of little importance in a context in which the stakes are high, the resources are limited (about one order of magnitude smaller than the NASA science budget), and the institutional framework is particularly complex (19 member states and a good number of competing scientific discipline to cope with). The second element to be stressed is the role of the Cassini-Huygens mission in shaping the relationship between ESA and NASA. The mutual recognition of scientific excellence and the awareness that the need for cooperation in space science goes in both directions form an important heritage of the Saturn-Titan venture. "Scientific research is difficult. It is hard to do it alone; it is even harder to do it together," wrote Hendrik van de Hulst, one of the founding fathers of ESRO, in 1961.[119] I think it is fair to conclude that implementing together Cassini-Huygens was certainly difficult, but its success helped to make the future easier.

Notes

* The research for this chapter was supported by the HSS/NASA Fellowship in the history of space science, by the University of Palermo, and by the NASA Space Science Mission Directorate. I am indebted to ESA's director general, Jean-Jacques Dordain, and former director of science and robotic exploration, David Southwood, for the authorization to perform archival research at the ESA Headquarters in Paris. I am grateful to the head of the records management section Nathalie Tinjod and the staff of the Science Directorate for their helpful support. I thank the ESA coordinator of solar system missions, Marcello Coradini, for his generous intellectual support, and Roger Bonnet, Agustin Chicarro, Marcello Fulchignoni, Daniel Gautier, and David Southwood for interviews on their experience in ESA's Scientific Programme. Unless otherwise specified, all documents cited in the notes have been located in the archives of the ESA Headquarters (coded documents) or in the working archives of the Science Directorate (letters, mission proposals, technical reports, etc.).

1. The scientific background and technical aspects of the mission are discussed in D. M. Harland, *Cassini at Saturn* (Berlin, Germany: Springer, 2007). A US-focused sociological analysis is in B. Groen and C. Hampden-Turner, *The Titans of Saturn* (Singapore: Marshall Cavendish; London: Cyan, 2005). See also L. J. Spilker, ed., *Passage to a Ringed World: The Cassini-Huygens Mission to Saturn and Titan* (Washington, DC: NASA SP-533, October 1997).
2. R. Bonnet, "Cassini-Huygens in the European Context," in *Titan: From Discovery to Encounter* (Noordwijk: European Space Agency SP-1278, 2004), pp. 201–209, on p. 206.

3. Other cases are discussed by the author in J. Krige, A. Russo, and L. Sebesta, *A History of the European Space Agency 1958–1987* (2 volumes) (Noordwijk: European Space Agency SP-1235, April 2000), thereafter *History of ESA*.
4. R. Lüst, "The [ESA member state] ministerial conference and beyond," *ESA Bulletin* 53 (February 1988), 9–13, on 9. Also "Cooperation between Europe and the United States in space," *ESA Bulletin* 50 (May 1987): 98–104.
5. The obvious reference is Krige, Russo, and Sebesta's *History of ESA*. Also K. Madders, *A New Force at a New Frontier* (Cambridge: Cambridge University Press, 1997). ESA's original member states were Belgium, Denmark, France, Germany, Italy, Netherlands, Spain, Sweden, Switzerland, and the United Kingdom. These were the founding states of ESRO in 1964. ELDO's membership included Belgium, France, Germany, Italy, Netherlands, the United Kingdom, and Australia.
6. Following a recent (2008) reform of the ESA executive, the new official denomination is director of science and robotic exploration (D/SRE).
7. *History of ESA*, vol. 2, p. 87. The quotation is from the ninth meeting of the Science Programme Board (October 23, 1974), the ESRO Council's delegate body for the Scientific Programme in the transition period from ESRO to ESA. It was eventually replaced by ESA's Science Programme Committee (SPC).
8. *History of ESA*, vol. 1, pp. 41–51, 209. The quotation is from the 28th meeting of the ESRO Launching Programme Advisory Committee (May 22–23, 1969).
9. *History of ESA*, vol. 1, pp. 224–32.
10. Ibid., p. 225. The quotation is from the report of the Geophysics Panel set up by the LPAC to advise on solar system science.
11. Ibid., pp. 240–44.
12. These feelings have been described to the authors by M. Coradini, a former student of Fulchignoni, and A. Chicarro, a former student of Masson.
13. *History of ESA*, vol. 2, pp. 158–74. Also N. Calder, *Giotto to the Comets* (London, UK: Presswork, 1992).
14. *History of ESA*, vol. 2, pp. 138–58. The quotation is from the SSAC's *Recommendations on the Development of Space Science in the 1980s* (Noordwijk: European Space Agency SP-SP-1015, December 1978), p. 33.
15. R. Bonnet, *Les horizons chimeriques* (Paris: Dunod, 1992).
16. *History of ESA*, vol. 2, 199–216; *European Space Science Horizon 2000* (Noordwijk: European Space Agency, ESA SP-1070, December 1984). Also R. Bonnet, "The New Mandatory Scientific Programme for ESA," *ESA Bulletin* 43 (August 1985): 8–13.
17. *Horizon 2000*, p. 10. Giotto was successfully launched by an Ariane vehicle on July 2, 1985, and achieved a spectacular rendezvous with Halley in March the following year. As the spacecraft survived the encounter with the comet's turbulent environment, it was eventually decided to extend the mission, redirecting Giotto to flyby comet Grigg-Skjellerup in July 1992.
18. The M-missions included in Horizon 2000 in 1985 were the Infrared Space Observatory (ISO), the astrometry mission Hipparcos, the European participation in the NASA Hubble Space Telescope, the heliospheric mission Ulysses and Giotto.
19. Coradini, interview with the author, March 8, 2007.
20. D. Leverington, *Babylon to Voyager and Beyond: A History of Planetary Astronomy* (Cambridge, UK: Cambridge University Press, 2003),

pp. 426–66. Also D. M. Hunten, "Jupiter," and T. Encrenaz, "The Planets beyond Jupiter," in J. A. M. Bleeker, J. Geiss, and M. E. C. Huber, eds., *The Century of Space Science* (Dordrecht: Kluwer Academic Publishers, 2001), vol. 2, pp. 1425–30 and 1431–50, respectively.

21. J. Comas Solà, *Astronomische Nachrichten*, 4290 (1908): 289–90.
22. G. Kuiper, "Titan: A Satellite with an Atmosphere," *Astrophysical Journal* 100 (1944): 378–83.
23. D. M. Hunten, "Titan's Atmosphere and Surface," in J. A. Burns, ed., *Planetary Satellites* (Tucson: University of Arizona Press, 1977), pp. 430–43; "A Titan Atmosphere with a Surface Temperature of 200K," in D. M. Hunten and D. Morrison, eds., *The Saturn System* (Washington, DC: NASA Conference Publication 2068, 1978), pp. 127–40.
24. N. Thomas, "The Satellites of the Outer Planets," in Bleeker, Geiss, and Huber, eds., *Century of Space Science*, pp. 1451–78; Harland, *Cassini at Saturn*, pp. 87–95.
25. D. L. Matson, L. J. Spilker, and J.-P. Lebreton, "The Cassini/Huygens Mission to the Saturnian System," *Space Science Reviews* 104 (2002): 1–58, on p. 14. The whole volume is devoted to the Cassini/Huygens mission.
26. F. Raulin, "Titan's Organic Chemistry and Exobiology," in *Huygens: Science, Payload and Mission* (Noordwijk: European Space Agency SP-1177, 1997), pp. 219–29; "Exo-Astrobiological Aspects of Europa and Titan: From Observations to Speculations," *Space Science Reviews* 116 (2005): 471–87.
27. T. Owen, F. Raulin, C. McKay, J. I. Lunine, J.-P. Lebreton, and D. L. Matson, "The Relevance of Titan and Cassini/Huygens to Pre-Biotic Chemistry and the Origin of Life on Earth," in *Huygens: Science, Payload and Mission*, pp. 231–33. For an overview of Titan science prior to the Cassini-Huygens mission, see A. Coustenis and F. Taylor, *Titan: The Earth-like Moon* (Singapore: World Scientific, 1999); R. Lorentz and J. Mitton, *Lifting Titan's Veil: Exploring the Giant Moon of Saturn* (Cambridge: Cambridge University Press, 2002). A more technical review is in *Proceedings: Symposium on Titan* (Toulouse, September 9–12, 1991) (Noordwijk: European Space Agency SP-338, April 1992).
28. W. Ip, D. Gautier, and T. Owen, "The Genesis of Cassini-Huygens," in *Titan: From Discovery to Encounter* (Noordwijk: European Space Agency SP-1278, 2004), pp. 211–27.
29. Ip had started his scientific career at the University of California in San Diego and arrived in Lindau in 1978. He is presently at the National Central University in Taiwan.
30. *Project Cassini: A Proposal to the European Space Agency for a Saturn Orbiter/Titan Probe Mission in Response to the Call for Mission Proposals Issued on 6th July 1982*, November 12, 1982. The author thanks professor Ip for providing him with a copy of this document.
31. A. Cassini, "Cassini at Saturn," in *Titan: From Discovery to Encounter*, pp. 31–41.
32. *History of ESA*, vol. 2, 193–95.
33. *United States and Western European Cooperation in Planetary Exploration.* Report of the Joint Working Group on Cooperation in Planetary Exploration (Washington, DC: National Academy Press, 1986).
34. *History of ESA*, vol. 2, pp. 180–85.
35. R. Bonnet and V. Manno, *International Cooperation in Space: The Example of the European Space Agency* (Cambridge: Harvard University Press, 1994),

pp. 98–108. The Ulysses spacecraft was launched in 1990, after a long delay due to the tragic accident of the shuttle Challenger in 1986. NASA contributed the launch on the shuttle Discovery, the Radio-isotope Thermoelectric Generator (RTG), the Deep Space Network (DSN), and 50 percent of the experiments.

36. R. W. Smith, "The Biggest Kind of Big Science: Astronomers and the Space Telescope," in P. Galison and B. Hevly, eds., *Big Science. The Growth of Large-Scale Research* (Stanford: Stanford University Press, 1992), p. 202.
37. Bonnet and Manno, *International Cooperation in Space,* pp. 102–103.
38. Ibid., p. 106. See also R. Bonnet, "European Space Science—In Retrospect and In Prospect," *ESA Bulletin* 81 (February 1995): 6–17.
39. SSAC, 32nd meeting (12–13/10/83), SSAC/MIN/32, 01/12/83.
40. Fulchignoni's group was represented in the original Cassini proposal by G. Valsecchi.
41. *Cassini: Saturn Orbiter and Titan Probe*, ESA/NASA assessment study, ESA SCI(85)1, August 1985, p. 1.
42. SSWG, 50th meeting (09/01/86), SOL(86)3, 30/04/86. The recommendation is reported in SOL(86)2, 09/01/86.
43. SSAC, 43rd meeting (10/01/86), SSAC/MIN/43, 03/02/86. All quotations in this paragraph are on pp. 3–5.
44. This decision had been taken by the SSAC earlier at that same meeting,
45. SSAC(86)2, 10/01/86.
46. SPC, 41st meeting (6–7/02/86), ESA/SPC/MIN/41, 17/03/86.
47. SSAC, 45th meeting (08/10/86), SSAC/MIN/45, 18/11/86, with attached SSAC(86)6, 20/10/86; SPC, 43rd meeting (17–18/11/86), ESA/SPC /MIN/43, 10/11/87. The updated study report, coded SCI(86)5, was published in December 1986.
48. SSAC, 46th meeting (05/02/87), SSAC/MIN/46, 27/03/87, with attached SSAC(87)2, 05/02/87.
49. SSAC, 47th meeting (29/04/87), SSAC/MIN/47, 22/06/87, with attached SSAC(87)8, 30/04/87; SPC 45th meeting (25–26/05/87), 24/07/87.
50. GRASP was originally designed to be flown on the European Eureca carrier onboard the Space Shuttle. Following the setback of the shuttle launching program, the assessment study on GRASP was redirected toward an Ariane launch option.
51. *Cassini: Saturn Orbiter and Titan Probe. Report on the Phase-A study*, ESA SCI(88)5, October 1988.
52. On the eve of this meeting, a synthetic presentation of all missions was published in *ESA Bulletin* 55 (August 1988): 10–40.
53. SSWG, 61st meeting (26/10/88), SOL(88)12, 01/02/89, p. 3.
54. SSWG, 55th meeting (04/02/87), SOL(87)5, 01/04/87, with attached SOL(87)2, 04/02/87; SSWG, 56th meeting (27–28/04/87), SOL(87)12, 04/06/87, with attached SOL(87)10, 28/04/87; SSAC, 47th meeting (29/04/87), SSAC/MIN/47, 22/06/87, with attached SSAC(87)8, 30/04/87. Also, *Vesta: A Mission to the Small Bodies of the Solar System. Report on the Phase-A study*, ESA SCI(88)6, October 1988.
55. SSWG, *Recommendation on Next Scientific Project*, SOL(88)11, 26/10/88. According to Gautier, Cassini prevailed by eleven votes against two for Vesta: Ip, Gautier and Owen, "The Genesis of Cassini-Huygens," p. 220.
56. AWG, *Recommendation for the next scientific project*, ASTRO(88)9, 26/10/88.

57. SSAC, 50th meeting (27/10/88), SSAC/MIN/50, 15/03/89.
58. SSAC, *Recommendation on the Choice of the Next Scientific Project*, SSAC(88)5, 27/10/88. According to Gautier, the SSAC selected Cassini by 5 votes to 2 (cf. Ip, Gautier, and Owen, "The Genesis of Cassini-Huygens," p. 220), but the minutes of the SSAC meeting report a unanimous approval of the recommendation (SSAC/MIN/50, p. 4). The role of Mezger and Puget in the final decision has been mentioned to the author by M. Coradini and D. Southwood.
59. SPC, *Selection of the Next Scientific Project (M1)*, ESA/SPC(88)31, 07/11/88, p. 6.
60. SPC, 50th meeting (24–25/11/88), ESA/SPC/MIN/50, 02/02/89, p. 2. A similar feeling regarding cooperation with NASA was expressed by the Italian delegation (p. 4).
61. Ibid., p. 5. It is worth recalling that Huygens also discovered Saturn's ring: A. van Helden, "Huygens, Titan, and Saturn's ring," in *Titan: From Discovery to Encounter*, pp. 11–29; Leverington, *Babylon to Voyager*, pp. 91–98; C. D. Andriesse, *Titan: Biography of Christiaan Huygens* (Utrecht: Utrecht Universiteit, 2003), pp. 117–43.
62. Ip, Gautier, and Owen, "The Genesis of Cassini-Huygens," p. 225.
63. M. Coradini, letter to the members of the SSWG, July 26, 1989.
64. R. Bonnet, "Cassini-Huygens in the European Context," in *Titan: From Discovery to Encounter*, pp. 201–209, on p. 206.
65. *ESA/NASA Memorandum of Understanding Cassini-Huygens mission*, ESA/SPC(90)20, 21/05/90 (signed on December 17, 1990). Also *Science Management Plan of the Huygens Part of the Cassini Mission*, ESA/SPC(89)17, rev. 1, 15/11/89.
66. ESA, *Announcement of Opportunity, Cassini Mission: Huygens Probe*, SCI(89)2, October 1989. NASA, *Announcement of Opportunity, Cassini Mission: Saturn Orbiter*, NASA A.O. No. OSSA-1–89, October 10, 1989.
67. ESA/SPC(90)52, rev. 1, 29/11/90. Also, Matson, Spilker, and Lebreton, "The Cassini-Huygens Mission to the Saturnian System"; Harland, *Cassini at Saturn*, pp. 175–91. One of the facility instruments was an ion mass spectrometer whose accommodation in the payload presented some difficulties. In the event, it was decided to include such an instrument and a specific AO was issued in May 1991. The results of the selection process were announced in February 1992.
68. M. Coradini, letter to SSWG members, February 15, 1990.
69. *Huygens Probe: Payload Selection*, SSAC(90)8, 06/08/90, Annex. Also J.-P. Lebreton and D. L. Matson, "The Huygens Probe: Science, Payload and Mission Overview," *Space Science Reviews* 104 (2002): 59–100; Harland, *Cassini at Saturn*, pp. 191–98.
70. SSWG, 67th meeting (23/07/90), SOL(90)11, 01/10/90, with attached SOL(90)9, 23/07/90; SSAC, 56th meeting (21/08/90), SSAC/MIN/56, 29/10/90, with attached SSAC(90)11, 21/08/90; SPC, 57th meeting (17–18/09/90), ESA/SPC/MIN/57, 23/10/90.
71. A. Lawler, "1992 Budget Curb Stunts NASA Growth," *Space News*, 2:34 (October 7–13, 1991): 1; D. Isabell and A. Lawler, "NASA Budget Funds All Sectors, Snarls Science Projects," ibid., p. 10
72. J. M. Luton, letter to NASA administrator R. H. Truly, November 7, 1991.

73. D. Southwood, letter to B. Moore III, October 28, 1991. The letter had been urged by the SSAC at its 61st meeting (22/10/91), SSAC/MIN/61, 18/02/92, p. 7.
74. SPC, 61st meeting (6–7/11/91), ESA/SPC/MIN/61, 17/01/92, p. 5.
75. SPC, 62nd meeting (25/26/02/92), ESA/SPC/MIN/66, 15/04/92, p. 4.
76. SSAC, 62nd meeting (17–18/02/92). SSAC/MIN/62, 07/04/92, p. 6.
77. Cf. topical articles in *Space News*, vol. 2 (1992), nos. 3 (January 27–February 2); 4 (February 3–9); 5 (February 10–16); 8 (March 2–8).
78. SSAC, 64th meeting (12/05/92), SSAC/MIN/64, 22/10/92, p. 10.
79. Ip, Gautier, and Owen, "The Genesis of Cassini-Huygens," p. 226.
80. A. Lawler, "NASA Science May Take Big Hit in Budget Crunch," *Space News*, 5:3 (January 17–23, 1994), p. 4.
81. *Space News*, 5:21 (May 23–29, 1994): 1. AXAF is the acronym for Advanced X-ray Astrophysics Facility. After launch (1999), the satellite was named Chandra X-ray Observatory, after the Indian physicist Subrahmanyan Chandrasekhar.
82. Groen and Hampden-Turner, *Titans of Saturn*, pp. 147–57. On Goldin's "faster, better, cheaper" approach, see H. E. McCurdy, *Faster, Better, Cheaper: Low Cost Innovation in the U.S. Space Program* (Baltimore, MD: Johns Hopkins University Press, 2001); P. J. Westwick, *Into the Black: JPL and the American Space Program, 1976–2004* (New Haven: Yale University Press, 2007), pp. 207–27; G. H. Riecke, *The Last of the Great Observatories: Spitzer and the Era of Faster, Better, Cheaper at NASA* (Tucson: University of Arizona Press, 2006).
83. L. Tucci, "Goldin Subjects Cassini to Cost Risk Reduction," *Space News*, 5:11 (March 14–20, 1994): 3.
84. L. Woltjer, telefax to R. Bonnet, April 2, 1994, distributed to SPC delegations as annex I to ESA/SPC(94)27, 31/05/94.
85. R. Bonnet, letter to D. Gautier, March 30, 1994. An echo of their worries is in the minutes of the 69th SSAC meeting (16/11/93), SSAC/MIN/69, 27/01/94, pp. 2–3.
86. R. Bonnet, letter to C. Sagan, March 11, 1994. R. Bonnet remembers Sagan calling him on the phone from California asking for help because NASA was trying to stop the mission. Cf. also Groen and Hampden-Turner, *Titans of Saturn*, p. 31.
87. "Hand off AXAF, Cassini," *Space News*, 5:24 (June 20–26, 1994): 16.
88. For the NASA/ASI cooperation on the Cassini mission, see B. Pernice, "The Cooperation between NASA and ASI on the Cassini Mission," *Il Nuovo Cimento*, 15C (6) (1992): 1133–36.
89. J.-M. Luton to A. Gore, June 13, 1994. The letter is also reported in Groen and Hampden-Turner, *The Titans of Saturn*, pp. 195–97.
90. M. Coradini, D. Gautier and R. Bonnet told the author of this diplomatic initiative.
91. A. Lawler, "Clinton Comes to NASA's Aid," *Space News*, 5:26 (July 3–10, 1994): 1; "Mikulski Panel gives NASA 14.4 B$," *Space News*, 5:28 (July 18–24, 1994): 1.
92. Bonnet, "Cassini-Huygens in the European Context," p. 208.
93. ESA press release N. 27–1997, July 16, 1997.

94. Telefax message from I. Pryke, head of ESA's Washington Office, to ESA Headquarters in Paris, September 9, 1997, with attached the FCPJ "Schedule of events" and other documents. Many documents related to the activity of the "Stop Cassini" movement are available at the website http://www.animatedsoftware.com/cassini/cassini.htm (accessed on April 24, 2012).
95. R. Bonnet, letter to "European scientists involved in the Huygens mission," July 18, 1997.
96. W.-H. Ip, letter to B. Clinton, August 11, 1997. An identical letter was sent by Ip to J. Gibbons, director of the White House Office of Science and Technology Policy.
97. J. H. Gibbons, letter to D. S. Goldin, October 3, 1997.
98. ESA press release N. 32–1997, October 15, 1997.
99. Email message from the Huygens Project to "Distribution," October 24, 1997. Unfortunately Hassan passed away in 1999 and could not celebrate the arrival at Saturn and Titan.
100. Cassini's journey (including scientific activity) toward and within the Saturnian system up to end 2006 is described in detail in Hartland, *Cassini at Saturn*, pp. 203–304.
101. Huygens Communications Link Enquiry Board Report (D.C.R. Link et al.), *Findings, Recommendations and Conclusions*, ESA, December 20, 2000, available at the website: http://klabs.org/richcontent/Reports/Failure_Reports/ESA_Cassini/huygens_enquiry_board.doc (accessed on April 24, 2012).
102. L. J. Deutsch, "Resolving the Cassini/Huygens Relay Radio Anomaly," Transactions of the IEEE Aerospace conference, Big Sky, Montana (USA), March 2002, available at the website: http://klabs.org/richcontent/Reports/Failure_Reports/ESA_Cassini/huygens_deutsh.pdf (accessed on April 24, 2012).
103. K. Clausen and L. Deutsch, *Huygens Recovery Task Force: Final Report*, HUY-RP-12241, ESTEC, Noordwijk, The Netherlands, July 27, 2001.
104. *Joint NASA/ESA Resolution on the Huygens Recovery Task*, signed by E. Weiler (June 21, 2001) and D. Southwood (June 29, 2001). ESA Press release n. 39–2001, July 2, 2001.
105. Details in ESA press releases NN. 36–2004 (July 1, 2004); 67–2004 (December 25, 2004); 3–2005 (January 14, 2005). Also J.-P. Lebreton, O. Witasse, C. Sollazzo, T. Blancquaert, P. Couzin, A.-M. Schipper, J. B. Jones, D. L. Matson, L. I. Gurvits, D. H. Atkinson, B. Kazeminejad, and M. Pérez-Ayúcar, "An Overview of the Descent and Landing of the Huygens Probe on Titan," *Nature* 438 (December 8, 2005): 758–64.
106. A. Abbott, "Titan Team Claims Just Deserts as Probe Hits Moon of *Crème Brûlée*," *Nature* 433 (January 20, 2005): 181.
107. Landing on Titan's surface occurred at about 11:38 UTC. Huygens continued thereafter to transmit data for at least 3 hours 14 minutes, as determined by the detection of the probe's signal by the Earth-based radio telescopes, but useful scientific data could only be relayed to Earth during its visibility by the Cassini orbiter.
108. "Europe Arrives at the New Frontier—The Huygens Landing on Titan," *ESA Bulletin* 121 (February 2005): 6–9; J. C. Zarnecki, M. R. Leese, B. Hathi, A. J. Ball, A. Hagermann, M. C. Towner, R. D. Lorenz, J. A. M. McDonnell, S. F. Green, M. R. Patel, T. J. Ringrose, P. D. Rosenberg, K.R. Atkinson,

M. D. Paton, M. Banaszkiewicz, B. C. Clark, F. Ferri, M. Fulchignoni, N. A. L. Ghafoor, G. Kargl, H. Svedhem, J. Delderfield, M. Grande, D. J. Parker, P. G. Challenor, and J. E. Geake, "A Soft Solid Surface on Titan as Revealed by the Huygens Surface Science Package," *Nature* 438 (December 8, 2005): 792–95.

109. A presentation of some important results with pertinent publications (by October 2006) is in Harland, *Cassini at Saturn*, pp. 244–344. For more recent accounts, see R. Lorentz and J. Mitton, *Titan unveiled: Saturn's Mysterious Moon Explored* (Princeton: Princeton University Press, 2008); A. Coustenis and F. Taylor, *Titan: Exploring an Earth-like World* (Singapore: World Scientific, 2008).
110. *Horizon 2000 Plus. European Space Science in the 21st Century*, ESA SP-1180, November 1994. A second (astronomy) cornerstone is dedicated to a mission for high precision astrometric observations (Gaia); in a more distant future there is a cornerstone mission aimed at detecting gravitational waves (Lisa).
111. The total cost of the Rosetta mission is close to 1 billion Euro, including the launch, the spacecraft, the scientific payload, and operations. The mass of the spacecraft (including propellant) is approximately 4 tons. ESA's foreseen investment in BepiColombo is about 665 million Euro, covering the development of the Planetary Orbiter, the launch, and the operations. The costs of the instruments funded by member state institutes is over 200 million Euro. In Mercury orbit, the Planetary Orbiter and the Magnetospheric Orbiter will weigh 1,150 and 275 kilograms, respectively.
112. *A Moon Programme: A European View*, ESA BR-101, May 1994; ESA/SPC(94)43, 21/10/94. The scientific case had been presented in *Mission to the Moon*, ESA SP-1150, June 1992.
113. L. Woltjer, *Europe's Quest for the Universe* (Les Ulis: EDP Sciences, 2006), pp. 167–68.
114. R. Bonnet, interview with the author, November 20, 2007.
115. A. Russo, "Europe's Path to Mars: The European Space Agency's Mars Express Mission," *Historical Studies in the Natural Sciences* 41:2 (2011): 123–78.
116. The 7-ton Mars-96 mission consisted in an orbital module, two small stations to be landed onto the surface and two penetrators into the Martian soil. Its scientific payload included instruments contributed by scientific groups in several western European countries, with ESA also being directly involved by the provision of a mass memory unit. The spacecraft was lost in a launch failure of the Proton-K vehicle on November 16, 1996.
117. *Cosmic Vision. Space Science for Europe 2015–2025*, ESA BR-247, October 2005.
118. P. Messina, B. Gardini, D. Saccotte, and S. Di Pippo, "The Aurora Programme: Europe's Framework for Space Exploration," *ESA Bulletin* 126 (May 2006): 11–15; J. Vago, B. Gardini, G. Kminek, P. Baglioni, G. Giafiglio, A. Santovincenzo, S. Bayón, and M. van Winnendael, "ExoMars: Searching for Life on the Red Planet," *ESA Bulletin* 126 (May 2006): 17–23.
119. H. van de Hulst, "International Space Cooperation," *Bulletin of Atomic Scientists* 17 (1961): 233–236, on 233.

Chapter 13

Pluto: The Problem Planet and its Scientists

David H. DeVorkin

From its discovery at the Lowell Observatory in February 1930 to the present time, Pluto has always been a problem. First, was it discovered by chance? Or was it predicted? That problem took over 60 years to resolve. Most recently, a debate flared over its status as a planet. If it is a planet, what kind of planet? That question has just barely been resolved and is a valuable study in the complex process of consensus formation.[1] In between, Pluto presented astronomers with many other problems, from explaining its lop-sided orbit to determining its origin. Was it an escaped moon of Neptune, or did it form as a planetary body in the original solar nebula? Recounting these problems—why they were raised and how they were resolved—reveals not only critical stages in twentieth-century astronomy when our conception of the nature and extent of the solar system itself changed in profound ways, but a period of time when we learned that systems of planets themselves circling other stars are not rare occurrences due to chance, but are common in the universe arising through processes intrinsic to the formation of stars themselves. Pluto, we now realize, is the first inhabitant to be detected in a vast region of the solar system that, in its aggregate of countless thousands to millions of cold little bodies orbiting in a belt around the sun, are directly visible from interstellar distances. Recounting these problems also illustrates how astronomers acquired and applied new tools in the twentieth century, and how these tools, along with their makers and patrons, changed disciplinary practice in astronomy.

Searching for Planet X

After the discovery of Neptune, Urbain Leverrier proudly proclaimed before the French Academy of Sciences that it would only be a matter of time and effort before Neptune's deviations from Newtonian law would reveal the presence of yet other planets.[2] The Boston Brahmin Percival Lowell was one among several mathematically adept practitioners who pursued this discovery, making a first attempt as early as 1905. Lowell had been a student of Benjamin Peirce at Harvard, and through Peirce became excited about the mathematical prediction of the existence of new planets using the anomalies in the motions of known planets and the statistics of cometary orbits. After proclaiming that such a body existed, Lowell activated searches for the ninth planet starting in 1905 from his eponymous observatory in Flagstaff, Arizona. He had photographs taken of likely fields at his observatory but also explored ways to refine predictions.[3] He predicted that his "Planet X" moved in an eccentric (0.202) orbit with a semimajor axis of 43 astronomical units and an inclination of 10 degrees.

There had been two searches for the planet subsequent to Lowell's predictions and death, but neither had been conclusive. With partial support of the Lowell family, Vesto Melvin Slipher, Lowell's successor as observatory director, mounted a new search in the late 1920s. Slipher appealed to the National Academy of Science's nascent National Research Fund (NRF) for support to hire observers and research assistants. The NRF astronomy committee chairman, Henry Norris Russell of Princeton, and a friend of the Lowell Observatory, assured Slipher that he favored a trans-Neptunian planet search: "[I] have given them a high place in my list…" but neglected to admit to him that "high place" was not the highest of his funding priorities; astronomy had greater needs.[4] The NRF never materialized, so Slipher secured supplemental funds from the Lowell Trust to hire a suitable assistant. Such a man was not too expensive.[5]

Slipher's plan was to photograph regions on the ecliptic where the suspected planet might be, using large photographic plates in a 13-inch widefield astrograph. But the instrument had to be thoroughly tested and the search system evaluated. The staff was capable of doing this, but Slipher again asked Russell for help. Russell, seeking a respite from the Princeton climate for his ailing colleague Raymond Smith Dugan, offered a trade in expertise: Dugan would visit Flagstaff for a season and a Lowell staff member, Carl Lampland, would take up residence at Princeton, broadening horizons all around.

Throughout the spring and summer of 1929, Clyde Tombaugh—their recently hired eager "young man from Kansas"—secured upward of 50 plate pairs (photographing the same area twice, separated in time by days or weeks) while a "blink" comparator was prepared.[6] Slipher well knew that the deep survey would yield much serendipitous data, such as new nebulae, interesting star clusters, and other classes of deep-space objects. As a survey instrument, he thought it was unparalleled, and he planned to use Lowell Observatory's 40-inch reflector to take spectra of some of the new objects that were being discovered. Slipher sent samples of Tombaugh's work to Lampland at Princeton, for his and Russell's examination, and they both agreed that even if no new planet were found, the survey was yielding useful data. On these grounds observing astronomers have justified long-shot searches many times since.

Tombaugh's unprocessed plates accumulated into a serious back-log during the summer of 1929, since Slipher had not secured anyone from the senior staff to perform the extremely tedious task of examining them, and did not get to them himself. Eventually, Tombaugh took up this part of the work as well, which was painstaking and often frustrating, requiring that an observer peer into a microscope eyepiece for hours at hand as small fields in the two aligned plates were brought into view and sequentially compared. The rate of "blinking" or imaging one and then the other plate in the pair was adjustable, so that, in theory, one would see a continuous field of stars through the eyepiece and anything that changed in the field would shift back and forth, indicating planetary motion of some sort. Or pulse in size, indicating a variable star. In practice, however, no two photographic plates were the same (see figure 13.1). Variations in transparency, in the stability of the air, and the temperature, could alter everything creating something of a stroboscopic background, mesmerizing, dulling, maddening. There could also be spurious objects that were flaws in the photographic emulsions. These would at first bring hope and excitement, then disappointment. Bad weather in the summer and fall gave Tombaugh time to "blink" his plates, and think through how to maximize the efficiency of his search system. Through the fall and into the winter, he began to examine the dense starfields where the ecliptic crossed the Milky Way, encountering sometimes

Figure 13.1 Astronomer Clyde W. Tombaugh at the door of the Pluto discovery telescope, Lowell Observatory, Flagstaff, Arizona. (Credit: NASA.)

hundreds of thousands of measurable stars on a single plate. Tedium that would have daunted even the most dedicated professional seemed not to affect Tombaugh as he plowed through the heavens, shifting between night duties taking new plate pairs when the weather allowed, and blinking his backlog on the Lowell comparator. For a young man who spent his free time making telescopes, this was god's work.[7]

The Lowell Object

The object he eventually found presented no disk, and was at least 14 times fainter than Lowell had predicted. But its very slow motion meant that it was in orbit beyond Neptune. Through February and into March, the Lowell staff was very quiet about their apparent discovery. The Lowell family was informed, as were close observatory friends.[8] Finally, on March 12, to coincide with Lowell's birthday the following day, Slipher, as was the custom, cabled Harvard College Observatory announcing the object. Harvard would then tell the rest of the astronomical world.

William Hoyt, the chronicler of Lowellian history and lore, makes the point that the Lowell staff faced a problem for which they had little expertise: calculating the orbital characteristics of the object.[9] Indeed, the discovery created an immediate problem Lowell staff could not handle: calculating its orbit and proving that it had been predicted by their founder and patron. Celestial mechanics, still the mark of respectability for an astronomer, was not a strength of the staff. Even V. M. Slipher was uncomfortable with orbital mechanics. When his old Indiana mentor J. A. Miller, now at Swarthmore, invited him to take a sabbatical semester to teach elementary astronomy and mathematics in his place, so Miller could escape to indulge his passion for solar eclipse chasing, Slipher turned him down. "I might be able to get by with the astronomical part of it" Slipher admitted, "but the mathematics would be more than I would dare attempt."[10] Slipher was known to take the morally superior attitude that his mission was to be with his telescopes, staying at his posting, but it is also clear that he knew the limits of his mathematical capabilities.[11]

Soon after Slipher's announcement, the distinguished celestial mechanician Ernest W. Brown of Yale applauded Lowell for the discovery, but knew it did not come from the discrepancies noted between the observed and predicted positions of Uranus and Neptune, the signature of an unseen gravitationally perturbing body. To mainstream mathematical astronomers, the faint object Clyde Tombaugh found could not have sufficient mass to be Lowell's predicted perturber.[12] The discovery had been "purely accidental."[13]

Brown pointed out that the planetary positions used by Lowell for his predictions were too crude, but he did not criticize Lowell's theory. The actual perturber, Brown mused, might yet be found.[14] These remarks by Brown, and others by Armin Leuschner at Berkeley, raised the collective blood pressure on Mars Hill. Harvard Observatory director Harlow Shapley was no help either, reporting excitedly the Planet X fervor at Harvard:

> I am in the midst of considerable perturbation arising from Planet X. An hour ago we first determined the photographic magnitude on Harvard plates. Two hours ago we received the first precise positions from the Yerkes Observatory. Three hours ago, and before and after, letters come and reporters come with inquiries and suggestions. Is the name to be Osiris, or Bacchus?[15]

Slipher knew he had to act fast if Lowell Observatory was to retain a role in the discovery, and in the naming of the new planet.[16] Requests for positions came from Leuschner at Berkeley, from the Naval Observatory, from Mount Wilson, and from Yerkes. Yielding to pressure from Shapley, Roger Lowell Putnam, the Lowell trustee, told the Lowell staff to release their data. As he pleaded with Putnam to resist such demands, for the sake of Lowell's name, Slipher also pleaded with Miller to drop everything and come to their rescue. Miller quickly packed up and arrived at Flagstaff on March 22 ready to compute the orbit of the new planet with the Sliphers and Lampland.[17]

Anyone who could compute an orbit, and could obtain accurate positions for the new planet, jumped into what was very much a blood sport among astronomers. As Miller and his charges labored away, tentative orbits were issued from Berkeley and Harvard, using positions that lacked the time span available from the still-embargoed Lowell data. Harvard, announcing an orbit computed in Cracow, Poland, came up with circular elements, whereas a Berkeley team of graduate students, Fred Whipple and Ernest Clare Bower, made no commitment to the all-important eccentricity or period of the orbiting object, stating only that it was at present some 41 astronomical units away from Earth, with an inclination to the ecliptic of 17 degrees. Although their professor, A. Leuschner, suggested that these parameters could fit the description of an asteroid or comet, and not a major planet, there were not yet enough data at hand to say anything definite.

From late March and well into June, astronomers worldwide with access to plate vaults scoured their records for prediscovery observations, and feverishly shared their positions via telegram. Whipple and Bower, now joined by N. U. Mayall and S. B. Nicholson at Mount Wilson, combined observations from Flagstaff with earlier ones from Yerkes and a critical observation from the Royal Observatory of Belgium at Uccle to produce the first reliable orbit, placing the remarkable object "some 40 au's from the Sun with a period of 250 years." During the past century, it had resided at maximum distance from planets like Neptune and Uranus.[18] By July, and the publication of a second and refined report from Berkeley, Pluto had been officially named and its status as a planet generally accepted.

By late 1931 and into early 1932, many of the nagging questions about Planet X's orbit were resolved, leaving its mass the most contested issue. Meanwhile, other questions loomed larger and larger: was it a true prediction or chance discovery? Much of the outcome to this question depended on how closely the orbit of Planet X agreed with Lowell's predictions, and if its mass agreed as well.

Pluto as Planet X?

Whereas the orbit was tolerably close, the predicted mass was far too high. E. W. Brown continued to argue forcefully that no one could predict anything definitely from the small residuals remaining in the Uranian orbit after Neptune had been accounted for. Certainly not Lowell's now tiny object, which, at around magnitude 15.4 as determined by Walter Baade at Mount Wilson, would have an estimated mass ranging from barely one-tenth to nine-tenths the Earth's mass. This was in any event only a small fraction of the mass required by Lowell's predicted perturber. Brown's criticisms, and those of others, did not lead to definite answers. A. C. D. Crommelin, a major voice in the British Astronomical Association and an old Lowell ally, backed the Lowell prediction based upon a separate set of calculations, and defended Lowell's predictions against Brown's conclusions. While Crommelin publicly answered Brown, Slipher privately stayed away from the fray because his arguments remained obscure to the Lowellians. It was here that Russell came to the rescue and provided them with the rhetoric of defense they were searching for, in a series of reviews in *Scientific American*.

In his monthly *Scientific American* column, which he used more often than not as a bully pulpit, Russell was always kind to the Lowell Observatory, even at the height of the controversy over life on Mars in the 'teens. Percival Lowell had claimed that his observatory site was superior to those elsewhere, allowing him to perceive the canals and water vapor when others using larger telescopes failed. Lowell's assertions, and the fact that he sought a public forum for his views, threatened elite observatory directors at Yerkes, Lick, and Mount Wilson. Russell, a theorist and therefore less dependent upon the quality of observations from his own observatory at Princeton, believed that such controversies were positive indicators of healthy debate within a self-regulating community of astronomers; only privately did Russell reveal his annoyance with the Boston Brahmin. By the 1920s, Lowell Observatory senior staff thought of Russell as an ally because of his visits and his friendly, highly influential column.

By April 1930, Slipher was in a panic, hoping Russell could set things right for the Lowellians.[19] Russell dismissed erroneous orbits claiming that they resulted from "bad data." For Russell, whose professional life was marked by the deft employment of approximate methods, there was but one consideration: "[I]t is quite incredible that the agreement can be due to accident," he argued, "the actual accordance is all that could be demanded by a severe critic."[20]

Back home in the fall, Russell hurried west to reassure his Lowell friends first but mainly to get to Mount Wilson, where the best astrometric data had been collected on the planet's varying position. Junior astronomers there, Seth B. Nicholson and N. U. Mayall, had found that Pluto's mass came out similar to Earth's only if they factored in an ancient 1795 observation by Lalande. Russell was elated, and without checking the calculations, wrote it up rapidly for *Scientific American*, even though Nicholson felt far from comfortable with his calculation and wanted it verified by their more critical neighbors to the north at Berkeley. Russell had no such reservations: "So all contradictions vanish and we may welcome Pluto to a fully accredited place among the major planets of our system."[21]

Nicholson well knew that without Lalande's observation, Pluto's mass calculated from the modern observations would be far lower. He sent his analysis to Brown, who immediately found flaws not only in his use of the old data, but in his analysis. Brown also saw Russell's *Scientific American* article at about the same time, and knew he had to act. He knew Russell well enough to know that there was nothing Nicholson could have done. Politically astute, Brown knew what was driving Russell, and shared his values. But Brown could not let something this critical, which hit right at the heart of his cherished mathematical methods, stand unanswered, even if it was in a popular journal. Over the years Russell had been able to convert his casual column on the night sky into "a monthly classification of the state of astronomy."[22] Brown acted accordingly.

Brown was one of few astronomers in the American astronomical community who could confront Russell. He was the leading architect of the standard lunar theory of the day, and had collaborated with Russell in the past in a position of mathematical mentorship. Trained at Christ's College, Cambridge, Brown had served as the president of the American Mathematical Society and throughout his later career held an abiding sense of high standards of practice in mathematical astronomy. Brown therefore needed something more than a simple retraction: "[W]ith your very extensive experience of least squares and your knowledge of the residuals, it was somewhat astonishing that you did not notice the discrepancy right away," he chided Russell. Nevertheless, Brown was not concerned with Russell's reputation; in fact it was Russell's reputation that was the problem: "The trouble is that journals like 'Nature' quote you from the 'Scientific American,' and when Nicholson's paper comes out, there will be general confusion as to what is the fact." The retraction had to be more than a mere statement of the fact, Brown continued, it had to be an admission that physical intuition and arguments influenced by issues of chance might be helpful techniques in astrophysical practice, but not in mathematical analysis:

> The trouble is, however, that there have been some bad howlers from the gravitation point of view. If it were a question of physical astronomy where a good deal of latitude may be permitted with doubtful data, it would not matter much. Where it is a question that can be settled by recognized methods or where it is a mere question of mathematics, I think it is not fair on the coming generation to leave the matter uncorrected.[23]

Brown's assessment nicely contrasts how he viewed the differences between traditional astronomy and the new quantitative astrophysics as Russell practiced it. He sensed a widening gap in modes of practice: those advocated by astrophysicists and those who held to the precision of mathematical formalism. Russell well understood; a few years hence, he addressed the American Astronomical Society as retiring president, calling his style of practice "a tissue of approximations."[24] Russell had long advocated statistical methods, computational shortcuts, and iterative schemes, basing many of his best-known and most powerful empirical correlations, discoveries, and techniques upon them. Although Russell met with much resistance from traditional astronomers for his computational methods early in his career, by the 1930s he was in command of the American community and his methods were the ones younger astronomers had to emulate, though one had to be crafty about it. Russell treated theory pragmatically, celebrating the heuristic power of his tenuous methods because, he argued later in life, they provided a useful framework upon which a stronger fabric might someday be woven. Russell always argued that "[m]athematical approximations are involved in all our theoretical discussions, since Nature is far too complex to analyze as a whole." In tune with other Americans like P. W. Bridgeman and John Clarke Slater, he preferred numerical methods to formal probability functions, like Gaussian statistics, because he feared that such practices removed physical insight, or physical control from the worker. What Brown knew only too well was that few had sufficient physical insight to engage Nature as Russell dared do. For Brown, Russell's methods in less capable hands would lead to a weakening of science.[25]

Faced with Brown's rebuke, Russell admitted to the error privately and appended a note to his next column that made the error appear as a typo in the carbon he received from Nicholson, which really did not help matters much. But Russell knew he had blundered and had let his affinity for the Lowellians and their patronage cloud his view. Though he agreed with Brown that he had to minimize the damage, he remained determined to uphold the Lowell discovery; though not a prediction, it was still a major event, one the Lowell family should well be proud of. In this Brown fully concurred.

Pluto's Fate

Both Russell and Brown knew and appreciated the power of patronage. In 1935, Russell helped A. Lawrence Lowell prepare a biography of his brother, writing a new set of essays on his observatory's contributions to astronomy, and even let him quote extensively from his *Scientific American* columns. Neither Brown nor Russell objected. One way or another, the Lowell family deserved the credit for the discovery. Above all, the family had to continue supporting the observatory and astronomy.[26]

The question of the accidental nature of the discovery of Pluto has lingered as an issue among dynamical astronomers. Gibson Reaves has reviewed more recent history,[27] which included extensive deep searches for additional planetary bodies. Finally an analysis by E. Myles Standish in the early 1990s utilizing revised masses of the Jovian planets from the Pioneer and Voyager flybys, pinpointed the Uranian anomaly as caused by imperfect knowledge of its mass, and not to the existence of a still unknown Planet X.[28]

Though it was clearly not predicted, Pluto's fate was indeed to be discovered, mainly because the assets of the Lowell family made such an extensive and open-ended search possible, leading to the spectacular discovery, years ahead of its time. This fact, evidently appreciated by both Brown and Russell, made the recognition of Lowell's prediction a matter of political necessity since it indeed had prompted a major discovery through the allocation of considerable resources.

The Boundaries of the Solar System

At the Lowell Observatory, V. M. Slipher realized that Pluto was probably not the Planet X predicted by Lowell. In the summer of 1930, he told Tombaugh to keep looking. Tombaugh went from being "in the most excited state of mind in my life" to, "in the course of several weeks, the feeling grew that the real Planet X was yet to be found."[29] So he continued looking, systematically photographing and blinking for the next 13 years. He recalled feeling at first, as others did, that Pluto represented a new class of object in the solar system. And if so, there had to be more of them. By 1938, Tombaugh had examined and reexamined some 65 percent of the entire visible sky from Lowell, searched through some 70 million star images discovering all sorts of things, from asteroids to new nebulae to huge accumulations of tiny faint galaxies. But no new planets appeared on the scale or distance of Pluto. By July 1943 he had completely covered the visible sky from the declination of Canopus (50 degrees south of the celestial equator) to Polaris. He blinked most of this area too, save for the northernmost regions where no one believed a planetary object could possibly lurk.[30] He found nothing he could call a planet.

In the decades since Pluto's discovery, estimates of its mass diminished several orders of magnitude, from Lowell's six earths, to a single earth in the late 1930s, to one-tenth earth by the 1940s through the early 1950s. Continued efforts by leading celestial mechanicians, including Dirk Brouwer, Gerald Clemence, and Wallace Eckert, to apply perturbation theory led to estimates on the order of one earth mass,[31] but photometric studies and estimates of reflectivity (albedo) and visual diameter by Gerard Kuiper using the world's largest optical telescope led to significantly lesser values, as will be covered later in this chapter. Between 1949 and 1950, direct measurements of its diameter still varied by a factor of 2, but after 1978, and the discovery of its companion, dubbed Charon, by US Naval Observatory astronomers, its physical dimensions became more tightly constrained to a combined system mass of barely 1/300th Earth and a diameter of four thousand kilometers.[32]

Soon after Pluto's discovery there was speculation about it being a prototype of a wholly new outer zone of the solar system. On March 28, 1930, the *New York Times* received a telegram from Vienna announcing that Hanns Hörbiger, a successful engineer who had developed and heavily promoted a theory that the universe consisted of an all-pervading medium of ice aether, believed the new planet to be an example of a vast zone of icy planetoids that only waited to be discovered by improved astronomical equipment. Hörbiger viewed Pluto, according to the news wire, as a vindication of his controversial theory known variously as the "World Ice Theory" or "Glacial Cosmogony."[33] Hörbiger also claimed that Pluto could not be the body that was perturbing the inner planets, and called for "close observation and exact calculation over years rather than months."[34] This short article drew at least one supporting letter the very next day, lauding how Hörbiger's theory was the only one that could offer "a more rational scientific explanation" for the new planet.[35]

Hörbiger's followers of his Glacial Cosmogony did not further exploit the link with Pluto made by the founder just before his death. But the link to the existence of an outer zone did not die. In August 1930, Whipple's former teacher at UCLA, Frederick C. Leonard, published a popular essay exploring the significance of the new planet, using Whipple's barely one-month-old graphic solutions. Leonard made much of the pronounced ellipticity of the orbit, and explained that this diagram did not do the oddities of the orbit justice: it was also highly inclined to the general orbital plane of the other planets. The variation of all known planets prior to Pluto was not more than 7 degrees Leonard pointed out; Pluto however deviated from the mean by 17 degrees. Leonard speculated that Pluto was somewhere between the sizes of Mercury and Mars, given its faintness. Pluto's existence not only expanded the size of the solar system, but it also led Leonard to further speculate, in a most prescient way, that there was no reason now not to believe that the Sun's "control extends far beyond the orbit of Pluto." Indeed, were there not

> probably similarly constituted members revolving around outside the orbit of Neptune? Indeed, it may ultimately be found that the solar system consists of a number of zones, or families, of planets, one within the other... Is it not likely that in Pluto there has come to light the first of a series of ultra-Neptunian bodies, the remaining members of which still await discovery but which are destined eventually to be detected?

Although he grasped the significance of Pluto's discovery, it would be many years before this connection would be made with any conviction. Indeed, even though in the 1930s astronomers remained preoccupied with classifying the object, very few ventured to speculate as did Leonard. Leuschner, for instance, in his June 1932 address to the Astronomical Society of the Pacific as its retiring president, described "the continuing romance" astronomers had with Pluto as they searched for its identity and cosmic status.

Dominating the course of deliberations, Leuschner observed, was the fact that "[t]he more intensively Pluto is studied, the smaller the value of its mass becomes" and hence, Pluto should still be considered an "object" and not a "planet." Leuschner echoed Hörbiger that any final verdict awaited further study, but if indeed Pluto's mass turned out to be very small, and then "as is the case with comets" it could not possibly sensibly affect the motions of the other planets, then Pluto would have to be classed as a minor planet, like Ceres. But this only raised another problem astronomers were wrestling with: when is an object a comet or a minor planet, "even if the object shows all the characteristics of the former and none of the latter?" Leuschner cited as example of this classification conundrum the curious case of "Neujmin's object," announced in 1913, whose orbit was cometary but appearance planetary. One early observation at Yerkes Observatory showed it to have a very faint coma so it was classed as a comet. But upon its reappearance and scrutiny at Mount Wilson in 1931 with the 100-inch, no coma or tail was ever found. Even so, Leuschner concluded, the object remained known as a comet and not a planet. But had it first been seen in 1931, it would have been classed as a planet. "I think the implications I have in mind are clear." Whatever its ultimate classification, Leuschner concluded, all credit was due the Lowell Observatory for the fact of the discovery itself.[36]

Tombaugh's failure to detect any additional members of this new zone, especially anything that could be deemed Planet X, the massive perturber predicted by Lowell, is in all likelihood only one of several factors that contributed to the hiatus in exploring the outer boundaries of the solar system and the realization that Pluto is the harbinger of a new class of objects in that realm. Several lines of investigation ultimately led to this realization and some detracted, but they all, more or less, were conducted knowing Pluto was out there and had to be explained. Among them are continuing refinements of meteor and cometary orbits and classifications of their various behaviors into families distinguished by their orbital dynamics, their sources, and how they informed theories of the formation of the solar system and of stars in general. The predictive powers of the former, as one of several continuing preoccupations of the remaining but dwindling faction of mathematical astronomers following Brown and Leuschner, had to somehow meet the approximate modeling of the astrophysical theorists and dynamicists motivated by the evolutionary considerations of the latter—followers of George Darwin including Henry Norris Russell.

Factors Inhibiting a Larger Solar System

Pluto's eccentric orbit stimulated efforts to explain the object as an escaped moon of Neptune and hence be explained as an accident. This would, for one thing, retain the dimensions of the known solar system. Russell suggested this as speculation in his influential set of lectures, *The Solar System and its Origin*, and his thoughts attracted the energetic and endlessly creative young British theorist Ray Lyttleton to spend an academic year with Russell in Princeton in the mid-1930s. Among many studies, including an effort to produce the present solar system arising from a binary star system experiencing a tidal disruption by a third star, a scenario also suggested by Russell, Lyttleton explored the possibility that Pluto was once a moon of Neptune. Indeed, Neptune's remaining moon certainly behaved oddly; Triton moves in a retrograde orbit, the largest moon to do so. Lyttleton had been attracted to the problem when he asked if Pluto might someday encounter the Neptune system, and came away convinced that it was originally a moon of Neptune and in an encounter with Triton had been ejected, leaving Triton in a retrograde orbit.[37] Lyttleton's ejection mechanism followed traditional lines in celestial mechanics and remained an intriguing idea among astronomers for decades.[38]

Another factor inhibiting speculation on a larger solar system was the general feeling of those who studied the velocities and orbits of meteors, both in swarms and sporadic, that many meteors encountered the Earth at hyperbolic velocities. This meant that they were not gravitationally bound to the solar system and were interlopers from interstellar space. The brilliant but feisty Estonian astronomer Ernst J. Öpik promoted this view based upon visual observations of meteor trails that he examined statistically.[39] In 1930 Öpik was invited to the Harvard College Observatory by its director, Harlow Shapley, to team up with S. L. Boothroyd of Cornell to test his views. They designed and executed a campaign to determine meteor heights, directions, and velocities using an elaborate set of visual and telescopic stations built near Flagstaff, Arizona. Feeling that visual methods were still more sensitive than photographic means to reach faint meteor trails, Öpik and Shapley created a set of stations separated by a baseline of some 38 kilometers and led teams of observers between October 1931 and July 1933 to amass a large amount of observations. Their results when analyzed by Öpik's methods supported his conclusions completely, leading the team to speculate that the great majority of all meteors were extrasolar and hence there was no invisible repository lurking within moderate distances just beyond the known planets.

Öpik was a man of firm and particularly bold beliefs. After Fred Whipple graduated from Berkeley and was hired by Shapley at Harvard, he encountered the Estonian polymath and at first was captured by his conclusions. Whipple was also fascinated by another Öpik notion: the orbits of both meteors and long-period comets suggested that they did come from a vast reservoir centered on the Sun but at nearly interstellar distances, and that stellar perturbations on this reservoir brought them into the planetary realm from time to time. Öpik believed that all stars possessed these circumstellar shells, and some meteors encountered on Earth could even have originated in those vastly distant places.[40] Öpik's genius at crafting statistical methods for analyzing families of meteor orbits as well as his grand speculations led Whipple to study meteor radiants starting in 1933. Whipple soon decided though that he could improve upon visual methods for tracking them by applying the technique of wide-field photography, which led in 1936 to significant technical improvements. Based upon the availability of small, reliable synchronous motors and stabilized AC current he created precision image choppers to place in front of fast photographic cameras that he felt would equal and surpass visual reconnaissance and also greatly improve earlier photographic attempts. As he recalled in an oral history, "The cycles were kept very well, by the power company. You had a very accurate timing device."[41]

With this innovation and Shapley's approval, in the mid-1930s Whipple set up triangulation stations mounting cameras on the Harvard campus and at the Agassiz Observing Station at Oak Ridge some 26 miles away, and was rewarded with sets of trail plates that he assumed would yield confirmation of Öpik's hyperbolic orbits. However, upon careful analysis, especially using only dual observations that could be accurately timed with a visual record, Whipple realized that all the orbits were elliptical. This surprise result initially drew Öpik's enmity and delayed publication for some years, but it also led Whipple to refine techniques of observation to improve knowledge of meteor orbits, especially their velocities and deceleration rates through the upper atmosphere of the Earth.[42]

Öpik resisted Whipple's conclusions for years, even after radio Doppler observations at Harvard, in Canada, and after the war at Jodrell Bank more than vindicated Whipple's conclusions. "He never forgave me until the fifties for tearing down the hyperbolic meteor theory. (Laughter) He finally wrote a very nice paper in the late fifties about my work and admitted he was wrong. But up until that time, he never forgave me. Finally he did when he was completely convinced."[43] Now the meteors had to come from someplace nearby, at intermediate distances somewhere beyond the planets but not as far as the stars.

Oort's Cloud, Whipple's Snowball, and Kuiper's Belt

In the immediate post–World War II era a broad range of astronomers, astrophysicists, geophysicists, and geochemists were motivated by new conceptual frameworks, new techniques of investigation, and new patterns of patronage.[44] Among astronomers, Gerard Kuiper, Jan Oort, and Fred Whipple led the charge. Kuiper, a highly creative specialist in binary star formation and evolution at the interface of theory and observation, applied new infrared formulations to planetary studies, and Oort, a leader in studies of galactic structure, dynamics, and evolution, applied radio technologies that lent powerful new perspectives to integrate stellar evolution and cosmogony. And although he migrated (or accreted) less than the others, Whipple brought a whole new perspective to meteor and cometary origins not only using traditional orbital means improved by technology, as we have seen, but also by revolutionizing views of the physical structure of cometary bodies through new studies of the interaction of cometary debris, in the form of meteor swarms, with the Earth's upper atmosphere, utilizing data from rocket-borne instrumentation.

A major outcome of this watershed period was that the solar system became a far larger and more complex place, a place where Pluto became a member of a class of objects and less an oddity. Encounter theories and interstellar origins were replaced by autonomous mechanisms producing comets and meteor swarms arising within the gravitational confines of the Sun, as defined by the original solar nebular cloud that contracted, condensed, and spun up into a disk. Although many of the elements of this combined view had been suggested earlier, first by Öpik and then by gifted speculators like Kenneth Edgeworth, it was the combined force of Oort's, Kuiper's and Whipple's contributions that changed the way astronomers thought about the solar system.

Öpik's vision of a vast reservoir of comets, for instance, did look superficially like Jan Oort's highly revised model in 1950. Indeed, when Whipple first read Oort's manuscript on the existence of a cloud of comets, Whipple initially felt that Oort's work did not go beyond Öpik's some 18 years prior, but as he examined the Dutchman's analysis he realized that it was a very solid contribution, more acceptable, more mathematically rigorous, dynamically sound, and far-reaching than Öpik's. Stimulated by a recent exhaustive analysis by a Leiden thesis student, A. J. J. van Woerkom, which convinced him that all known comets are part of the solar system, Oort showed that the cloud itself could not resist the systematic perturbative effect of passing stars and that the effect would systematically influence cometary orbits in the cloud. These random deviations from the systematic perturbations would cause comets to lose orbital energy and enter the realm of the planets. For these reasons Whipple credited Oort with the firm establishment of this new realm of the solar system, the skeletal frame of the original interstellar-scale cloud that formed the sun and its system of planets.[45]

The establishment of the Oort Cloud, as it is today called, was a major factor leading astronomers to think more about what was going on in the outer reaches of the solar family. The question now was, what if anything existed between the solar family of planets and the Oort Cloud? Neither Oort, nor Whipple, nor anyone else associated this new zone with Pluto.

Underlying, and parallel with, Oort's efforts, and those under Oort at Leiden Observatory, and Lyttleton's continuing effort to promote encounter mechanisms was the deeper question of the origin of the solar system. Interrupted by the war, the formerly dominant theory of a two-star tidal encounter and the hot origin of the planetary system was on the wane and no longer in serious play. Another Russell graduate student in the late 1930s, Lyman Spitzer, had shown convincingly that hot filaments arising from tidal collisions would dissipate and not condense, and his continued studies after the war, bolstered by Russell, laid the encounter theory to rest. In its place, variations on the nebular theory arose once again, the most promising by Carl F. von Weizsäcker that envisioned planetary formation to take place within vortices in a collapsing solar nebula.[46]

Nebular theories, however, required all members of the solar system to have originated in a single process, or at most a stepwise continuous set of processes. In his first formulations of his theory, Oort had envisioned his vast cloud to be the reservoir out of which both comets and asteroids arose, and that there was a long-term replenishment process acting that variously had removed asteroids from the well-known belt between the orbits of Mars and Jupiter, a process due mostly to perturbations by Jupiter, and a reverse process now acting through which perturbations by passing stars decreased the perihelion distances of comets in the cloud and caused them to enter the planetary realm. Whipple was never comfortable with this scenario, since he was just then showing, based upon meteor trail data and rocket-borne in situ measurements of atmospheric properties, that comets were fundamentally different from asteroids. Writing to Oort in August 1949, Whipple felt that "the arguments are fundamentally secure" as there are indeed many correlations between comets and meteors, but he did not agree with a common origin for comets and minor planets: "I am convinced from all of the evidence from physical studies of meteors, that we are really dealing with two different types of objects."[47]

Whipple was just then putting together a major statement that described comets as icy conglomerates of dust and rock. It had long been suspected that meteor swarms were remnants of periodic comets or had been ejected from comets still active, like comet Swift-Tuttle and the Perseids, or the Leonids being associated with the faint comet Tempel-Tuttle. As Whipple noted to Oort in November 1949, swarm radiants were typically found in the ecliptic, "very suspicious in itself." He and a Harvard graduate student, Salah Hamid, had been studying secular perturbations on the Perseid meteor stream trying out various models to see how each would affect the "rate of expulsion of meteoric particles from comets." He was also collecting information on the spectra to be expected on the basis of his comet model and was searching for every scrap of evidence available.[48]

Whipple and Oort were in frequent contact during these years discussing their parallel interests in physical models for comets and hints coming from the different classes of meteoritic phenomena: those thought to derive from asteroidal origins and those from comets. Whipple's view of comets as highly fragile bodies was based upon continuing observations from his meteor cameras, after the war greatly expanded with a new network of wide-angle high-speed cameras in Arizona, the first Super-Schmidts, as well as in Massachusetts.[49] Those with trajectories typical of comets broke up far quicker and evidently easier upon entering the earth's atmosphere than those of asteroidal origin. He sent his drafts to Oort for comment, and Oort responded positively.[50]

Whipple was working along several parallel fronts toward establishing his icy conglomerate model. He was already well along developing his model when van Woerkom's work appeared in December 1948, and Whipple immediately saw it as a major confirmation of the ancient nature of comets and the necessity of viewing capture perturbations by Jupiter and Saturn as a mechanism for the replenishment of comets populating the inner solar system. His most direct evidence, however, came from observing disintegration rates and secular acceleration rates for short-period comets like Encke and his team's recent observation that the Giacobinid meteor trails ended higher in the atmosphere than expected, indicating their fragility.[51]

Whipple, who had always hoped to excel in a mainstream problem area in astronomy, now realized that his niche subject was getting far more attention from mainstream astronomers. To establish priority, but moreover to be sure he had touched upon the many specialties that informed his subject, he mimeographed his working paper and sent it around to a wide circle of readers. Oort responded quickly and with enthusiasm, calling his "A Comet Model" an "inspiring" work.[52] He felt that Whipple was definitely on the mark regarding how the rotational dynamics of an icy body would divest itself of fragments: "It may well be that your 'iceberg' theory is correct, but in order to be convinced I should want first to see some good evidence that comets and large meteorites are racially different."

In 1937, Öpik first suggested that interplanetary dust particles would survive frictional and collisional heating upon entry in the high atmosphere because they would radiate heat faster than larger bodies, and thus avoid vaporization. In 1947, from evidence of space densities of dust from zodiacal light observations, H. van de Hulst, a former Oort student at Leiden, estimated that the Earth sweeps up anywhere between a thousand and three thousand tons of dust per day. Core sampling of deep ocean beds and arctic ice fields was revealing a significant amount of dust that had compositions characteristic of nickel-iron meteoritic bodies. And, most intriguing were acoustic and mechanical collisional impact studies conducted by navy scientists by placing sensors on the skins of captured German V2 missiles flown from White Sands, New Mexico, by Army Ordnance. This effort had Whipple's enthusiastic endorsement, as he was then a member of a panel of scientists and project managers from military laboratories learning how to utilize rockets for studies of the upper atmosphere.[53] By 1950, Whipple was calling these particles micro-meteorites, and by the end of 1951 his cometary ice model was strengthened by continued V2 studies of the properties of the upper atmosphere in the range where meteor trails were observed. These in situ density measurements confirmed that the parent colliding bodies had average densities less than water. As Peter Millman summarized the correlation in a 1952 review, "Thus we find a physical difference between the material of planetary origin that falls [and is recovered] as meteorites, and that of cometary origin that appears as meteor shows and sporadic visual meteors."[54]

The third major contribution leading to the modern conception of the solar system, linking it as well to star formation, was Gerard Kuiper's demonstration that a vast flattened disk of material exists beyond the realm of the major planets extending out to the Oort Cloud itself. Like the contributions by Oort and Whipple, there were precursors envisioning what came to be known as the Kuiper Belt. We have seen elements in Leonard's speculations, and of course in Hörbiger's vision. The idea was most certainly in the air.

Kenneth Essex Edgeworth, electrical engineer and scion of an "archetypal gentleman literary and scientific" family, spent his later life deeply fascinated with problems of cosmogony, from the fission of rotating bodies to star formation and to the origin and development of the solar system.[55] He envisioned the formation of two families of planetary bodies in a collapsing and flattened nebular disc to be due to a successive hierarchy of condensations caused by what he called "viscous friction" between interacting layers of the disc. The inner family of bodies formed the terrestrial planets, centered on Venus and Earth, tapering off beyond Mars to the asteroid belt. Similarly, the Jovian planet family tapered off into a vast disc of frozen bodies, the comets. Edgeworth presented his general theory of viscous formation of planetary bodies to the Royal Astronomical Society in January 1948, drawing mild interest and light criticism from theorists such as Fred Hoyle and E. A. Milne, as well as the talented Royal Observatory astronomer Robert d'Escourt Atkinson. Milne had been given a copy of his paper by the RAS secretary to be sure it was worthy of discussion, and evidently felt it was in that it "achieved a great deal with simple mathematical methods." But Milne politely pointed out that some of his initial assumptions needed serious rethinking, such as the initial size and nature of the protoplanetary cloud, which he required to be larger than the present-day Galaxy.[56] In a 1968 review of theories of solar system formation, I. P. Williams and A. W. Cremin recounted Edgeworth's efforts, noting that Sir Harold Jeffreys regarded his viscous mechanism to be "hopeful" but very difficult to verify mathematically, and regarding its significance, the authors conclude that it differentiates the two families by "postulate rather than by calculation."[57]

The concept, however, of a lingering field of frozen debris beyond the planets was soon to take hold in the work of Gerard Kuiper. Kuiper viewed planetary formation as a special case of binary star formation and evolution, which was his original interest dating from the 1930s when he explored the properties of close binary systems and searched for evidence of binaries in proper motion studies. In the winter of 1943–1944, already engaged in wartime work at the Radio Research Laboratory (RRL) at Harvard, Kuiper took brief leave for his allotted observing time on the McDonald 82-inch reflector to obtain low dispersion spectra of the ten largest planetary satellites, as well as Pluto, utilizing panchromatic and newly formulated but very slow near-infrared Kodak emulsions. Kuiper brought his plates back to the RRL and devoted his personal time there to analysis. In rapid order Kuiper found that Titan, the largest moon of Saturn, had an atmosphere of similar composition to its parent planet.[58]

Well before publication, Kuiper sent word of his discovery far and wide. Henry Norris Russell responded enthusiastically in January 1944 "Your discovery is of great cosmogonic interest [as it] provides really new material bearing on the history of the solar system . . . " Russell seized upon the simple fact that the presence of an atmosphere, especially the identification of methane, indicated that Titan could never have been very hot and so could not have condensed out of a hot gas. This was another nail in the coffin of both encounter and eruption theories. Russell also applauded Kuiper's opportunistic style. With access to a very large telescope but other pressures bearing down upon him, he looked for a limited problem that required only finite observing time. Kuiper was the man for the job: "The main point is knowing what to try and you have done this before repeatedly."[59]

In his published paper, Kuiper admitted that "[t]he case of Pluto is puzzling." His observational data did not show an atmosphere, and without knowledge of its reflectivity, he could not say much about its mass or radius. It might be no larger than Neptune's larger moon Triton, Kuiper speculated, and until someone was able to measure its angular diameter, nothing could be said about the existence of an atmosphere.[60] Russell suggested that Kuiper follow up this work by gaining observing time on Caltech's 200-inch telescope, then under construction and due to the war still some years in the future. After repeated visual attempts with a unique photometric "disk meter" at McDonald using the 82-inch telescope, Kuiper would follow up this suggestion in 1950 with new diameter estimates at Palomar. Carrying his disk meter to California he made a series of visual measurements at the prime focus (at a magnification power of 1140X) with the aid of the observer Milton Humason. They once again reduced Pluto's size and nature: "You certainly keep on giving us a series of most convincing surprises . . . ," Russell responded in April. Pluto's angular diameter was less than 0.20 seconds of arc, making it not larger than .46 earth diameters and less than one-tenth the earth's mass.[61]

In this paper Kuiper made no mention of the possible bearing his new determination of Pluto's size, and estimate of its mass, might have on his developing theory of the origin of the solar system and the existence of a belt of debris beyond Pluto. But he was indeed exploring both just at that time, trying to determine how such a field of small frozen bodies might be produced and maintained. Speculation by various workers including van Woerkom, Whipple, Oort, and Kuiper in the recent past over how such a field could be produced ranged widely, and Kuiper's new mass determination did not rule Pluto out as a perturber that brought cometary bodies into the inner solar system. One way or another Pluto had to be accounted for in any new general scenario that explained the known bodies in the solar system. Kuiper, in an almost continuous progression of studies in the early 1950s, came the closest.

In some 15 related studies that appeared as journal articles, compilations, and review essays between 1949 and 1955 Kuiper explored both the dynamics and the physical character of planetary atmospheres, surface features, asteroidal and cometary origins, and the nature of the interplanetary medium, amassing a consistent and convincing amount of observational data in support of a coherent theory of the origin of the planetary system that built upon how regions of varying gravitational instability in the collapsing solar nebula produced both the major planets, and zones of residual material. As did Whipple, Kuiper developed observational programs, including a three-year photometric asteroid survey starting in 1949 at McDonald to improve knowledge of colors, rotational properties, and statistics that led to a better understanding of asteroidal bodies as fragments of collisional processes rather than as "original condensations" thus confirming the view that the asteroid belt is the product of an oblique collision of two small planetary bodies.[62]

Kuiper had been stimulated to put his thoughts together when Otto Struve organized an edited collection of essays by Yerkes and McDonald astronomers to commemorate the fiftieth anniversary of the Yerkes Observatory in 1947. J. Allen Hynek was the editor, and preparations and article development took several years. Hynek's assignment was to convince his authors, mostly present or former members of the Yerkes staff, to survey the past half-century of their specialties, identify major problems of the present, and aim to a readership level of a "first-year graduate student, well versed in fundamentals but by no means a specialist."[63] The essays that resulted, from leaders in astrophysics ranging from Bengt Strömgren and S. Chandrasekhar, to W. W. Morgan, Lawrence Aller, Otto Struve, Jesse Greenstein, and, among the few who were not Yerkes alumnae, Cecilia Payne-Gaposchkin, constitute a watershed assessment of the state of astrophysics immediately post World War II. Most authors blended either dynamical or spectroscopic observational astronomy with experimental or theoretical atomic, quantum, and nuclear physics or hydrodynamics to demonstrate the power of the new synthesis. Kuiper's efforts were exemplary in this regard, effectively combining Russell's astrophysics with Brown's dynamics.

By the time Kuiper began to focus on his essay for Hynek, he was becoming convinced from both recent theoretical work by his Yerkes colleague S. Chandrasekhar and experimental work in fluid dynamics that he had the tools to differentiate between laminar and turbulent flow conditions, and that as a result he felt he could dismantle von Weizsäcker's theory of planetary formation through vortex action: "[T]he primary vortices cannot arise at all since they are a region of laminar flow in a medium," he told Lowell Observatory's Carl Lampland in September 1949.[64] Kuiper hoped he was within a few weeks of finishing his own theory, which utilized a two-step process: turbulent eddies in the collapsing nebula produced initial density fluctuations and these would last long enough to further collapse into protoplanets. He retained the primary vortex motion of von Weizsäcker's theory, and then found a way to accelerate collapse before other forces could dissipate the vortex: "[W]hat is needed," Kuiper wrote, "to get the planets under way is *time*."[65]

Kuiper found, in the light of modern studies of turbulence, that von Weizsäcker's vortex eddies had lifetimes far too short to develop into major protoplanets. He therefore looked for a way to keep the eddies together long enough to collect enough material to build planets. The timescale he found was of the order of a hundred thousand years or more, and he found he could meet this timescale if the density of the cloud exceeded a certain limit, that of the well-known tidal disruption condition known as the Roche limit. Kuiper called this critical density the Roche density, where "the self-gravitation of a cloud of gas will be equal to the solar tides" or greater, at any given distance from the sun. He then moved outward in the solar system deriving conditions for planetary formation, and represented these in various ways, including a comprehensive profile of the distribution of mass in the solar nebula. He made much of the fact that this mass profile fit neatly the known frequency distribution for separations of binary stars of all classes, and that the mean positions of the major planets in the solar system fit the maximum of that distribution. Next, Kuiper observed that the slope of the mass distribution beyond Uranus and Neptune was very steep, creating densities far too low for planetary formation. This suggested, he explained, "why no trans-Plutonian planets have been found." Bluntly, they never formed.[66]

Kuiper's model left the origin of Pluto "still somewhat in doubt" and it continued to attract his attention periodically throughout the 1950s. Given its excessively small mass, it was still possible that it could be an escaped moon of Neptune.[67] But Kuiper was more interested in suggesting that other "small" planets existed beyond Pluto, more akin to asteroids in mass and dimension. What did likely lie beyond Pluto, Kuiper concluded, in the range of some 38–50 AU, was a debris field of "snowballs, with silicate and metallic particles suspended in them" with a combined mass equivalent to Oort's cloud of comets. In Kuiper's view, this field of nascent cometary nuclei could be perturbed by Pluto, bringing them periodically into the inner solar system. He felt this was a preferable model for the origin of the short-period comets than Oort's initial idea of ejection from the asteroid belt. Since the observed dissipation of comets into meteors and zodiacal light was an observed fact, "evidently resulting from the evaporation of the ices…[i]t would be puzzling if they were asteroidal bodies."[68]

Kuiper first submitted his paper to Hynek in November 1949 and sent it around for comment in February 1950. As the Hynek edition moved through the press, Kuiper redrafted his basic findings into a series of papers for the *Publications of the National Academy of Sciences*, where he was more explicit about the role of Pluto as a disperser of the original disk of cometary bodies between 38 and 50 AU. Further, his discussion clarified points about his differences with the Oort idea that comets were ejecta from the asteroid belt.

The work of Oort, Whipple, and Kuiper expanded the scope of the solar system, and provided new options for considering the origins and characters of its many components, from the major planets and their satellites, to the asteroids, comets, meteors, and especially Pluto. They also helped to unify views of the origins of comets and the connections between comets and meteors. By no means were astronomers and geophysicists of common mind as to the quantitative details of the new views; there was still widespread concern over the difficulty of accounting for the fact that the Sun comprised by far the dominant mass in the system, but the planets harbored 98 percent of the angular momentum. This puzzle was raised heatedly and repeatedly by Fred Hoyle, at first partly in defense of Lyttleton's interstellar cloud encounter theories, but by the end of the decade more in terms of his own views of how angular momentum could be transferred from the Sun to the condensing planetary disk through magnetic torqueing.[69] They differed also over details of compositional differences and if planets formed by accretion or condensation. Further, Kuiper himself became entangled in a continuing debate with Harold Urey and other geochemists and geophysicists over the temperature history of the Moon and the origin of its visible features that prevented the full acceptance of any one theory across disciplines.[70]

Throughout this entire time period, however, the origin and character of Pluto remained enigmatic. It was not fully comfortable in either ejection theories or in models of independent planetary formation. It was still thought to be massive enough to be blamed for the production of cometary visits into the inner solar system, but was small enough to be consistent with the idea that as one moved outward from the Jovian region, planetary masses fell off precipitously. Kuiper expressed it best by analogy, arguing that the difference we see between the Earth-like planets and the Jovians is "not because it originated closer to the sun but primarily because proto-Earth was less massive than proto-Jupiter. The existence of a planet like Pluto outside Neptune becomes at once 'acceptable.'"[71] Pluto's existence as a solitary body was reinforced throughout the 1950s, first and foremost by the lack of observational evidence of anything else lurking out there, and second by the success of evolutionary models like Kuiper's. At the very least though, one can say that by end of the 1950s, the solar system itself was a much larger place, and a place that could be studied by many new and increasingly more powerful techniques.

Searching for Kuiper's Belt and Pluto's Place

Given the debate over Pluto's status in recent years, there is no surprise that a growing literature exists; science writers and participants have poignantly captured the complexity of the highly diverse and loosely knit field of workers involved in the searches in the past 20–30 years.[72] Most of them recount the process through which Pluto went from being a uniquely curious body, one of a kind and a problem, to being the prototype of a whole new class of planetary bodies in the solar system. Others offer insights into social aspects of the debate, especially how different factions approached the issue and were guided by differing "cosmological" views.[73] But all demonstrate, most implicitly but a few explicitly, how the application of new critical technologies finally closed the gap in comprehension caused by the premature discovery of Pluto.[74]

The application of high-speed computers initiated the process. In 1964, long runs on the powerful Naval Ordnance Research Calculator at Dahlgren showed Pluto's orbit to be in stable three/two mean-motion resonance with Neptune over millions of years, throwing into doubt theories of its origin by interstellar capture or Neptunian ejection.[75] And in the same year Fred Whipple, armed with a new IBM-7094 computer, reexamined his old problem planet, and suggested that Pluto was far less massive than previously supposed: slight measured perturbations on Neptune that had been detected were due not to the planet, but could be explained by a vast belt of cometary bodies 40–50 AU from the Sun.[76] As much as he wanted this conclusion to stand, Whipple engaged his Harvard-Smithsonian colleagues Brian Marsden and S. E. Hamid to test it using a closer analysis of perturbations on the orbits of well-known comets. Even though they could find no evidence for the perturbative effort of a ring of comet material, Marsden keenly recalls that "Whipple always persisted with it that there was something there."[77]

But computers were soon to play a different role. Beyond Tombaugh's valiant efforts, no major search was mounted to specifically find other Plutos until the last three decades. As usual, multiple reasons exist. Pluto's diameter and mass were constantly overestimated, until the discovery by James Christy and Robert Harrington in 1978 that photographic images of Pluto over the past several decades often showed an asymmetric lump that, they realized, orbited the centroid of the image just like an unresolved moon might do, and in a period equal to Pluto's rotational period of six days. Pluto's resulting mass dropped again by a factor of 10, from 1 earth mass in the 1940s through 1960s, to .1 in the late 1960s and 1970s, to less than .002 by 1980 with the discovery of Charon. And subsequent studies of its brightness and color rendered it a silicate poor ice-rock mixture, a dirty snowball, in other words, maybe a nascent Whippleian monster comet.[78]

After 1980, Pluto's diminished stature brought it well into the realm of bodies existing beyond its orbit that had been qualitatively envisioned by Kuiper. On those few occasions where speculation reached print, however, Kuiper and others suspected that it would probably take a telescope the size of Palomar to find anything the size of Pluto in this outer region, and that would be a chance discovery. But the lack of effort to do so could have been due also to changing priorities and personalities in astronomy. Tom Gehrels, who had trained under Kuiper at the Yerkes Observatory in the 1950s, and had followed his mentor to the University of Arizona, pointed out in 1971 that the main obstacle to deeper surveys, whether they be for planets, planetesimals, or in his case, for fainter asteroidal bodies that might someday strike earth, was "not in the availability of funds or telescopes, but in the lack of dedicated personnel such as [George] Van Biesbroeck and the van Houtens to execute the enormous task of blinking, identification, etc."[79] Indeed, Gehrels knew the tedium of blinking. He had participated in one of the last of the great asteroid campaigns of the 1950s led by Kuiper, which depended mainly on the countless hours required to blink 1,094 8×10 inch photographic plate pairs, and consequently several years for analysis.[80] Gehrels had also teamed up with van Houten and others for a follow-on effort: a multi-institutional faint asteroid survey, the "Palomar-Leiden Survey," that took decades to fully analyze a set of plates taken by Gehrels with the Palomar Schmidt during observations made between two consecutive dark runs in 1960.[81]

All along clues emerged that bodies of some sort must lie in solar-centered orbital motion beyond Neptune. Comets were the first and most lasting of them. In the late 1970s and 1980s other discoveries added to the clues. The Voyager encounters with the satellite systems of Jupiter and Saturn between 1979 and 1981 accelerated the growing conviction that the smaller bodies of the outer solar system were icy and numerous. It had long been suspected that the densities of the satellites decreased with distance from their planetary captors, but precise measures from the Voyager flybys sharpened focus considerably, showing that water ices were a major constituent of even the larger of these bodies. The Voyagers also found that virtually every satellite was heavily cratered and although the majority of the cratering was from a bombardment phase early in the history of the formation of the solar system, cratering was still significant even on bodies where geological forces resurfaced the body in recent history.[82]

The flood of new data, however, did not fit into a neat simple picture, a fact that frustrated reductionists like Fred Whipple, who, with Brian Marsden and S. E. Hamid, tried to verify the existence of Kuiper's Belt by searching for perturbations of known comets that had been tracked over several orbits and had aphelia greater than 35 AU. In the late 1960s and early 1970s they failed to find evidence for these perturbations and so placed a highly restrictive upper limit on the amount of material that could exist beyond that distance. But in the 1980s, improved computer simulations of the dynamical behavior of observed short-period comets confirmed that they originated from a belt beyond Pluto but nearer than the Oort Cloud. In a highly cited article from 1980, Julio Fernandez from the National Astronomical Observatory in Madrid refined Kuiper's model using statistical methods analyzing the influence of close encounters between a hypothesized belt of comets in a region between 35 and 50 AU from the Sun. He showed how those comets lost orbital energy and came under the influence of Neptune, ending up as a family of cometary bodies with properties characterized by known short-period comets.[83]

Stimulated by this sort of work, in the late 1980s, Martin Duncan, Thomas Quinn, and Scott Tremaine consumed "several months of CPU time on a dedicated Sun-3 microcomputer" to show that their numerical simulations could match the distribution of orbital inclinations for observed Neptune-crossing comets. The predominance of low-inclination trajectories fit the concept of a disk-shaped source, leading them to conclude that short-period comets "arise from a cometary belt in the outer solar system" between 35 and 50 AU from the Sun. John Davies in retrospect recalled that he and his colleagues felt then that it would be only a matter of time: "All that remained was to find some."[84]

The first discovery of a wandering icy object as puzzling as Pluto came inadvertently during a patrol program at Palomar that had been searching deep space for decades (and generally avoiding the Milky Way and ecliptic regions) for supernovae. Since the 1950s, Caltech/Palomar staff led mainly by the irascible genius Fritz Zwicky conducted a continuous search for supernovae using the Palomar Schmidt telescopes, in 1959 monitoring some 4,000 galaxies in 64 fields and cooperating with other observatories to improve the statistics of these cosmologically significant events. Charles T. Kowal joined the team in the early 1960s as a research assistant, contributing to the monitoring and noting objects of particular interest as a by-product of this powerful celestial surveillance program. By 1970 Kowal began reporting fast-moving stellar-appearing objects from these searches, near asteroidal bodies. He also began searching for and finding new outlying satellites of Jupiter.

Because of his success finding rapidly moving faint objects, Kowal was able to establish a "Solar System Survey" program at Palomar using the large Schmidts. Comets and asteroids were detected, and in October 1977, an IAU *Circular* announced "Slow-Moving Object Kowal" based upon observations taken by both Kowal and Tom Gehrels over eight days of time that revealed a stellar object whose motion was "scarcely greater than that of Uranus...extraordinarily slow for an object so close to opposition." In subsequent days and weeks additional observations by at least five staffers recruited to assist yielded no definite orbit, but it lay at an estimated distance of some 14–17 AU, placing it between the orbits of Saturn and Uranus.[85] Further observations and reductions well into 1978 by Kowal, assisted by expert orbit calculators such as Brian Marsden (figure 13.2) and William Liller and numerous others, yielded a "completely unprecedented" orbit with aphelion near Uranus and perihelion near Saturn, in a 3:5 resonance with Saturn. By the spring of 1978, Kowal named the new object Chiron, and the discovery stimulated new searches for additional distant objects in the solar system.[86]

Like Pluto, however, at a time when manual searching was still dominant, Chiron remained unique for well over a decade, being joined only in 1992 with a discovery from Gehrels's partially automated Spacewatch system using electronic image detection and computer-assisted blinking. On January 9, 1992, the Spacewatch moving-object detection system announced that it had found an object moving slower than an asteroid. Once the team determined its orbit, they named it Pholus, and its similarities to Chiron established a new family known today as Centaurs for their hybrid orbital

Figure 13.2 Brian Marsden, critic of Pluto's planetary status, was longtime director of the Central Bureau for Astronomical Telegrams at the Smithsonian Astrophysical Observatory and its associated Minor Planet Center, both sanctioned by the International Astronomical Union. (Courtesy Smithsonian Astrophysical Observatory, photograph by Hal Dormin.)

characteristics, possessing both cometary and asteroidal properties. By then, continued scrutiny of Chiron as it moved toward perihelion and into warmer climes also revealed evidence that it was outgassing and some reports even suggested a coma.

Voyager's encounters with the satellites of the outer planets, the dynamic studies by Fernandez, Duncan, and so on, the chance discovery of Chiron, and the emergence of computer-assisted means of detecting anomalous motion among faint stellar objects with medium-sized telescopes, all were factors attracting new patrons and new teams to the search. Typical of the new teams was one headed by Dave Jewitt and Jane Luu in the late 1980s who were attracted more by the realization that cratering dominated resurfacing processes throughout the solar system than by an effort to test Kuiper's model. Along with other teams from the US Naval Observatory and at the University of Texas, they all adopted electronic methods to perform their searches, and, as CCD chips became larger and more capable, as John Davies has pointed out recently, the pace of discovery picked up proportionately.[87] Davies followed each team's progress as they employed larger chips, from 385×576 pixels, then to 800×800 (as in HST's first generation WFPC camera), then 1024×1024 and by 1992 to 2048×2048. They designed new and more efficient digital search routines that could be tailored to pick up not only anything that moved, but they could select objects that moved only within a certain spatial range.

Jewitt and Luu optimized their searches, for instance, to select objects moving with Centaur-like orbits, and on August 30, 1992, found their first candidate. After three days of analysis they announced the existence of the twenty-third magnitude object, designated 1992 QB_1. Brian Marsden determined, however, that it was somewhere between 37 and 59 AU, rather large for a Centaur, but the exact nature of the orbit was unknown. Meanwhile they and their competitors were finding other objects, and the discoveries became a continuous stream of objects with similar photometric and astrometric properties. As Davies recounts the situation, "Paradoxically, it was not long before the problem was not so much discovering new objects, but keeping track of the ones that had already been found."[88]

Through the mid-1990s, with digital advances in miniaturization allowing for the introduction of mosaic arrays of 2048×2048 chips to create focal planes that rivaled and surpassed large photographic plates, more teams formed, more telescope time became available to track as well as to discover, and the flow of discoveries of new objects increased to the point where statistics on their orbital behaviors revealed distinct families of new planetary bodies. One of those families was called "Plutinos" because their orbits were similar to Pluto's. Jewitt and Luu coined the term to describe the family of bodies that "reside in or near the 3:2 mean motion resonance with Neptune, as does Pluto." From the list of some 13 objects of this class, including Pluto, they concluded that Pluto is distinguished from the others only by its size, "rather than as an independent (but orbitally eccentric) planet."[89] In less than a decade, searches by more and more teams revealed larger and larger bodies, pushing for what many of them thought was the inevitable. Along the way, in March 2004, after almost three years of searching and cataloguing objects detected with a large-format CCD array on Palomar's 48-inch Schmidt with a search program optimized for detecting large slow-moving bodies, Mike Brown of Caltech, Chad Trujillo of the Gemini Observatories, and David Rabinowitz from Yale announced the existence of a body they dubbed "Sedna" lying some three times further from the Sun than Pluto. They surmised that Sedna was the largest Trans-Neptunian object yet discovered, still smaller than Pluto, but moving at a rate at the low end of their search range. After adjusting their lower limit downward, they found another large object, even brighter. They called it "Xena" at first, but it was soon renamed "Eris" according to convention.[90] Its photometric and spectral properties, determined by the eight-meter Gemini reflector and confirmed by HST, indicated that, like Pluto, it had a methane atmosphere, and, most exciting, a diameter larger than Pluto's.[91] Pluto at last had company. But now, what was it to be called?

In their various reviews, scientists revealed how the emergence and elaboration of a new technology, in this case the digital detection, scanning, and isolation of objects by their behavior, brought this once quiet, borderline, and definitely very frustrating field into the mainstream. The more recent studies, post 2005 at least, pointed the direction toward what has been termed "The Battle of Prague." They also indicate that the battle had been brewing for quite some time.

The Public Pluto

John Davies described in 2001, but did not assign much real importance to, naming conventions, whether they determined what to name a new planetary body, or what to name the Kuiper Belt. Similarly, but referring more to legacy naming conventions, David Jewitt has observed that "sufficiently vague statements that had no impact when they were made can be retroactively interpreted in almost any way you like." Even though he participates with others in using the term "Kuiper Belt" he feels that it was not the source of his stimulus for embarking on his searches, and, apparently, what Kuiper predicted was not what was precisely found to exist. Alternatively, Fernandez was explicitly influenced by Kuiper's 1951 work, and apparently believed he had, more or less, vindicated it.[92] As it may have been with Kuiper, so it was with Pluto.

Writing before Eris was detected and its mass estimated, Davies reflected lingering resistance to calling Pluto a Kuiper Belt object (KBO) based upon the fact that Pluto was so much larger than others located in the same region, and even had its very own moon. But Davies took very seriously the fact that unlike any other naming issue in astronomy, the question of Pluto's planetary status garnered considerable media and public attention, far beyond any questions about the role such bodies play in the formation and evolution of planetary systems.

As early as 1992, the discovery of 1992 QB_1 by Jewitt and Luu brought the *Boston Globe* to Brian Marsden's door, asking for commentary about Pluto's status: whether it too was a KBO or still a planet.[93] Media interest continued to be punctuated by these detections through the 1990s, but there was no media acceleration of popular concern until a young astronomer recently hired at the American Museum of Natural History as its ninth planetarium director was faced with the challenge of developing new and dramatic exhibits in a new planetarium facility that would justify the $230 million pledged by Frederick P. and Sandra P. Rose.

As Neil DeGrasse Tyson relates the situation in his memoir, *The Pluto Files*, his task was not to "face-lift the existing facility, but to invent something entirely new."[94] He knew the exhibitry had to be spectacular, and thereby expensive, and therefore he knew, as do all museum professionals, that what he set in place had to last a good long time. Tyson therefore set about assessing the "shelf life" of the various factoids the public would look for in his museum. Solid astronomical evidence rated the most expensive and permanent displays, those on less solid ground could be covered in less expensive ways to make them easier to change.

In 1998, Tyson hired the astronomer and creative specialist Steven Soter to curate the exhibits and Soter raised the issue of how to treat Pluto, handing him a recent article from *The Atlantic* on the subject. Thus prompted, Tyson wrote an essay for popular consumption indicating his personal desire to "defend Pluto's honor" but also his professional responsibility to "vote—with a heavy heart—for demotion." Tyson thus facilitated seeing this change as a downgrade, even though he was quick to add that in so doing, Pluto "went from being the runt of the planets to the undisputed King of the Kuiper belt. Pluto is now the 'big man' on a celestial campus."[95] Nicely chosen words, but, as Tyson soon learned in a "flow of letters to me and to the magazine" they garnered strong emotional replies, demonstrating to him how delicate the situation was. Also aware that the IAU had just issued a press release clarifying that it was not, nor did it have any intention of, changing Pluto's status as a planet, Tyson decided to inform his exhibitry decisions with "a panel debate on Pluto's status, inviting the world's leading thinkers on the subject to duke it out, on stage, for our benefit and for the benefit of the interested public."[96]

The February 1998 *Atlantic* essay by David H. Freedman had centered on Marsden's efforts to redefine Pluto's classification and clearly indicated that there was an ongoing debate over "demoting" Pluto within the astronomical community. Freedman also interviewed a leading member of the IAU's Working Group for Planetary System Nomenclature, "the only route to an official downgrading," who emphatically denied any plans to demote it from planetary status. Hotly contesting any change were people like the popular amateur astronomy writer and comet discoverer David H. Levy, whom Freedman quoted as arguing that "it's a mistake to frame the debate in terms of technical definitions, because something more important than precision is at stake. 'This isn't about science, or things,' [Levy] says. 'It's about people.'"[97]

After reading Freedman's essay, and no doubt keenly aware of the sensitivities, Tyson knew he had to address the situation as well, hoping that his convened panel would somehow inform his decision and lend it gravitas. His panel was well chosen, including the main figures in Freedman's essay, David Levy and Brian Marsden, along with Jane Luu, Alan Stern, and Michael A'Hearn, a highly respected leader among planetary scientists and a specialist in the physical properties of cometary bodies. Even though Tyson claimed that "[y]ou couldn't get more expert experts" than these, it was also clear that all were planetary scientists specializing in planetary studies or orbital dynamics. Also worthy of note was Tyson's choice for a title for the evening: "Pluto's Last Stand: A Panel of Experts Discuss and Debate the Classification of the Solar System's Smallest Planet."[98] Rhetorically, given the obvious sensitivities, he acted as if he had no intention of quieting the situation.

Some eight hundred people packed the auditorium on the evening of May 24, 1999, and were treated to a divided panel: "one for uncompromising iceballhood, two for dual status, two for planethood."[99] In order, these were Luu, then Marsden and A'Hearn, and finally Stern and Levy. The last two appealed strongly to cultural values, although Stern also invoked physical characteristics based upon the consequences of self-gravitation ("roundness" defining the lower end of the planetary mass range, and hydrogen fusion defining the upper end). The first panelist clearly preferred science over sentiment, as Tyson characterized Luu's position. The two advocating dual status were also those most senior and most visible in the mainstream profession, and in positions of making judgments through its institutions. They appreciated that the astronomical community was not monolithic, and therefore unlikely to accept any single definition. As A'Hearn testified soon after at the 2000 meetings of the IAU in Manchester, England, considered from the viewpoint of its orbit, or its dynamical characteristics, Pluto behaved like a TransNeptunian object, but considered as a physical object, it looked and behaved like a planet. And since recent suggestions to merely catalogue it along with the trans-Neptunian objects (TNO) "led to considerable controversy within the community," he concluded that "the only sensible approach is to use dual classification."[100]

As the climax to the evening, Tyson polled the audience, and according to his record, they only mildly preferred planet status over kicking Pluto "out of the planet club." He did not record if he asked the audience its opinion on assigning dual status, concluding only that the evening convinced his staff that it was really all just a laughing matter, of interest only for "reasons of nostalgia"; yet he marked the event as "The night Pluto fell from grace."[101]

In his memoir, Tyson details the fallout from his decision, reporting that it "would disrupt my life for years to come."[102] He does not tell us if this disruption was welcome, but, indeed, throughout his recounting it is clear that he and many others had great fun gaining public notoriety responding to what, they all knew, or hoped, would ultimately be good press for astronomy, and for the American Museum of Natural History. Surely there were unpleasant or tense moments in the flood of mail and public derision, but when a noted and highly respected geophysicist and former NASA high official offered his mild critique, he classed Tyson's institution as "The Science Citadel of the Capital of the World."[103]

The extent to which Tyson's efforts helped to inflame the matter, or to confuse it in the public mind, are beyond the scope of this essay. They did result, however, in exposing the normally private "trading zone" astronomers long employed to hammer out differences.[104] His memoir, along with others now appearing, helps to preserve the record of a rich and varied debate, among astronomers, the media, and the public. Tyson did not invent the emotional issue of "demotion" but clearly neither he nor others were able or even willing to redirect attention to what many astronomers felt was the real issue and indeed, the real meaning behind the new knowledge gathered over the years. As Tyson ironically testified in response to continuing media pressure in February 2001, this new knowledge redefined the nature of the solar system itself, revealing important new realms. And though most of it still remains uncharted, Tyson recently concluded, we can now name it. It's "called the Kuiper Belt, of which, Pluto reigns as king."[105]

The "Battle of Prague"

Among the Tyson panel members, as the most experienced and invested in the production of a rapidly growing population of known trans-Neptunian objects, Luu felt it was only a matter of time before Pluto was matched or even exceeded: "What if we find objects fairly close in size to Pluto—maybe even bigger, or maybe a bit smaller—will these objects also be called planets, or what?"[106]

Luu's question was echoed in debates over the next several years, as more and larger trans-Neptunians were detected. The announcement of Eris in July 2005 by Michael Brown and his team finally put an end even to Pluto's kingly status (figure 13.3). But well before then, preparing for the next triennial general assembly of the International Astronomical Union to be held in Prague in August 2006, the IAU felt it had to prepare a decisive public statement. In March 2004 the chair of the Planetary Systems Division of the IAU appointed a 19-member panel to meet in virtual space to work out the definition of a planet. Here was where all specialist interests in the subject finally had the opportunity to debate. This time it was not so much a clash over standards of practice and precision as it was in 1930, for the discipline overall had matured considerably. Now, however, it was a clash over inclusiveness, because the many specialties in the discipline that now were engaged in the matter dealt with overlapping interests and concerns that competed for attention and for resources, and each had their preferred definition of a planet. Unfortunately, no single definition emerged, and the committee issued a final report in November 2005 that was more a summary of the range of opinions and a series of individual proposals, none given more weight than any other.[107]

Figure 13.3 This comparison of the Earth and Moon with three Kuiper Belt bodies in the outer solar system demonstrates graphically the reason for assigning Pluto a new status. (Credit: NASA/JPL-Caltech/R.Hurt(SSC-Caltech.)

The process itself has been the heated subject of many blogs, chats, listservs, articles, essays, and books.[108] As the general assembly drew nearer, closure was still elusive, especially as Sedna and Eris came on the scene and Pluto was no longer even "king." The IAU executive committee convened a small panel of seven people as a new "Planet Definition Committee" chaired by the noted astronomer/historian Owen Gingerich, and joined by author Dava Sobel, solar system specialists Richard Binzel, Andre Brahic, Junichi Watanabe, Iwan Williams, and by the president-elect of the IAU, Catherine Cesarsky. Gingerich had examined the history of Pluto's discovery and theories of its origin in a 1959 article in *Scientific American*, and notably, four of the remaining six members were also active and successful as popular interpreters of astronomy and its lore; four were solar system specialists, and one a popular writer of astronomical history. This mixture of talents, consciously selected to be "historically and culturally sensitive," reveals the importance to the IAU of achieving a consensus that would be understandable and acceptable to the wider public. In the words of the committee chair, they sought "a scientific, but culturally sensitive, definition."[109] The charge now was explicit: decide on the definition of a planet, and do so by July 2006, in time for deliberations and approval by the general assembly in August.

As is very well presented in the recent popular literature, the committee's deliberations did result in a draft statement that was hotly debated at the IAU general assembly. It went through several revisions, its elements were subjected to rump sessions and open floor votes, and finally on the last day of the two-week conclave, a consensus emerged.[110] What was approved was significantly different from what the committee proposed, and in fact was a reflection of the diverse nature of the modern astronomical community.

In summary, the committee originally chose to define a planet by its physical characteristics: what it looks like, how it got that way. Thus IAU attendees learned through a press release to the media on August 16, the second day of the meetings, that "our Solar System will include 12 planets, with more to come: eight classical planets that dominate the system, three planets in a new and growing category of 'plutons'—Pluto-like objects—and Ceres. Pluto remains a planet and is the prototype for the new category of 'plutons.' "[111] The press release spoke of the shapes of planets—"a large round body"—and of their mass range, and that they were bodies in orbit around a star, were not stars themselves, nor satellites of another planet. There were two types of planets: those previously known to exist with periods less than two hundred years, and "plutons," those with periods longer than two hundred years and highly inclined, eccentric orbits. This definition left open the possibility that more planets would be added in the future. Beyond the known planets, including Pluto as the first Pluton, two more qualified: Charon, Pluto's giant moon (considered as partner with Pluto in a double planet), and Brown's 2003 UB_{313} or Eris. The asteroid Ceres was included as a cisjovian minor planet.

The August 16, 2006, press release rationalized the distinction between pluton and classical planet saying that "[a]ll of these distinguishing characteristics for plutons are scientifically interesting in that they suggest a different origin from the classical planets." But the committee's effort to name this new class of object in a manner that placed Pluto as its prototype, defining the class, did not satisfy members of the IAU that truly dynamic features had been adequately considered in the formal definition. Much debate ensued.

As Boyle, Schilling, and others emphasize, and I can personally attest to, the proceedings were far from smooth and harmonious.[112] On Wednesday, the day of the press release, I heard mainly mild joking and snickers in the hallways, but then a few astronomers of my acquaintance started expressing concern. The media was pressing for conclusions, asking questions that were neither friendly nor easy to answer. What started at first as a series of quiet interviews with microphones and cameras became loud gesticulating debates under hot lights. By late Thursday it was no longer a laughing matter. As Boyle related the situation: "Most of the astronomers in Prague felt they had to approve something, even if the process or the result was flawed."[113] For me, however, the issue was really if the perceived need for solidarity among astronomers trumped any attempt to provide a rational definition. This came clear when more than one astronomer hotly expressed their preference to me that a decision was critical, explicitly fearing that "If we don't do it, someone will do it for us."

Those who felt most compromised by the draft of the Planet Definition Committee, the dynamicists and the extrasolar planetary systems people, made their opposition clear at the first opportunity.[114] This led to additional sessions and revised text and to a decision to define a planet as a body massive enough to have cleared its orbital path of other bodies, through accretion or ejection. Pluto was not one of these, so it and its cohorts would be classed as "dwarf-planet." For balance, and to redress the possibility of a hierarchy again creeping in to the debate, Gingerich suggested another revision that would in effect reinstate and refine the term "classical," which was inferred in the original press release of August 16. All planetary bodies would have hyphenated names: those planets left that were large enough to clean out their respective portions of the solar system would be formally called "classical planets." Thus the term "planet" would include these two major categories in equal standing. Gingerich's revision was roundly rejected, so in the end, a clumsy compromise was reached that continued to define a planet in terms of the first two original criteria, but added the third dynamic criterion and provided recognition that Pluto was the prototype of all dwarf planets with trans-Neptunian orbital properties, and the linguistic oddity that dwarf planets were no longer "planets." In an essay for *Daedalus* written soon after the meeting, Gingerich felt that in the heat of the controversy, in the media room and on the voting floor of the IAU, one critical issue was lost to its members: sensitivity to public reaction. "It behooves us to pay attention to public relations" Gingerich lamented, keenly aware that even the president of the IAU and a member of his committee, Catherine Cesarsky, was unable to convince the membership.[115]

In the weeks and months following the Prague meeting, a flood of news articles and web-based discussions predicted another showdown and a reversal in three years, at the next general assembly in Rio de Janeiro. As Jewitt and Luu observed in the same issue of *Daedalus*, "[O]ne cannot buy the level of public interest that has been triggered by the planethood debate."[116] But with Gingerich and others they hoped that, flawed as the process was, it would still result in a better understanding by the public of "what science is about."

In the interim, there were many attempts to clarify the picture. A "Great Planet Debate" was staged at the Applied Physics Laboratory in Maryland, site of the Mission Operations Center for the New Horizons mission to Pluto, launched in January 2006 and due to fly by Pluto in July 2015. The IAU's Planetary Systems Division formed working groups on nomenclature to decide how to name newly discovered objects, and to decide on the lower mass limit to "dwarf-planet" status.

Whatever efforts there actually were for reopening the issue at Rio, however, it did not come to pass. As in all matters scientific, the case will never be closed, but Pluto still exists, as a provocative historical object that stimulated many questions about the nature of its discovery, about its origin and destiny, and about its class. During the course of asking these questions, a new realm to our solar system was postulated and, through a variety of stimuli, finally detected. It is now being elucidated, and our solar system itself is now known to be one of a class of systems that commonly populate the stars in the sky. Alan Stern said it best in 1996: Pluto is merely "the most easily detectable . . . member of an enormous ensemble,"[117] an ensemble that is, indeed, detectable over interstellar distances.

Conclusions

Since its discovery, Pluto went from being a unique, solitary body not classified easily with any other objects, members, or zones in the solar system, to being an exemplar of a wholly new class of object occupying a vast new region of the solar system. There were many steps and stages in the journey to this realization, most of which were not sequential but ran in parallel. The earliest stages involved how different astronomers, with differing backgrounds and toolkits, viewed the discovery itself. This debate dovetailed with efforts to determine Pluto's size and mass, and paralleled other lines of effort to ascertain the limits of the solar system itself, its origin, evolution, fate, and place in the dynamical hierarchy of the galaxy.

Recounting the many problems Pluto posed for astronomers helps to test the nature of scientific discovery generally: that the probability of discovering, and of comprehending, something entirely new is determined by the detection and recognition of the existence of some trait that, as Martin Harwit has suggested in his book *Cosmic Discovery*, "would label such a newly observed object or set of events as strikingly different from anything previously discovered."[118] Pluto started out being strikingly different, but due to forces of patronage and pride overriding standards of practice, and then of tradition and emotion, remained classified as a planet until a crisis of confidence finally led to its reclassification.

Finally, Pluto repeatedly provided an example of one of the paths to astronomical discovery. The mission-oriented focus fostered by the Lowell Observatory in the 1920s that led to the detection of Pluto in the first place, certainly, technically, ahead of its time, was definitely a large and unusual investment for that day. It was not as large as other projects like Hale's efforts in California, but it was unique in its nature: it was concentrated to a specific end, as opposed to the open-ended exploratory programs observatories typically fostered. In like manner, astronomers such as Kowal and Brown began discovering other Plutos in abundance only when provided with adequate incentives: technologies, funding, sufficient observing time, and an increased probability of success.

When *New Horizons* passes through the Pluto/Charon system and whizzes on by in its probe of the Kuiper Belt, no doubt many new questions will be raised as astronomers interrogate the data stream. One thing is certain: Pluto will become revealed in sufficient detail from the battery of instruments aboard the craft and will, once again, present a whole new set of problems to chew on for generations to come.

Notes

1. Lisa R. Messeri, "The Problem with Pluto: Conflicting Cosmologies and the Classification of Planets," *Social Studies of Science* 40 (2010): 187–214, explores factors that guided astronomers' reactions to the demotion. The appearance of her work, as I was revising this present chapter, which grew out of a talk given over seven years ago, has had very positive influence on the revision. I am indebted to readers of that essay as well as an earlier draft of the present form; the comments kindly provided by Owen Gingerich, Brian Marsden, Neil Tyson, Goetz Höppe, and Michael Neufeld were of great help. Archival material from the Lowell Observatory as well as Princeton University is acknowledged, as well as oral history material from the American Institute of Physics and the National Air and Space Museum. This work was supported in part by a grant from the NASA history office. Abbreviations used in the notes include: LowA–Lowell Observatory Archives; HUA–Harvard University Archives; VMS–V. M. Slipher; PUL/HNR–Princeton University Library, Henry Norris Russell papers; AIP–American Institute of Physics, Center for History of Physics; KP/UAZ–Kuiper Papers, University of Arizona.
2. Understandably there is a substantial literature on the discovery of Neptune. An early accessible history is Morton Grosser, *The Discovery of Neptune* (Cambridge, MA: Harvard University Press, 1962; reprinted, Dover Publications, 1979). The Leverrier quote is from Owen Gingerich, "The Solar System beyond Neptune," *Scientific American* 200 (April 1959): 86–100, on 86.
3. W. G. Hoyt, *Planets X and Pluto* (Tucson: University of Arizona Press, 1980), pp. 84–85. The best biography of Lowell is: David Strauss, *Percival Lowell—The Culture and Science of a Boston Brahmin* (Cambridge, MA: Harvard University Press, 2001).
4. V. M. Slipher to Russell, May 12, 1928; Putnam to Slipher, May 28, 1928; LowA/VMS/RLP. Russell, "Summary of Requests for Grants For Astronomy," attached to Russell to Hale, May 28, 1928; PUL/HNR, Box 19, folder 21. On the still-born National Research Fund, see Lance E. Davis and Daniel J. Kevles, "The National Research Fund: A Case Study in the Industrial Support of Academic Science," *Minerva* 12, 2 (April 1974): 207–20.
5. Hoyt, Planets X and Pluto, p. 175.
6. V. M. Slipher to Putnam, March 9, 1929. On Tombaugh, see Slipher to Putnam, February 7, 1929. LowA/VMS, and Hoyt, *Planets X and Pluto*, chapter 9, p. 182, which also discusses predicted magnitudes and exposure limits.
7. Tombaugh's character is well explored in David H. Levy, *Clyde Tombaugh: Discoverer of Planet Pluto* (Cambridge, MA: Sky Publishing, 2006).
8. Slipher to Miller, March 8, 1930. LowA/VMS.
9. This sequence of events has been taken from Hoyt, *Planets X and Pluto*, chapters 9 and 10.
10. Miller to Slipher, February 27, 1925; Slipher to Miller, March 8, 1925; LowA/VMS.
11. David H. DeVorkin, *Henry Norris Russell* (Princeton, NJ: Princeton University Press, 2000), p. 307.
12. See note 4.

13. E. W. Brown, "On the Predictions of Tran-Neptunian Planets from the Perturbations of Uranus," *Proceedings of the National Academy of Sciences* 16 (1930): 364.
14. Brown to Putnam, March 17, 1930: LowA/Putnam Pluto Folder.
15. Shapley to Slipher, March 20, 1930. LowA/VMS.
16. On naming conventions in planetary astronomy, governed by International Astronomical Union rules, see http://www.iau.org/public_press/themes/naming/ (accessed January 29, 2010).
17. Hoyt, *Planets X and Pluto*, pp. 201–204.
18. Whipple and Bower PASP 42 (1930): 239.
19. Russell, 1930, "Planet X," *Scientific American*, pp. 21–22; Russell, 1930, "How Pluto's Orbit was Figured Out," p. 364.
20. HNRjr, 18. MRE #1, p. 32. Russell, 1930, "More About Pluto," p. 446.
21. Russell, 1931, "Refining Pluto's Orbit," p. 91.
22. Jesse Greenstein to Russell, December 26 [1940], PUL/HNR.
23. Brown to Russell, February 7, 1931. PUL/HNR.
24. Russell, 1938, "Address of Retiring President," pp. 112–13.
25. Ibid.; Russell to John C. Cobb, March 2, 1931; PUL/HNR; "computational approach" from Schwarzschild Oral History Interview (OHI), June 18, 1982, pp. 16–17. On Bridgman and Slater, see S. Schweber, 1990. Russell to some extent shared Bridgeman's skepticism regarding the attainability of certainty. Walter, 1990, pp. 173–75.
26. Brown to Putnam, March 17, 1930. Pluto folder, LowA/RLP.
27. Reaves, 1997, p. 20.
28. E. Myles Standish, Jr., "Planet X: No Dynamical Evidence in the Optical Observations," *Astronomical Journal* 105 (1993): 2000–2006.
29. Clyde Tombaugh and Patrick Moore, *Out of the Darkness: The Planet Pluto* (Harrisburg, PA: Stackpole Books, 1972), pp. 127, 151.
30. Ibid., pp. 146, 164–72.
31. Brouwer, D. (1940). "Comparison of Newcomb's Tables of Neptune with an Orbit Obtained by Numerical Integration, and Discussion of the Perturbation by Pluto," [abstract] *Publications of the American Astronomical Society* 10 (1940): 7–8. Henry Norris Russell, "Bleak Black Ball of Rock," *Scientific American* 163 (1940): 18–19. This work is nicely situated in a thesis on early scientific machine computation. Allan Olley, "Just a Beginning: Computers and Celestial Mechanics in the Work of Wallace J. Eckert," University of Toronto, 2010.
32. J. W. Christy and R. Harrington, "The Discovery and Orbit of Charon," *Icarus* 44 (October 1980): 38–40; and "1978 P 1," Circular No. 3509 Central Bureau for Astronomical Telegrams.
33. See Robert Bowen, *Universal Ice: Science and Ideology in the Nazi State* (London: Belhaven, 1993), for a general description of these theories, but consider the remarks of a critical review [Michael Neufeld, "Hörbigerism," *Science* 262 (December 1993), 2069–2070] concerning the social implications of the title.
34. "Links New Planet to Icy Cosmos Idea," *New York Times*, March 29, 1930, p. 8.
35. Gustav Lindenthal, "Origin of the Solar System," *New York Times*, March 30, 1930, p. E5.
36. Armin Leuschner, 1932, "The Astronomical Romance of Pluto," *PASP* 44, 260 (August 1932): 197–214, from 197 to 98, 210, 213.

37. Raymond J. Lyttleton, "On the Possible Results of an Encounter of Pluto with the Neptunian System," *MNRAS* 36 (1937): 108–15; "The Origin of the Solar System," *Monthly Notices of the Royal Astronomical Society* 96 (1928): 559–68; H. N. Russell, *The Solar System and its Origin* (New York: Macmillan, 1935). At the time of Lyttleton's work, Neptune had only one moon. Nereid was discovered by Kuiper in 1949.
38. See, for instance, Owen Gingerich, "The Solar System beyond Neptune," *Scientific American* 200 (1959): 86–100; Thomas B. McCord, "Dynamical Evolution of the Neptunian System," *Astronomical Journal* 71 (1966): 585–90.
39. E. Öpik, "On the Fundamental Problem of Meteor Statistics," *Harvard College Observatory Circular* 355 (1930): 1–12
40. E. Öpik, "Note on Stellar Perturbations of Nearly Parabolic Orbits," *Proceedings of the American Academy of Arts and Sciences* 67 (1932): 169. Whipple OHI 1977, pp. 46–55.
41. Whipple 1977 OHI, pp. 46–50, on 50; "Incentive of a Bold Hypothesis."
42. Fred L. Whipple "Upper Atmosphere Densities and Temperatures from Meteor Observations," *Popular Astronomy* 47 (1939): 419–25; "The Incentive of a Bold Hypothesis: Hyperbolic Meteors and Comets," in R. Berendzen, ed., *Education in and History of Modern astronomy* [*NY Acad. Sci.* 198 (1972): 219–24], reprinted in *The Collected Contributions of Fred L. Whipple* (Washington, DC: Smithsonian Institution Press, 1972), pp. 3–17.
43. Whipple OHI, p. 53. See: Richard A. Jarrell, "Canadian Meteor Science: The First Phase, 1933–1990," *Journal of Astronomical History and Heritage* 12 No. 3 (2009): 224–234.
44. Ronald E. Doel, *Solar System Astronomy in America: Communities, Patronage and Interdisciplinary Research, 1920–1960* (Cambridge, UK: Cambridge University Press, 1996), especially chapters 3 and 4.
45. A. J. J. van Woerkom, "On the Origin of Comets," *Bull. Astron. Inst. Neth.* 10 (1948): 445–72; Jan H. Oort, "The Structure of the Cloud of Comets Surrounding the Solar System and a Hypothesis Concerning its Origin," *Bull. Astron. Inst. Neth.* 11 (1950): 91–110.
46. Doel, Solar System Astronomy in America, pp. 116–17.
47. Whipple to Oort. August 16, 1949, Box 6 Folder "O 1940–1950" HUG 4876. HUA.
48. HUG 4876 Box 6 Folder "O 1940–1950" 19491102: November 2, 1949, Whipple to Oort. HUA.
49. Fred L. Whipple, "Photographic Orbits of Sporadic Meteors," [abstract] *Astronomical Journal* 52 (1946): 50–51; "The Orbits of Meteors Photographed at Two Stations," [abstract] *Astronomical Journal* 54 (1948): 53. On the Super-Schmidts, see Teasel Muir-Harmony, David H. DeVorkin, and Peter Abrahams, "Wide-Field Photographic Telescopes: The Yale, Harvard and Harvard/Smithsonian Meteor and Satellite Camera Networks" (IUHPS Conference Proceedings, in press).
50. Oort-Whipple correspondence, 1949–1950. Whipple Papers, HUG 4876, Box 6 folder "O" Harvard University Archives (hereinafter HUA).
51. Whipple Comet Model I Encke p. 780.
52. Oort to Whipple. January 2, 1950. HUA.
53. David H. DeVorkin, *Science with a Vengeance* (New York: Springer Verlag, 1996); Peter Millman, "A Size Classification of Meteoritic Material Encountered by the Earth," *JRASC* 46 (1952): 79–82.

54. Millman, "Size Classification of Meteoritic Material," 80–81.
55. J. McFarland, "Kenneth Essex Edgeworth—Victorian Polymath and Founder of the Kuiper Belt?" *Vistas in Astronomy* 40 (1996): 343–354, on 343.
56. "Meeting of the Royal Astronomical Society," *Observatory* 68, 9 (1948): 10–12.
57. I. P. Williams and A. W. Cremin, "A Survey of Theories Relating to the Origin of the Solar System," *QJRAS* 9 (1968): 40–62.
58. G. P. Kuiper, "Titan: A Satellite with an Atmosphere," *ApJ* 100 (1944): 378–83. Kuiper explicitly mentioned that the Eastman 1N emulsions were slow. This is the first mention of this product in any publication scanned by the Astrophysics Data Service.
59. Russell to Kuiper, January 17, 1944, reviewed in detail in Doel, *Solar System Astronomy in America*, pp. 46–47.
60. Kuiper, "Titan," 380.
61. Russell to Kuiper, April 3, 1950, KP/UAZ; G. P. Kuiper, "The Diameter of Pluto," *PASP* 62 (1950): 133–37. Kuiper's work is also discussed in Owen Gingerich, "The Solar System Beyond Neptune," *Scientific American* 200 (1959): 86–100.
62. Ingrid Groeneveld and Gerard P. Kuiper, "Photometric Studies of Asteroids. I," *ApJ* 120 (1956): 200–219, on 219.
63. J. A. Hynek, ed., *Astrophysics: A Topical Symposium* (New York: McGraw-Hill, 1951), p. v.
64. Kuiper to Lampland, September 26, 1949. KP/UAZ.
65. G. P. Kuiper, "On the Origin of the Solar System," chapter 8, in Hynek, ed., *Astrophysics*, pp. 357–424, on 377; emphasis in the original.
66. Ibid., pp. 377, 400.
67. G. P. Kuiper, "Further Studies on the Origin of Pluto," *ApJ* 125 (1957): 287–89.
68. Kuiper, "On the Origin of the Solar System," J. A. Hynek, ed., *Astrophysics: A Topical Symposium* (New York: McGraw-Hill, 1951), pp. 400, 402.
69. Fred Hoyle, "The Origin of the Solar Nebula," *Quarterly Journal of the Royal Astronomical Society* 1 (1960): 28–55.
70. M. M. Woolfson, "The Evolution of the Solar System," *Rep. Prog. Phys.* 32 (1969): 135–85; S. G. Brush, "The Impact of Modern Physics on Theories of the Origin of the Solar System," *Bulletin of the American Astronomical Society* 16 (1984): 547; "Theories of the Origin of the Solar System, 1956–1985," *Reviews of Modern Physics* 62, 1 (January 1990): 43–112; Doel, *Solar System Astronomy in America*, pp. 148–50.
71. Kuiper, "On the Evolution of the Protoplanets," *PNAS* 37 (1951): 9.
72. A. J. Whyte. *The Planet Pluto* (New York: Pergamon Press, 1980); Mark Littmann. *Planets Beyond: Discovering the Outer Solar System* (New York: Wiley, 1988); S. Alan Stern, "The Historical Development and Status of Kuiper Disk Studies," in T. W. Rettig and J. M. Hahn, eds., *Completing the Inventory of the Solar System*. ASP Conference Series 107 (1996): 209–32; John Davies, *Beyond Pluto: Exploring the Outer Limits of the Solar System* (Cambridge, UK: Cambridge University Press, 2001); Govert Schilling, *The Hunt for Planet X* (New York: Springer, Copernicus Book, 2009); Neil deGrasse Tyson, *The Pluto Files: The Rise and Fall of America's Favorite Planet* (New York: W.W. Norton and Co., 2009).
73. Alan Boss, *The Crowded Universe* (New York: Basic Books, 2007); Tyson, *The Pluto Files*; Alan Boyle, *The Case for Pluto* (New York: Wiley, 2010).

Messeri makes the case for how these conflicts led to a "forced consensus" over a formal definition of planet, and Pluto's ultimate reclassification.

74. The use of the term "premature" is a contention based partly on the discussion here that Pluto was discovered but not predicted, and as we now know was discovered long before its true role in the solar system could be discerned based upon observations. Also a contention at this time is that the rise of wide-field electronic area detection techniques in astronomy and the parallel development of automated means of comparing fields over time to look for changes made it economically feasible and therefore desirable to search for and find objects of the class Pluto represents. This supports to some degree Martin Harwit's conclusion that "[t]he most important observational discoveries result from substantial technological innovation in observational astronomy." See Martin Harwit, *Cosmic Discovery* (New York: Basic Books, 1981), p. 18.
75. C. J. Cohen and E. C. Hubbard, *Science* 145 (1964): 1336; reported in Ann Ewing, "Pluto Theories Crumble," *Science News Letter* 86 (October 3, 1964): 213.
76. Fred L. Whipple, "Evidence for a Comet Belt Beyond Neptune," *Proceedings of the National Academy of Sciences* 51 (May 15, 1964): 711–18.
77. Marsden OHI, October 17, 2005, pp. 41–42; S. E. Hamid, B. G. Marsden, and F. L. Whipple, "Influence of a Comet Belt Beyond Neptune on the Motions of Periodic Comets," *Astronomical Journal* 73 (1968): 727–29.
78. Pluto's diminishing stature prompted Alan Dressler and C. T. Russell to humorously suggest that Pluto was in fact initially massive enough to be Lowell's perturber, but that its purported accelerating loss of mass caused by its inexorable trek toward perihelion (as comets do) will cause it to disappear completely by 1984. Alan Dressler and C. T. Russell, *FORUM* , "From the Ridiculous to the Sublime: The Pending Disappearance of Pluto," *EOS* 61, 44 (October 28, 1980): 690. For additional irony, see also R. L. Duncombe and P. K. Seidelman, "A History of the Determination of Pluto's Mass," *Icarus* 44 (October 1980): 12–18.
79. Tom Gehrels, "Future Work," in Tom Gehrels, ed., *Physical Studies of Minor Planets, Proceedings of IAU Colloq. 12* (Washington, DC: NASA SP-267, 1971), pp. 653–62, on p. 654.
80. G. P. Kuiper, Y. Fujita, T. Gehrels, I. Groeneveld, J. Kent, G. van Biesbroeck, and C. J. van Houten, "Survey of Asteroids," *Astrophysical Journal Supplement* 3 (1958): 289–334 and tables.
81. C. I. Van Houten, I. Van Houten-Groeneveld, P. Herget, and T. Gehrels, "The Palomar-Leiden Survey of Faint Minor Planets," *Astro and Astrophy Suppl* 2 (1970): 339–448.
82. A good summary of the contributions of Voyager imagery to planetary science is Ronald A. Schorn, *Planetary Astronomy from Ancient Times to the Third Millennium* (College Station: Texas A&M University Press, 1998), pp. 274–79.
83. Julio A. Fernandez, "On the Existence of a Comet Belt beyond Neptune," *Monthly Notices of the Royal Astronomical Society* 192 (1980): 481–91.
84. Davies, *Beyond Pluto*, p. 46; Brush, "Theories of the Origin of the Solar System"; Martin Duncan, Thomas Quinn, and Scott Tremaine, "The Origin of Short-Period Comets," *Astrophysical Journal* 328 (May 15, 1988): L69–L73, quotes from L69; L70. Based upon a search using the Astrophysics

Data Service, this is the first paper that describes the zone explicitly as the "Kuiper Belt," L72.

85. Marsden, ed., "Central Bureau for Astronomical Telegrams Circular," Nos. 3129 and 3130.
86. C. T. Kowal, William Liller, and Brian Marsden "The Discovery and Orbit of 1977 UB," *Bull. Am. Astron. Soc.* 10 (1978): 481; C. T. Kowal, "Chiron," in Tom Gehrels, ed. *Asteroids* (Tucson: University of Arizona Press, 1979), pp. 436–39.
87. Davies, *Beyond Pluto*, pp. 48–70. The influence of patronage on problem choice in this area needs further examination, especially insofar as it can be seen as promoting or exploiting new technologies, or responding to emerging mission priorities for funding in bodies like NASA. See Harwit, *Cosmic Discovery.*
88. Davies, *Beyond Pluto*, p. 74. Comments, Brian Marsden to the author, August 26, 2010.
89. David Jewitt, and Jane Luu, "Reflection Spectrum of the Kuiper Belt Object 1993 SC," *Astronomical Journal* 111 (1996): 499–503; Davies, *Beyond Pluto,* p. 84.
90. On naming conventions in planetary astronomy, governed by International Astronomical Union rules, see http://www.iau.org/public_press/themes/naming/ (accessed January 29, 2010).
91. Mike Brown, Chad Trujillo, and David Rabinowitz, "Discovery of a Candidate Inner Oort Cloud Planetoid," *ApJ Letters* (August 2004); *ApJ* 635 (2005), L97–L100; Alan P. Boss, *The Crowded Universe: The Search for Living Planets* (New York: Basic Books 2009), pp. 115–17.
92. David Jewitt, "The Discovery of the Kuiper Belt," *Astronomy Beat* No. 48 (May 3, 2010), www.astrosociety.org., p. 4; Fernandez, "On the Existence of a Comet Belt beyond Neptune."
93. Davies, *Beyond Pluto*, pp. 202–205.
94. Neil DeGrasse Tyson, *The Pluto Files* (New York: W.W. Norton and Co., 2009), pp. 61–62.
95. Ibid., p. 65.
96. Ibid., p. 69. Tyson reproduces the press release, and Schilling, *Hunt for Planet X*, pp. 238–40, provides clarifying background on what caused it to be issued; specifically that it was the result of considerable internal debate within the IAU sparked by Marsden.
97. David H. Freedman, "When is a Planet Not a Planet?" *The Atlantic* (February 1998).
98. Tyson, *Pluto Files*, p. 69.
99. Ibid., p. 75.
100. Michael F. A'Hearn, "Pluto: An Edgeworth-Kuiper Belt Object and/or a Planet?" *The TransNeptunian Population*, 24th meeting of the IAU, Joint Discussion 4, August 2000, Manchester, England—meeting abstract.
101. Tyson, *Pluto Files*, p. 75.
102. Ibid., p. 81.
103. Ibid., p. 108.
104. Messeri , in "The Problem with Pluto," deftly applies this metaphor to explore the Pluto debate and its effect on the professional community.
105. Tyson, *Pluto Files*, p. 174.
106. Ibid., p. 71.

107. Schilling, *Hunt for Planet X*, pp. 242–43. In the past decade, the IAU reorganized into a set of major divisions, and the original commissions, over 40 of them, were clustered together. IAU Division III, Planetary Systems, consists of 7 commissions.
108. One of the most recent compilations (Boyle, *Case for Pluto*) contains an extensive discussion of the proceedings at Prague, as well as a very extensive and useful bibliography including books, articles, and Internet sources. See also Tyson, *Pluto Files*; Schilling, *Hunt for Planet X*.
109. I am indebted to Goetz Hoeppe for a discussion that led to this realization. It is confirmed in Owen Gingerich, "Planetary Perils in Prague," *Daedalus* 136, 1 (2007): 137–40, quote from 137, and in email conversation with Owen Gingerich, quotations from both sources.
110. Detailed coverage can be found in Schilling, *Hunt for Planet X*, chapters 26 and 27; and in Boyle, *Case for Pluto*, chapter 9, and the reference material included therein.
111. "The IAU draft definition of 'planet' and 'plutons,'" IAU Press release IAU0601: http://www.iau.org/public_press/news/detail/iau0601/ (accessed January 29, 2010).
112. I arrived in Prague on the second day, the day of the announcement. Much of what they report I sensed in the hallways and meeting rooms. The press release did worry me, especially the use of the term "pluton," and so I expressed my concerns to Owen Gingerich. But I was not aware of the deeper issues that were brewing among the groups of dynamicists and extrasolar planetary specialists who were organizing rebuttals. I was aware of a growing low-level panic.
113. Boyle, *Case for Pluto*, p. 133.
114. Ibid., pp. 127–28; Schilling, *Hunt for Planet X*, chapter 27.
115. Owen Gingerich, "Planetary Perils in Prague," *Daedalus* 136, 1 (2007): 137–40, on 140. Gingerich also related the linguistic oddity.
116. David Jewitt and Jane X. Luu, "Pluto, Perception & Planetary Politics," *Daedalus* 136, 1 (Winter 2007), 132–36, on 136.
117. Stern, "The Historical Development and Status of Kuiper Disk Studies," p. 229.
118. Harwit, *Cosmic Discovery*, p. 198.

Chapter 14

Transcendence and Meaning in Solar System Exploration

William E. Burrows

James A. Michener, who wrote *Tales of the South Pacific* and many other stories of adventure, believed that exploration is essential to the human condition, both physically and spiritually.

> We risk great peril if we kill off this spirit of adventure, for we cannot predict how and in what seemingly related fields it will manifest itself. A nation that loses its forward thrust is in danger, and one of the most effective ways to retain that thrust is to keep exploring possibilities. The sense of exploration is intimately bound up with human resolve, and for a nation to believe that it is still committed to forward motion is to ensure its continuance.[1]

The prolific and endlessly imaginative novelist was apparently referring to two kinds of exploration: the physical kind, as when Voyager 2 went on the Grand Tour, and the imaginative one, which is the human process that conceived the mission and took a hard look at whether it was feasible.

Adm. Richard E. Byrd, who opened the Antarctic to the rest of the world, understandably was deeply intrigued by exploration as it affected the human spirit. "In that instant," he would recall when he saw the vast region of ice at the bottom the world for the first time, "I could feel no doubt of man's oneness with the universe...It was a feeling that transcended reason; that went to the heart of man's despair and found it groundless. The universe was a cosmos, not a chaos; man was rightfully a part of that cosmos as were the day and the night."[2] Those words speak to the oldest and most profound questions engaging humanity: Who are we and why are we here?

The feeling that transcended reason is the sense of being an integral part of a process of learning about the world by exploring it endlessly, as did those who came before us, and those who will follow. The urge to explore, for the sake of adventure, learning, or survival, is a fundamental human trait. It begins with the infant who looks at his immediate world and then instinctively starts to explore it, first by progressively touching all of his body and n time, taking the measure of his cradle, his crib, and the world beyond Exploration is implicitly about extending reach for the sake of increas- owledge and safety, and it is both intellectual and spiritual. It is the rea- e Vikings sailed to the New World; Apollo 11 sailed to the Moon; and ger 2 sailed to four of the outer planets. Transcendence for the men and n in the space program comes with the sure knowledge that they are an al part of a historic process that began when their ancestors took to the d that will continue—barring a natural or man-made catastrophe—as descendants successively explore and then colonize the Moon, Mars, elsewhere in this solar system, and beyond to infinity.

Exploration feeds our insatiable craving for knowledge of both the outer world and the one within ourselves. It by definition transforms the unknown into the known, and that transformation has been the central preoccupation of serious thinkers, religious and secular, from time immemorial. And as on Earth, exploration is inevitably followed by establishing a permanent presence in a place that is habitable. That is the plan for the Moon. NASA has correctly decided that its overarching mission in space—what would have been called manifest destiny in a bygone era—should be the continued exploration of the solar system.

Planning the colony, let alone establishing it, will be a transcendental event in the lives of those who do it. That is to say, they will know—as did Columbus, Magellan, Cortez, Ponce de Leon, Byrd, Perry, and other notable explorers—that they are part of a great and infinitely important historic process (figure 14.1). For the scientists and engineers who are now planning the Moon colony, transcendence comes with knowing that they are the decisive force that links the men who landed on the Moon with the men and women who will inhabit it and then Mars, perhaps one of more Jovian moons, and beyond. They are well aware that the logical, established, sequence of progressively more challenging missions that got us to the Moon and all of the planets except Pluto (if, indeed, Pluto still is a planet)—flyby, orbit, robotic landers, and finally the landing of their fellow humans—will be repeated by their successors at Mars, elsewhere in this solar system, and ultimately well beyond it. For the extended team of solar system explorers, self-definition and transcendence come with the certain knowledge that they are an integral part of that grand design. And they know that as the future unfolds, that design will prove to be of monumental importance for this civilization for spiritual reasons as well as for intellectual, economic, and other ones, certainly including survival.

The more reflective scientists, engineers, managers, and others who are planning this outward expansion believe that having planned a noble future, they are inseparably, indelibly, part of it, and will remain so long after they have ceased to exist physically. On what is perhaps its most basic level, it is the feeling people get when they plan to build a home for themselves in which they will not only live in relative comfort and safety, but will raise and nurture their children, and then leave it to their children and perhaps generations of children to come.

Figure 14.1 The 1842 voyage of the French vessels *Astrolabe* and *Zelee* toward Antarctica is depicted from *Voyage au pole sud et dans l'Oceanie* (1842) as it encounters a ice field. Exploration had ranged to the poles before the International Polar Year of 1882–1883, but never before was it so well coordinated. (Credit: National Oceanic and Atmospheric Administration, NOAA Library Collection, Washington, DC.)

That thought occurred to me on the morning of May 26, 2008—Memorial Day, as it happened—when I read in the newspaper that the Phoenix Mars Lander had made a perfect landing in the northern polar region of the red planet two days earlier. The spacecraft traveled 422 million miles in nine months to become the first visitor from Earth to land on Mars using braking rockets to slow its descent since a pair of Viking spacecraft accomplished that exquisitely intricate maneuver to help celebrate the nation's bicentennial back in 1976. So while Memorial Day justifiably commemorated the men and women who have given their lives in their nation's armed forces, it had an additional meaning for me. It commemorated the increasingly sophisticated and productive robotic expeditions to Mars that began with Mariner 4's wonderfully audacious flyby on November 28, 1964, making it the first visitor from Earth to reach it. And that, in turn, happened two years after Mariner 2 became the first probe to fly by Venus, which also made it the first emissary from Earth to reach another planet. That feat, the subsequent encounter with Mars, the encounters on the Grand Tour, the landing of men on the Moon, and other triumphs of exploration vividly demonstrated that, contrary to popular myth, science is deeply emotional. The standing applause, backslapping, and handing out of cigars—appropriately also a rite of birth—by shirt-sleeved flight controllers at JPL and the Johnson Space Center when an exploration mission or other daring feat is successfully completed is dramatic testimony to that. Mariner 4's triumph at Mars was nothing short of delicious. It decisively beat the Soviets, whose desperation to remain competitive with the United States, or at least stay in the race, evidently led to a crash program in both senses of the term. The bitterly painful record spoke for itself:

> On October 10, 1960, a Mars flyby spacecraft failed to achieve Earth orbit . . . Four days later another Mars flyby probe failed to make it to Earth orbit . . . On February 4, 1961, a Venus flyby probe failed to leave Earth orbit . . . Eight days later there was a Venus flyby communication failure 14 million miles from the destination . . . On August 25, 1962, yet another Venus flyby spacecraft failed to leave Earth orbit . . . And there was another failure on September 1 . . . Eleven days later another spacecraft supposedly headed for an encounter with Venus failed to leave Earth orbit . . . On October 24, 1962, a Mars flyby spacecraft failed to leave Earth orbit . . . Eight days later there was a Mars flyby communication failure 69 million miles from the destination . . . Three days after that another spacecraft that was supposed to fly by Mars never made it out of Earth orbit . . . On January 4, 1963, a lunar soft landing was abruptly aborted because the spacecraft did not make it out of Earth orbit . . . On April 2, 1963, a lunar soft lander missed its destination by 5,300 miles . . .[3]

And so it went, as Kurt Vonnegut Jr. might have put it with weary philosophical acceptance.

To be sure, the Russians were trying desperately to turn space spectaculars into propaganda victories to "one-up" their economically and scientifically better-off rival. But it was more than competition that drove the Russians. They were the abidingly proud descendants of Konstantin E. Tsiolkovsky, the self-educated rural schoolteacher who conceptualized the use of liquid fuel for rockets, which made a decisive difference in lifting capacity over the powder that had been used by the Chinese for at least a millennium. In a manner of speaking, it was Tsiolkovsky who made the Space Age possible, and every school-going child in the Soviet Union, let alone the enthusiasts in the rocket societies and engineering establishments, knew it. They in effect stood on Tsiolkovsky's shoulders, and on them tried mightily to transcend from this world to a higher one, both spiritually and physically, with him. And they succeeded on October 4, 1957, when one of Sergei Korolyov's R-7 rockets flung the first Sputnik into an orbit around Earth.

The US space program's science return, from both Earth-orbiters and the far-flung explorers that have scouted the Moon and all but one of the other planets, is immense and well documented. Voyager 1's inspection of Jupiter and Saturn and its twin's 12-year Grand Tour—the greatest feat of exploration in history, in my opinion—provided a cornucopia of information about the neighborhood in which we live. All four of the Grand Tour encounters, beginning at Jupiter in 1977 and ending at Neptune in the summer of 1989, returned enough information to fill a small library.

The late Merton E. Davies of the Rand Corporation, who did pioneering work on the US reconnaissance satellite program at its inception and was on the Voyager imaging team, once declared that "[t]he joy of exploration is finding answers for which there are no questions."[4] That has happened repeatedly during solar system missions. John R. Casani, a JPL engineer who was the project manager for the Galileo mission to Jupiter, recalls that to his knowledge, no one thought about the planet having rings until Voyager 2 got there in early March 1979 and found them (though rings had been spotted around Uranus and Neptune by telescope). It also never seems to have occurred to anyone that active volcanoes existed anywhere in the solar system except on Earth, or that moons varied, often dramatically, until Voyagers 1 and 2 spotted heat and smoke plumes coming out of holes on Europa and Io, two of the Jovian moons, and scrutinized Callisto's heavily cratered surface of ice and rock and Ganymede's "crazy quilt" pattern of grooved terrain and craters.

E. Myles Standish, an astronomer and retired principal member of JPL's technical staff who was responsible for providing the Voyager navigation team with accurate positions of the planets at all times, which is called establishing their ephemerides, wholeheartedly agrees. "We thought all the moons in the solar system looked like our moon" before they were inspected up close, he says. It therefore came as a surprise that Jupiter's moons, which were first spotted by Galileo in 1610, are all different. The Voyager mission also revealed that the relatively placid surface of Jupiter that shows through telescopes here is actually a vast kaleidoscope of swirling storms. The Great Red Spot is a raging storm that is three times the size of Earth and rotates once every six days. Voyager 1 discovered lightning in the giant planet's cloudtops and also reported the presence of a ring in 1979.

And it got better. Both Voyagers sent imagery home that revealed Saturn also had moons. And not only did it have its own distinctly different moons, but the scientists who pored over the imagery quickly concluded, much to their surprise, that the moons were shepherding all of the untold millions of pebbles, rocks, and boulders that comprised the rings. Like drill sergeants, they were maintaining the integrity of the long, curving lines. Then, getting back to Davies, there was another answer for which no one had raised a question. The rings were not perfect circles. "What we hadn't anticipated was 'misbehaving rings' that weren't on nice, pure, circular orbits," Torrence V. Johnson, the chief scientist of JPL's Solar System Exploration Programs Directorate, added. "The braided, twisted, whatever, F ring defied easy explanation with classical dynamics." He remembered Richard Feynman visiting the imaging team during the encounter and discussing various structural possibilities for the rings. The celebrated Caltech physicist did not offer a suggestion as to why the ring pattern was the way it was. Johnson's reaction again betrayed the very human side of science.

> I was somewhat relieved, actually, that he didn't have a ready solution either, that would make us look dumb for not having thought of it. The F ring

structure can now be easily reproduced with complex computer models of the gravitational interactions of the "shepherd" satellites with the ring particles, but it took a lot of asking new questions to get there.[5]

Solar system exploration has been deeply emotional for T. V. Johnson, who very much wanted to be out there but finally, reluctantly, accepted the fact that it was not going to happen. He was not going to be an astronaut.

> I'm part of the generation of scientists strongly affected by the legacy of Sputnik and "hard science" science fiction by the likes of Heinlein and Asimov. Space exploration has always seemed to me just one of the most fascinating things to do. As I went to college in the '60s, I was disappointed somewhat as I gradually realized that I wasn't likely to be taking personal trips to the planets *a la* my science fiction heroes. I stopped worrying about this as I got into working on planetary missions, starting with Voyager. Our spacecraft and observations have effectively become extensions of our senses, gathering data from distant worlds that we can't visit physically. The very language that people working on these missions use reflects this: "We" got into orbit yesterday. Next week, "we're" flying by Titan, etc. I don't know if I have felt transcended to a better place emotionally, but I sure have felt that I have been privileged for thirty-plus years to be on the bridge of a real *Starship Enterprise*, receiving reports from our probes and scanners throughout the solar system. And now, through things like the internet, we can share this experience with people worldwide.

The identification of those who control the missions with the machines that carry them out, as Johnson explained, has been obvious to anyone who listened to the minute-by-minute narrations of encounters in the mission control center at JPL and, for human missions, at the Johnson Space Center. Being in the mission control room or with the world news media in von Karman Auditorium at JPL during a planetary encounter would show how closely the men and women who participate in the missions identify with their spacecraft. During every planetary encounter, certainly including those on the Grand Tour, the studiously unemotional voice on the public address system that describes what is happening has often referred to the spacecraft in the first-person plural: "We," as in "We are thirty-eight hours from closest approach to Neptune." That is because the human so closely identifies with the spacecraft that they in effect become the same thing to him: he and the machine are a single, completely unified entity composed of metal, plastic, silicon, blood, bones, ears, eyes, and a brain. A reporter once raised that subject with John Casani, and the imaginative, quick-witted engineer's answer remains in JPL's rich store of folklore: "No, we don't anthropomorphize our spacecraft," he explained. Then, after a pause, he added: "They don't like it..."[6]

For Johnson, Casani, and thousands of their colleagues in NASA and in foreign space agencies, the exploration of the solar system has been and remains a transcendental experience because they deeply believe that they are an integral part of a long and profoundly important historical process—one that is infinitely larger than the dimensions of their own lives. The urge to explore for the sake of adventure and understanding is a uniquely human trait. It is the urge that sent Europeans across the Atlantic to the New World, and the Chinese across the Pacific bound for the same destination. It is the same process, the same relentless drive toward infinity that must have inspired those who built the Mayan and Egyptian pyramids, the Acropolis, and the monuments, temples, and tombs in the Valleys of the Kings and Queens in Luxor.

Unlike those formidable quests, though, the exploration of distant worlds requires the act of flying. That, too, is a primordial human longing. The urge to fly—call it bird envy—goes back to the early Chinese powder rocket builders and to Daedalus and Icarus. It runs through the science fiction that effectively began with *From the Earth to the Moon* and continues in *Star Wars*, *Star Trek*, and *Battlestar Galactica* (figure 14.2). It is the essence of who Tsiolkovsky and his innumerable heirs, including Wernher von Braun, Robert Goddard, Mikhail Tikhonravov, and Korolyov, were. They were driven to ride rockets to distant worlds, not use them to obliterate whole cities. Tsiolkovsky, who had a deeply depressing childhood and was later influenced by a Moscow mystic named Nikolai Fyodorov, also sought transcendence. For him, it was a resolute belief that his fellow humans would one day inhabit worlds away from this one—infinitely better worlds—and that

Figure 14.2 NASA artist's conception of the Voyager spacecraft. It shows the Voyager spacecraft as it travels outward from Earth on its outer planetary tour. (Credit: NASA/JPL.)

piritually, if not physically. He had this to say to the
ication called *The Science Review*:

al aspects of the problem of ascending into space by
ice... My mathematical conclusions, based on scien-
times over, show that with such devices it is possible
se of the heavens, and perhaps to found a settlement
e earth's atmosphere... People will take advantage of
all over the face of the earth but all over the face of

Humanity's destiny to spread throughout the solar system is accepted by the true believers with a fervor that is religious. And as is the case with religion they, too, are convinced that they are part of a process that will continue indefinitely. Their reward will be to go to a heaven populated not by angels strumming harps, but by homesteaders developing new frontier worlds away from the old one. And like other enthusiasts, including sports fans and certain kinds of college alumni, they are most comfortable with each other. That is why they join space organizations, which include the American Astronautical Society, the National Space Society, the American Institute of Aeronautics and Astronautics, the California Space Authority, the Federation of Galaxy Explorers, the Florida Space Authority, Global Space Travelers, the Mars Society, the Moon Society, the NASA Alumni League, the National Coalition of Spaceport States, the Planetary Society, ProSpace, the Space Access Society, the Space Generation Foundation, the Space Studies Institute, and the X Prize Foundation. All of them in turn formed the Space Exploration Alliance in June 2004 to support President Bush's mandate to establish the lunar base and eventually strike out for Mars.

The X Prize Foundation made headlines in October 2004 when it awarded the Ansari X Prize, worth $10 million, to the developers of the first reusable private spacecraft that made it to the edge of space—arbitrarily defined as one hundred kilometers—twice in two weeks. The winner, called SpaceShipOne, was carried into the air over the Mojave Desert by a twin-boomed jet aircraft and released, whereupon its rocket kicked in and it shot almost straight up to an altitude barely short of seventy miles. The purpose of the daring stunt, which was conceived by Burt Rutan, who made a fortune with his Virgin Atlantic and Virgin Galactic airlines, was to stimulate private space ventures as alternatives to exclusive NASA involvement in civilian space operations. That, in turn, was supposed to encourage the commercial development of space, including tourism, by innovative, low-cost approaches. Ironically, it was the former Soviet Union that led the way to commercial space tourism in 2001 and 2002, when it helped two earthly multimillionaires transcend to space for $20 million each. The first was a New York–born California investment manager and former aerospace engineer named Dennis Tito, who was launched in Soyuz TM-32 on April 28, 2001, and spent eight days in orbit. The second was a South African entrepreneur named Mark Shuttleworth,

Figure 14.3 Human and machine on Mars, artist conception, after 2040. When the first humans venture across the surface of Mars, they may be inclined to collect artifacts from early spacefaring years. Here an astronaut retrieves the *Sojourner* rover, which arrived on that planet in 1997. (Credit: NASA/Pat Rawlings.)

who went into orbit on Soyuz TM-34 on April 25, 2002. Two days later, he was transferred to the International Space Station, where he logged eight days, some of them participating in biology experiments. He returned home on May 5, floating down onto the Russian steppe under parachutes, as cosmonauts do.

The true believers have not the slightest doubt that the exploits of SpaceShipOne's test pilot, Brian Binnie, and Dennis Tito and Mark Shuttleworth are indelible signposts on the road to ultimate transcendence. The fact that their vision is for humanity to spread out in space and remain there forever is by definition transcendent (figure 14.3). The National Space Society, which claims more than twelve thousand members in more than fifty chapters in the United States and around the world (and supporters who are not technically members) is explicit on the subject:

> The Vision of NSS is people living and working in thriving communities beyond the Earth, and the use of the vast resources of space for the dramatic betterment of humanity. The Mission of NSS is to promote social, economic, technological, and political change in order to expand civilization beyond

> Earth, to settle space and to use the resulting resources to build a hopeful and prosperous future for humanity. Accordingly, we support steps toward this goal, including human spaceflight, commercial space development, space exploration, space applications, space resource utilization, robotic precursors, defense against asteroids, relevant science, and space settlement oriented education.[8]

The rationale for exploring and colonizing space is stated explicitly: The survival of humanity and the biosphere that nurtures it. And the situation is described as dire but not hopeless. "The human species is encountering increased natural, man-made, and extraterrestrial threats, including disease, resource depletion, pollution, urban violence, terrorism, nuclear war, asteroids, and comets." The statement continued:

> Space technology provides both means to monitor threats to life on Earth and ways to help curtail them. Space industrialization and settlement provide safety valves to relieve the pressures that cause Earth-bound threats. They also provide escape routes in case of catastrophic man-made or extraterrestrial threats. Humanity has inherited the stewardship of the planet Earth...It will therefore need the vast resources of outer space to reverse the damage it has caused to the Earth biosphere, and ultimately enhance all life on Earth.[9]

That is not the work of wild-eyed, romantic, space freaks whose imaginations dwell in a pulp fiction universe inhabited by the Dark Force and superheroes wielding light swords and ray guns. The members of the National Space Society, the American Astronautical Society, the American Institute of Aeronautics and Astronautics, the Planetary Society, and others tend to be level-headed scholars, scientists, lawyers, physicians, and business people who believe that exploring and then colonizing space is simply rational because it will help Earth in many ways and also provide a hedge against a worldwide catastrophe such as a major asteroid or comet impact, a massive volcanic eruption, a pandemic, or a thermonuclear war that, in addition to causing widespread death and destruction, would cloud the atmosphere with so much debris for so long that it would start what Carl Sagan and four other scientists called a long, devastating nuclear winter.[10] They are the sort of citizens who sit on community planning boards and participate in urban development projects. Here, in part, is what the NSS website said about exploring and exploiting space in June 2008:

> When the first person landed on the Moon in 1969 after only eight short years of intense effort, the National Aeronautics and Space Administration (NASA) proved that we could do nearly anything we put our minds and resources to that is consistent with the laws of physics.
>
> A few years later, Princeton physicist Gerard O'Neill and others showed that large orbital space settlements would fall within the laws of physics. Dr. O'Neill's analysis strongly suggested that asteroids and lunar mines could supply the materials, the Sun could provide the energy, and that our

technology had nearly reached the point where we could build space settlements. These communities could be placed almost anywhere in the solar system.

In 1990, Robert Zubrin and David Baker described a program called Mars Direct, an innovative approach to begin the settlement of Mars. Zubrin's 1996 book *The Case for Mars* went on to outline a long term program to bring Mars to life with a vibrant human civilization. While certainly difficult, every step in this program is also achievable within the laws of physics.

Many plans for space settlement have been proposed—in orbit, on the Moon, on Mars, the asteroids, or elsewhere. All are extremely difficult and expensive, but not much more difficult and expensive than things we have already done. After all, construction of today's civilization was a mighty task indeed. However, if we are going to spend an enormous amount of time, effort, and money on something, we'd better know why.

There are many reasons to move to space: growth, wealth, energy, survival, spiritual development, knowledge, diversity, to solve serious Earthly problems, to fulfill a sense of destiny and responsibility, and even to have fun. All of these boil down to a simple fact: *A future with space settlement is vastly better than one without it.*

This flows from another simple fact: *There are far, far more resources in space than on Earth.* For example:

- The largest asteroid, Ceres, has enough material to build orbital space settlements with a total living area well over a hundred times the land area of the earth.
- One smallish asteroid, 3,554 Amun, has about $20 trillion worth of metals. There are tens of thousands of asteroids.
- The energy available for space settlements exceeds 2 billion times the total energy currently used by humanity.

There are potentially profit-making industries: space tourism, space solar power, space materials, and others that can pave the path to the first self-sustaining space settlements. As the website noted:

Furthermore, we more-or-less know how to exploit these resources without hurting anyone, oppressing anyone, or harming any living organism for the simple reason that there aren't any living things there—it's just rock and radiation, both of which are usable (and valuable) resources. We can bring life into space at great advantage to those who dare try, as well as to humanity as a whole.

Arthur C. Clarke, inventor of the concept of using geosynchronous orbit for communication satellites, once wrote that new ideas like this pass through three stages:

- Stage 1: "It can't be done."
- Stage 2: "It probably can be done, but it isn't worth doing."
- Stage 3: "I knew it was a good idea all along!"

Arthur C. Clarke, the patron saint of space exploration, was right. His third book, *The Exploration of Space*, was a primer on the subject and published in 1951, has been a primer for space exploration ever since. We knew it was a good idea all along.

Notes

1. Roger D. Launius and Howard E. McCurdy, *Imagining Space* (San Francisco: Chronicle Books, 2001), p. 22.
2. "Brave Blue World," *Nature*, February 1, 2007, p. 459.
3. William E. Burrows, *Exploring Space: Voyages in the Solar System and Beyond* (New York: Random House, 1990), pp. 420–21.
4. The comment was made to the author.
5. Email from Johnson to the author, June 3, 2008.
6. Email from Johnson to the author, June 2, 2008.
7. Evgeny Riabchikov, *Russians in Space* (Garden City: Doubleday, 1971), p. 99.
8. NSS Statement of Philosophy, p. 1, undated.
9. Ibid.
10. Carl Sagan, *Pale Blue Dot* (New York: Random House, 1994), pp. 227–28.

Index

Printed in the United States of America